Semiconducting Chalcogenide Glass I
Glass Formation, Structure, and Stimulated Transformations in Chalcogenide Glasses

SEMICONDUCTORS
AND SEMIMETALS
Volume 78

Semiconductors and Semimetals

A Treatise

Edited by R.K. Willardson
CONSULTING PHYSICIST
12722 EAST 23RD AVENUE
SPOKANE, WA 99216-0327
USA

Eicke R. Weber
DEPARTMENT OF MATERIALS
SCIENCE AND MINERAL
ENGINEERING
UNIVERSITY OF CALIFORNIA
AT BERKELEY
BERKELEY, CA 94720
USA

Semiconducting Chalcogenide Glass I

Glass Formation, Structure, and Stimulated Transformations in Chalcogenide Glasses

SEMICONDUCTORS
AND SEMIMETALS

Volume 78

ROBERT FAIRMAN

Beaverton, OR, USA

BORIS USHKOV

JSC ELMA Ltd
Moscow, Russia

ELSEVIER
ACADEMIC
PRESS

Amsterdam • Boston • London • New York • Oxford • Paris
San Diego • San Francisco • Singapore • Sydney • Tokyo

ELSEVIER B.V.
Sara Burgerhartstraat 25
P.O. 211, 1000 AE Amsterdam
The Netherlands

ELSEVIER Inc.
525 B Street, Suite 1900
San Diego, CA 92101-4495
USA

ELSEVIER Ltd
The Boulevard, Langford Lane
Kidlington, Oxford OX5 1GB
UK

ELSEVIER Ltd
84 Theobalds Road
London WC1X 8RR
UK

1st. edition 2004

Library of Congress Cataloging in Publication Data
A catalog record is available from the Library of Congress.

British Library Cataloguing in Publication Data
A catalogue record is available from the British Library.

ISBN:0-12-752187-9
ISSN:0080-8784 (series)

♾ The paper used in this publication meets the requirements of ANSI/NISO Z39.48-1992 (Permanence of Paper). Printed in Great Britian

In memory of N.A. Gorjunova and B.T. Kolomiets,
who discovered chalcogenide vitreous semiconductors

Contents

List of Contributors

VICTOR S. MINAEV (1), *JSC "Elma", Research Institute of Material Science and Technology, Zelenograd, 124460, Moscow, Russia*

SERGEY P. TIMOSHENKOV (1), *Moscow Institute of Electronic Engineering (Technical University), Zelenograd, 124498, Moscow, Russia*

ANATOLIY POPOV (51), *Moscow Power Engineering Institute (Technical University), 14 Krasnokazarmennaya st., Moscow, 111250, Russia*

VALERY A. FUNTIKOV (97), *Kaliningrad State University, Universitetskaya Street, 2 Kaliningrad, 236040, Russia*

VICTOR S. MINAEV (139), *Kaliningrad State University, Universitetskaya Street, 2 Kaliningrad, 236040, Russia*

MIHAI POPESCU (181), *National Institute of Materials Physics, Str Atomistilor, 105 bis, P O Box MG7, Bucharest-Magurele (Ilfov), Romania*

OLEG I. SHPOTYUK (215), *Lviv Scientific Research Institute of Materials of SRC "Carat", 202, Stryjska Str., Lviv, UA-79031, Ukraine; Institute of Physics of Pedagogical University, 13/15, al. Armii Krajowej, Czestochowa, 42201, Poland*

List of Contributors

[illegible]

[illegible]

[illegible]

[illegible]

[illegible]

[illegible]

[illegible]

[illegible]

Preface

At the present, there are few individual or collective monographs written by Eastern Europe's scientists known to Western readers regarding the physical and structural-chemical phenomena observed in chalcogenide vitreous semiconductors (CVS), and the processes that take place under external influence.

This collective monograph, written by well-known East European scientists in the chalcogenide glass field, continues the tradition of Russian scientists from Ioffe's Physical and Technical Institute (St. Petersburg) who discovered the semiconductor properties of chalcogenide glass in 1955 and initiated fundamental research: chemist N.A. Gorjunova and physicist B.G. Kolomiets.

Chalcogenide glasses, and in particular, chalcogenide semiconductor glasses (CSG), are remarkable for their unique properties that are insignificant or even absent in crystal semiconductors: radiation resistance, reversible electric switching effect and memory, photo-structural transformation, an absence of impurities influence and synthesis from super pure materials, and simplicity of technology.

The book begins with a chapter that covers the problem of glass formation in chalcogenide systems. Existing criteria and concepts of glass formation are considered, and a systematic review of glass formation in binary chalcogenide systems of I-VII groups of the Periodic Table is presented. In addition, new (inversion) regularities in the periodic alteration of glass formation ability of binary and multi-component chalcogenide alloys are described, which open the possibility for forecasting glass formation in systems where glass formation is yet unknown.

Along with generally accepted concepts of atomic structure of glass and chalcogenide glass in particular, this monograph also considers the problems of glass structural modification under the influence of external effects, as well as the concept of eutectoid structure of glass. As a conclusion of the structural section, a new concept of polymeric polymorphous-crystalloid structure of glass is presented, proceeding from the work of prominent Russian investigators of glass, E.A. Poray-Koshits in particular. The new concept presented here combines different concepts of structure in a single consistent concept, and throughout the monograph, new views on glass structure are illustrated by examples of chalcogenide glass structures.

Chalcogenide glasses and their structure are very sensitive to external impacts, particularly photo and radiation impacts which can significantly alter the structure and properties of vitreous semiconductors. A separate chapter is devoted to each of these impacts.

Some problems discussed in this book are considered by authors from opposing positions, and different explanations are given to some processes in CVS. Photo-structural transformations, in particular, are explained by some authors by defect generation processes, and by other authors as structural transformation of polymorphous nature. The future will show which point of view is closer to the truth, or perhaps these different points of view will unite in a single strong system that will explain all the aspects of the structure of the glassy state.

This volume does not cover all problems connected with investigations of chalcogenide vitreous semiconductors. It is planned to publish future volumes describing various properties of CVS, their electronic phenomena, as well as a wide range of prospective applications for these materials.

In conclusion, we would like to express our gratitude to managers of JSC Elma ("Electronic Materials") and JSC Research Institute of Material Science and Technology for their help in collecting this group of authors, and in overcoming the technical obstacles that are inevitable in realizing these kind of projects.

V.S. Minaev

CHAPTER 1

GLASS-FORMATION IN CHALCOGENIDE SYSTEMS AND PERIODIC SYSTEM

V. S. Minaev

JSC "ELMA", RESEARCH INSTITUTE OF MATERIAL SCIENCE AND TECHNOLOGY, ZELENOGRAD, 124460 MOSCOW, RUSSIA

S. P. Timoshenkov

MOSCOW INSTITUTE OF ELECTRONIC ENGINEERING (TECHNICAL UNIVERSITY), ZELENOGRAD, 124498 MOSCOW, RUSSIA

1. Introduction

Existing theories, concepts, criterions, semi-empirical rules, and models of glass-formation can be divided into three groups: (1) structural–chemical, (2) kinetic, and (3) thermodynamic. As Uhlman (1977) noted, the differences between these groups are rather indistinct. Very often, concepts overlap from one group to another.

For example, Rawson (1967) did not distinguish the thermo dynamic group as separate from the others, regardless of variances in chemical bond energy and the energy of the system at crystallization (melting) temperature. It can be said that by doing so, Rawson has actually introduced the thermo dynamic (energetic) aspect in his structural–chemical criterion of glass-formation. At the same time, he has also stated that an acceptable theory of glass-formation cannot be created solely on the basis of one of the aspects.

Tammann (1935) was among the first scientists trying to characterize the glass-formation process, and his approach combined thermodynamic and kinetic descriptions of the process together with the first structural ideas related to glass structure and chemical bonding between constituent atoms. Even now, the harmonic combination of the most important elements of each of the three groups of theories[1] into a three-in-one concept and, in the ideal case, in a single, physically chemical founded integrated formula that can be applied to the prognosis of new chemically different glass-forming systems remains unresolved. Although, new studies (Chapter 4) indicate further improvement and mutual consolidation of the above-mentioned aspects of glass-formation. The application of systematic unification of these theories and conceptions to chalcogenide glasses is still waiting to be resolved, as well as its application to other glass groups and to glasses in general.

[1] Rawson (1967) said that many of these theories were too elementary and limited, and had not deserved the names 'theories'.

ISBN 0-12-752187-9
ISSN 0080-8784

In this chapter, we would like to consider glass-formation both from the standpoint of its physical–chemical essence, and from the standpoint of practical tasks in the field of creation of new chalcogenide glass-forming materials.

In this chapter, the body of study will be reviewed in conjunction with the problem of unification of all three main aspects of glass-formation—structural–chemical, energetic (thermodynamic), and kinetic—into a single concept. The consideration will be carried out in the most general, and at the same time, in a rather simplified form. However, discussion of some of the problems connected with the unification will not be possible here. One of the reasons for this lies in the fact that at present, although the thermodynamic and kinetic aspects are detailed sufficiently, the structural–chemical aspect of glass-formation is not as well defined. Chapter 4 of this collective work is devoted to the analysis and deeper understanding of structural–chemical features of glass-formation and glass structure.

The second standpoint of glass-formation presented in this chapter is connected with purely practical tasks.

Chalcogenide glasses are used in various fields of technology where their different properties are employed (discussed in corresponding chapters of this book). As the range of clearly defined properties of these materials becomes wider, it leads to a greater potential for their use in specific technical applications in devices, circuits, and systems.

According to the fundamental Kurnakov–Tananaev's rule of the physical–chemical analysis, as described by Tananaev (1972), a property of a substance is a function of its chemical composition, structure, and dispersivity. In this chapter, we would like to show to those who seek a foundation in expanding the range of certain properties of chalcogenide glasses, based upon the alteration of their chemical composition[2], i.e., to discover the location of chalcogenide glasses, the location of the main chemical elements that take part in glass-formation on the 'geographical map' of the Mendeleev's periodic system of elements.

Furthermore, we would like to demonstrate periodical regularities of glass-formation, considering them as the periodical property of elements that present the same type of chalcogenide systems 'chalcogen–non-chalcogen,' where elements are sequentially replaced with elements of the same subgroup of the periodic table with larger (or lesser) atomic numbers, changing correspondingly glass-forming ability (GFA) and properties of glass.

In seeking glasses with desired properties, it is extremely important to forecast new glass-forming compositions. In our opinion, there are two ways to solve the glass-formation prognosis problem in the absence of a unified concept of glass-formation that connects its structural–chemical, kinetic, and thermodynamic aspects.

The first method is related to revealing and using the above-mentioned periodical regularities of glass-formation. It allows the qualitative evaluation of GFA in simple (two- or three-component) systems. The second method—the quantitative determination of GFA—is more complicated. The task of this chapter is to advance along both these ways.

2. Main Regularity of Glass-Formation in Chalcogenide Systems and Its Infringements

Pioneers of glassy semiconductors, Goryunova and Kolomiets (1958, 1960) were the first to reveal the regularity that GFA, as determined by the size of the glass-formation

[2] Alterations of glass properties dependant on its structure are considered in Chapters 2 and 4.

region in two- and three-component chalcogenide alloys, decreases with replacing of one of the components of 4th (Ge, Sn), 5th (As, Sb, Bi), or 6th (S, Se, Te) main subgroups of the periodic table by an element with a greater atomic number.

The cause for such a decrease in GFA is the increase in the metallization degree of covalent bonds due to the increase in the element's atomic number.

Approximately, the same conclusion was also made by Hilton, Jones and Brau (1966), who compared regions of glass-formation in ternary systems and took them as a measure of GFA. Hilton lined up elements of VI, V, and IV groups with decreasing tendency of glass-formation: S > Se > Te, As > P > Sb, Si > Ge > Sn.

Based on the fact that up to 9 at.% B, 3 at.% Ga, and 1 at.% In can be added to vitreous arsenic selenides, Borisova (1972) came to the conclusion that GFA in the III group of the periodic table also decreases with the increase in atomic numbers of elements. The exclusion from the III group is thallium, with which significantly wider glass-formation regions were obtained in ternary systems with arsenic selenides and sulfides.

Despite the anomalous behavior of phosphorus and thallium, as well as some other elements which will be discussed later, a decrease in GFA in alloys with progressively higher atomic numbers among components of main subgroups of the periodic table is one of the main regularities of glass-formation in chalcogenide systems. Therefore, to predict qualitatively the relative GFA of glasses in a given system with an unknown glass-formation region (a ternary system, for instance), one should consider GFA expressed as the size of the glass-formation region in other systems of the same type where one of elements of the system under investigation is sequentially replaced by elements of the same subgroup with larger or lesser atomic numbers, when these glass-formation regions are known in the systems.

Unfortunately, the matter turns out to be more complicated in practice. Comparisons even in similar binary systems have revealed several violations in the projected regular decrease of the GFA with higher atomic numbers, and this led to additional research to determine the root cause of such variances, as well as to seek additional periodical regularities in binary chalcogenide systems. Such works were carried out by Minaev (1977–1979, 1980a,b, 1985a,b, 1991) in the late 1970s and 1980s and remained practically unknown to foreign readers.

In these works, the author managed to reveal the inversion nature of glass-formation in binary chalcogenide systems for several individual elements and even groups of elements of the periodic table connected with the secondary periodicity of elemental properties. These new regularities consistently violate earlier discovered regularities, and they are connected with the increased atomic number, as described below.

But even new regularities, giving a general picture of the glass-formation of chalcogenides in the periodic table, contain only qualitative agreement.

During the 1980s, the problem of quantitative determination of the GFA (Minaev, 1980a) was set. To solve the problem, it is useful to apply the experience of investigators who developed various theories and concepts of glass-formation and who analyzed various factors of the glass formation.

3. Criteria of Glass-Formation. Factors Affecting Glass-Formation

An analysis of structural–chemical concepts of glass-formation (Frankenheim's crystallite concept (1835), Lebedev's concept (1921, 1924), Zachariazen's disorder network (1932), the kinetic theory of glass-formation of Stavely, Turnbull, Cohen (1952, 1961) reviewed in detail by Rawson (1967)) does not suggest the possibility of a concrete quantitative prediction method for the GFA of substances.

Let us also consider the various standpoints regarding glass-formation and related effects as they were known in the 1980s.

Goldschmidt (1926) proposed empirical criteria for glass-formation, in which the ratio of the radii of cations and anions in glass-forming oxides lies in the region 0.2–0.4, typical for anion locations in vertexes of tetrahedrons.

Based on the concept of disordered locations of atoms, which must remain unchanged after cooling of melt and formation of glass that is incompatible with exact data of lengths and angles, Smekal (1951) suggested the idea that the presence of 'composed' chemical bonds is necessary for glass formation.

Stanworth (1952) has shown in oxides there is a correlation between the tendency of glass-formation and the degree of ionicity or covalencity of the bond. The quantitative expression from Stanworth has used values of electronegativities by Pauling (1970). The differences in electronegativities of elements and the degree of ionicity (covalencity) of adjustment bonds can be evaluated. Based on his criteria, Stanworth (1952) predicted the existence of tellurite (TeO_2-based) glasses.

In works of Myuller (1940) and his followers (Myuller, Baydakov and Borisova, 1962), detailed investigations of chemical bonds in glasses and glass-forming liquids began. In these works, they wrote that the type of the main structural unit and the nature of the chemical bond were of great significance in the formation of the glassy state. The disposition of certain substances towards glass-formation was connected by Myuller to the predominance of directional bonds with the reduced radius of action, which in the first turn were powerful covalent bonds. Important roles are played by valences of elements that determine trigonal and tetrahedral configurations of chemical bonding. Covalent bonds in the atomic network at moderate temperatures cause a reduction of the vibrational amplitude of atoms, when compared to the vibrational amplitude of ions in the ionic lattice. In Myuller's opinion, the cause of high viscosity and the increased activation energy of the atomic re-grouping, as observed in substances disposed towards the glass-formation, lie in this difference.

As for chalcogenide glasses, Leningrad's scientists Kolomiets and Goryunova (1955a, b), Myuller (1965), Kokorina (1971) and Borisova (1972) connected the glass-formation in chalcogenide systems with elements of main subgroups of III–V groups of the periodic table as having a predominance of directional localized bonds from shared electron pairs—covalent bonds in which the portion of ionicity determined from electronegativities of elements is in the range 3–10%.

Subsequent investigations have shown that glasses are formed in systems that have a more significant portion of ionicity of chemical bonds as well. For example, in investigations of the Cs–Te system's equilibrium diagram by Chuntonov, Kuznetsov, Fedorov and Yatsenko (1982) and the Cs–Se system's equilibrium diagram by Fedorov,

Chuntonov, Kuznetsov, Bolshakova and Yatsenko (1985) revealed that the equilibrium in these systems is established with difficulty due to their disposition to glass-formation. In the system $Cs_2S-Sb_2S_3$, the glass-formation region includes 100% Cs_2S composition as well (Salov et al., 1971).

These data indicate that glass-formation can be characterized not only by 'pure' covalent (S–S, Se–Se) or predominant covalent (As–S, P–Se) bonds, but also by covalent–ion bonds with the ionicity degree equal to ≈55% for Cs–S or equal to ≈40% for Cs–Te, judging by the dependence of the bond's ionicity degree on the difference of electronegativities of elements forming the chemical bond, as established by Pauling (1970). To compare with oxide glass-forming systems, it should be noted that in such glass formers as B_2O_3 and SiO_2, ion portions of chemical bonds can be evaluated according to Pauling (1970) as ≈45 and ≈51%, respectively.

Even greater ionicity is possessed by halide glasses, for example the glass former BeF_2, in which as Rawson (1967) indicates, the bond Be–F is presented by the approximately 80% ion component.

Thus, the concept of the exclusive role of the covalent bond in glass-formation in considered systems must be revised. For glass-forming chalcogenide systems, the covalent–ionic chemical bond is, as a rule, typical with the predominant role of the covalent component. There are some exceptions, however: glasses of the Cs–S system. Only in glass-forming chalcogens (sulfur and selenium) are chemical bonds 100% covalent, and in chalcogen glasses of the S–Se system they possess some ionic components (the electronegativity of sulfur is 2.5, selenium 2.4). It must also be remembered that chalcogen's chains in chalcogen and chalcogenide glasses are interconnected by van der Vaals bonds.

So, the most generalized point of view of Smekal concerning necessity of the presence of 'composite' bonds for glass-formation is completely applicable to chalcogenide glasses as well.

The most important feature of glass-formation is the polymerization of structural fragments of which the glass is built. The polymeric structure of glass was revealed in the second half of the 19th century when Mendeleev (1864) stated that 'the glass structure is polymeric.' This concept has been given new practical and theoretical confirmations in works of Sosman (1927), Zachariasen (1932), Kobeko (1952), Tarasov (1953) and Myuller (1960, 1965). The polymerization of glasses is the most important part of the polymeric-crystallite concept of the glass structure generalized by Poray-Koshits (1959).

The necessary and sufficient condition of glass-formation is considered by Kokorina (1971) as follows:

- the presence of localized paired electrons bonds in the structure;
- the construction of the main polymeric network from endless polymeric complexes;
- the connection of structural complexes only through a single bridge bond, i.e., the presence of bonds in the structure that can be called swivel bonds.

Winter (1955), in his turn, connected the GFA with the number of p-electrons in the external atom shell per one atom. The p-electrons criterion of Winter concludes in the fact that for glass-formation, the most favorable number of p-electrons per atom is four. The minimum number of p-electrons for glass-formation is two.

Sun's criterion (1947) of the bond strength is based on the idea that the stronger the bonds between atoms, the easier is the glass-formation. Actually, the intensity of the process of atomic re-grouping during crystallization of material, which is accompanied by rupture of individual bonds and formation of new inter-atomic bonds, is dependant on the strength of bonds. Therefore, the glass-formation ability is connected with the increased strength of chemical bonds. The strength of the chemical bond 'metal–oxide' is determined by dividing the oxide dissociation energy by the number of oxygen atoms surrounding the atom in the crystal or glass, i.e., by the coordination number (CN).

Rawson (1956) has modified the Sun's criterion through the introduction of the component taking into account 'the liquidus temperature effect' at glass-formation. Rawson has connected the glass-formation process not only with the bond strength, but also with the thermal energy that is present in the system and required for the bond rupture. The measure of this energy is the melting temperature (for an elementary substance or a compound) or the liquidus temperature (for a multi-component system) in Kelvin degrees. The Rawson's glass-formation criterion is the ratio of the bond strength to the melting temperature. The criterion allows for a sharper frontier between glass-forming and non-glass-forming oxides. Rawson has not applied his criterion to multi-component glasses, although he showed that many systems exist where no component forms glasses, but in two-component systems in the range of low liquidus temperature glasses are formed. The liquidus temperature effect explains also existence of conditional glass formers, which in principle can form glasses, but only at conditions that are more favorable. Such a condition is the reduction of the liquidus temperature due to the presence of the second oxide and, therefore, the reduction of the thermal energy facilitating the glass-formation owing to its insufficiency for rupturing of existing bonds and forming other bonds in the process of the atom re-grouping leading to crystallization.

In connection with this, Rawson indicates that investigations of phase diagrams make the understanding of glass-formation processes in two- and ternary-component systems significantly easier.

Even before Rawson, Kumanin and Mukhin (1947) came to almost the same conclusions, but from another position, crystallization tendency: in glass-forming systems, in the region of crystallization of a certain chemical compound (in general, for compounds with congruent melting), there is a progressive reduction in the crystallization tendency of glasses when their compositions are moved away from the compound composition (i.e., with the liquidus temperature reduction—V.M.). The crystallization tendency reaches the minimum in regions of the cooperative crystallization of this compound together with compounds of other chemical compositions.

The generally accepted physical–chemical factor of glass-formation (beginning from Tammann (1903, 1935), then Kumanin and Mukhin (1947), Rawson (1956) and others) is the presence of low-temperature eutectic points on phase diagrams. In eutectic points, the action of the liquidus temperature effect, proposed by Rawson, usually becomes the most apparent.

The problem of connection of the glass-formation process with the phase diagram appearance was described in the work of Dembovsky (1978) by the example of chalcogenide systems.

Turnbull and Cohen (1959) have suggested the evaluation criterion of disposition to glass-formation with the reduced thermodynamical crystallization temperature

$$\theta_c = kT_c/h \tag{1}$$

where k is the Boltsman's constant, T_c the equilibrium crystallization temperature and h the evaporation thermal energy per molecule or a kinetic unit (it characterizes the bond strength in a substance). In substances of the same type θ_c is lesser for greater dispositions to glass formation.

Evaluation of the glass-formation ability by the differential-thermal analysis method was proposed in the work of Hruby (1972)

$$\text{GFA} = \frac{T_c - T_g}{T_m - T_c} \tag{2}$$

where T_c, T_g, and T_m are the temperatures of glass crystallization, glass-transition, and glass melting, respectively.

In the last two cases, the prediction capabilities of the criteria are limited by the necessity to obtain experimental data for already synthesized alloys.

From the point of view of Funtikov (1987) and his electronic configuration model, the disposition to glass-formation and the properties of chalcogenide glasses depend on features of electronic configurations of initial atoms. He analyzed the maximum content X of various elements in glass-forming alloys of the As–Se–X systems and concluded that the GFA depends periodically on their atomic number. At that for elements of III row (3rd period), V row (4th period), and VII row (5th period) the character of the dependence is the same and has the minimum for elements of III group and the maximum for VI group (Fig. 1). The creation of GFA is greatly influenced by stable electronic configurations d^0, d^{10}, f^0, and f^{14}.

Dembovsky (1977) and Dembovsky and Ilizarov (1978) introduced the number of valence electrons (VE) of an element in the formula for GFA that they derived in the framework of the empiric theory of glass-formation in chalcogenide glasses

$$\text{GFA} = \gamma(A + E)(\text{VE} - K)/2 \tag{3}$$

where $\gamma = \sum_i T_i X_i / T_{\text{liq}}$, A is the number of atoms of different types, E the number of structural nodes, K the CN, T_i the melting point of the i component, X_i the mole fraction of the i component, and T_{liq} the liquidus temperature of the alloy.

The GFA value for glass-forming alloys is 4.0 ± 1.0. Differences between values calculated by the author of the empiric theory in accordance with this formula and experimentally determined areas of the glass-formation of 20 binary and ternary chalcogenide systems are in the region 5–10 at.%.

We fail to find data, based on this formula, concerning predictions of new glass-forming systems or new regions of glass-formation in known systems.

In some of above-mentioned concepts and criteria, the CN of atoms constituent in the composition of discussed glasses is present in explicit or implicit form (the Zakhariasen concept, criteria of Sun, Sun–Rawson, the theory of Dembovsky, the criterion of Winter).

According to Ovshinsky (1976), the important parameter that determines stability of non-crystalline materials and total constraint in them is the covalent connectivity of their

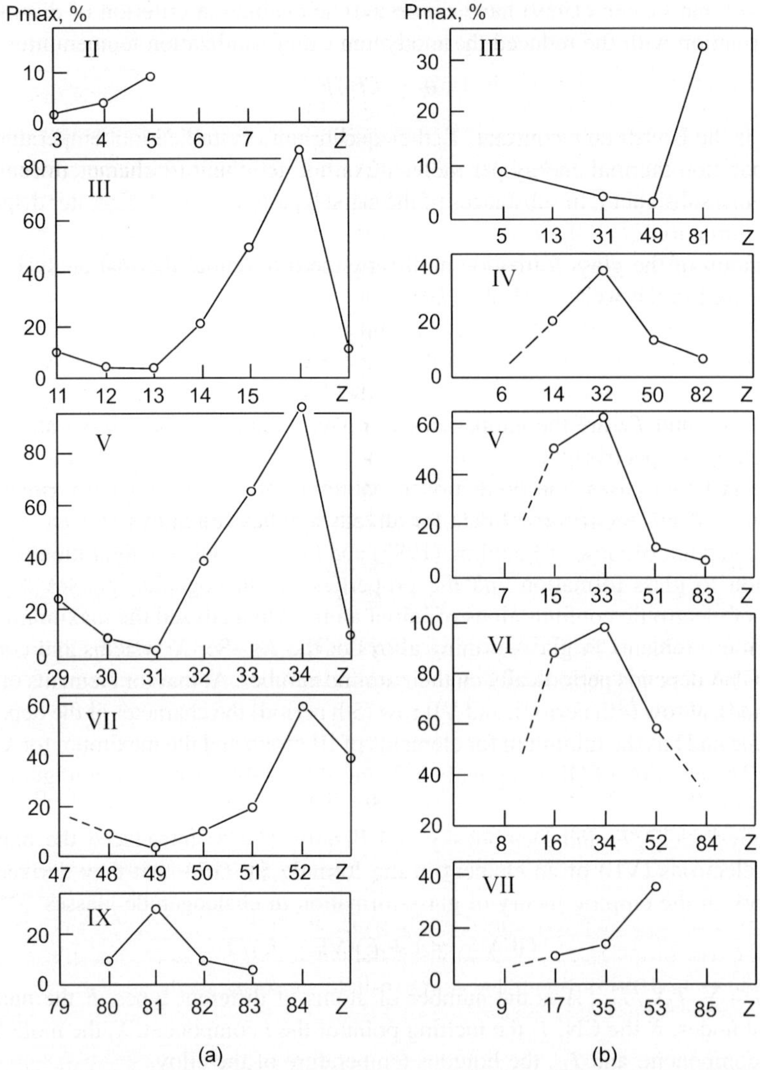

FIG. 1. The dependence of the maximal content P_{max} of elements (X) in the glass-forming system X–As–Se upon their atomic number Z in the periodic table (Funtikov, 1987): (a)—in rows (II, III, V, VII, IX); (b)—in groups (III, IV, V, VI, VII) of the periodic table.

atomic network. The connectivity is determined by the number of neighboring atoms with which the 'average' atom has covalent bonds, or the average covalent coordination number (CCN). Most of the atoms, constituent non-crystalline semiconductors, are located in IV, V, VI, and VII groups of the periodic table, and in accordance with the rule '8-*N*', where *N* is the number of the group, have valences and CCN equal to 4, 3, 2, and 1, respectively.

Boolchand, Bresser, Georgiev, Wang and Wells (2001) in their works have clearly shown the role and the influence of the CN and the connectivity of substances on their various properties, comparing germanium and selenium that have close values of the chemical bonds strength (according to Pauling (1970)): Ge–Ge, 37.6 Kcal mol^{-1} and Se–Se, 44 Kcal mol^{-1}; for germanium CN = 4 and for selenium CN = 2. As a result, thermal (the melting temperature and the heat of fusion), elastic (the Young's modulus), and plastic (the hardness) behavior of crystalline germanium (CN = 4) strikingly differs from that of trigonal selenium (CN = 2) due to the significant difference of the connectivity in them. In non-crystalline substances, for example in the binary glass-forming system Ge_xSe_{1-x}, the network connectivity or the average CN $r = 2(1 + x)$ are continuously changed depending on the composition-causing changes of the glass- transition temperature $T_g(x)$ and bulk elastic constants, which are progressively raised with increasing r and the degree of the cross-linking.

Phillips (1979) has proposed the idea of a correlation between the alloy's GFA, its average CN r, and the number of mechanical-bonding constraints which each atom undergoes as the result of the action of inter-atomic forces in accordance with the model of the valence-force field, and the number of degrees of freedom per atom. Having carried out some calculations and simplifications connected with the correlation of the bond-stretching α and the bond-bending β interactions in binary alloys, Phillips has concluded that

$$n_c = 1/2r^2 \tag{4}$$

The optimum value of the GFA, according to Phillips, should correspond to the situation where the number of mechanical-bonding constraints is equal to the number of degrees of freedom per atom

$$n_c = n_d \tag{5}$$

For systems in the 3D space, $n_d = 3$. It means that the most favorable average CN for glass-formation should be

$$r = \surd 2nc = \surd 2{\cdot}3 \approx 2.45 \tag{6}$$

Phillips' idea has been developed in the work of Thorpe (1983) who has come to the conclusion that the average CN describing the constraint-free network with the optimal GFA is the so-called Phillips–Thorpe mean-field rigidity threshold

$$r_c = 2.4 \tag{7}$$

Above this threshold, which was corrected by Boolchand and Thorpe (1994) and Boolchand et al. (2001), there is the stressed rigid phase; below the threshold is the intermediate unstressed rigid phase, and then the floppy phase.

4. Structural–Energetic Concept of Glass-Formation in Chalcogenide Systems

Predictability of some phenomenon or fact is always connected with the problem of preliminary establishment of main regularities leading to the origin of such phenomenon or fact. Glass-formation phenomenon is not an exception to the rule. However, an examination of existing publications on theories and practices of glass-formation (Section 3) does not suggest any rules that could help determine, at least approximately, glass-formation regions in unexplored two-, three- and more component systems. Attentive consideration of theories (criteria) of Goldshmidt, Zachariasen, Lebedev, Smekal, Steanworth, Winter, Phillips, and others show that all of them explain glass-formation to one extent or another, and formulate in more or less generalized form the conditions in which glass is formed, but they do not provide a guiding thread for the prediction of glass-formation regions in concrete chalcogenide systems or other systems. Moreover, they do not give even approximate coordinates where would-be regions could be searched for.

One exception appears to be the 'empiric glass-formation theory' of Dembovsky and Ilizarov (1978), when considered with the above-mentioned formula of GFA. However, neither the authors themselves nor others have used it for prediction of new glass-forming systems. It is possible that the explanation lies in the rather complicated formula and the potentially ambiguous qualitative interpretation of 'structural nodes' in multi-component compositions.

The second exception is, of course, the Sun–Rawson criterion (Rawson, 1956, 1967), which is applicable (and being applied!) for calculations of the GFA of individual oxides.

In accordance with three groups of theories explaining causes of glass-formation (theories emphasizing peculiarities of the structure, theories considering kinetics of the liquid crystallization, and theories paying attention to thermodynamical aspects of the glass-formation), two main factors of the potential glass-formation can be distinguished following Rawson (1958, 1967): the structural–chemical factor, considering mutual locations of atoms and the strength of chemical bonds, and the energetic factor whose measure is the liquidus (melting) temperature. The third factor—kinetical—operates only in a state when the first two factors create within the substance, conditions that are suitable for the glass-formation phenomenon to originate. The kinetical factor is the factor of the practical glass-formation. Its usage allows obtaining glass-formation regions different in size, depending on kinetics (the cooling rate of melt).

Structural–chemical and energetic factors of glass-formation, together with the condition of the relaxation of GFA with increase in atomic numbers of elements (Goryunova and Kolomiets, 1958, 1960), have been considered as a starting point for the development of the structural–energetic concept of glass-formation presented below, which has allowed the prediction of the existence of glasses in scores of chalcogenide systems, and to experimentally confirm the existence of semiconductor glasses in more than 20 of them (Minaev, 1980a, 1991).

When transitioning from theoretical consideration to the practice of glassmaking, the structural–energetic concept could not exclude from consideration the kinetic factor as well.

4.1. Glass-Formation and Phase Diagrams of Chalcogenide Glasses

The glass-formation in binary- and ternary- chalcogenide systems is directly connected with structures of corresponding phase diagrams.

An analysis of more than 60 phase diagrams of binary chalcogenide systems and data on glass-formation in these systems (Minaev, 1979, 1980c, 1981a, 1982a, 1985a, 1987b, 1988, 1991) allows a classification of the diagrams into four types based on the likelihood of obtaining glasses from their corresponding systems (Minaev, 1982a).

Glass-forming phase diagrams of binary systems are usually diagrams with low-temperature eutectics in the range adjoining chalcogen (Fig. 2-(1)). Such systems are Al–Te, Ge–Se, Si–Te, As–S, P–Se, Cs–Te, and others.

Also glass-forming diagrams are diagrams with the phase segregation in the region, adjoining the chalcogen, and a rather low-temperature eutectic, neighboring this region (Fig. 2-(2)). Such systems are Cs–S, K–Se, Tl–S, Tl–Se, Sb–S, and others.

Glass-forming diagrams are also diagrams of two-chalcogen systems. Diagrams S–Se and S–Te are of the eutectic type, and the diagram Se–Te is characterized by a continuous sequence of solid solutions (Vinogradova, 1984).

Non-glass-forming diagrams are diagrams with a sharp rise of the liquidus temperature in the range closely adjoining the chalcogen (Fig. 2-(3)), very often followed by the phase segregation (Fig. 2-(4)).

An additional analysis of glass-formation in more than 100 ternary chalcogenide systems (Minaev, 1982a, 1987a, 1991) has shown a genetic relation of glass-formation regions depicted with the phase diagrams of the glass-formation type of binary systems contained in these ternary systems. It was for this reason that this simple classification of binary glass-forming and non-glass-forming phase diagrams was considered as a foundation for the classification of ternary glass-forming systems as well (Minaev, 1982a, 1991). The main classification factor here is the above-mentioned composition of phase diagrams.

Glasses in ternary systems are formed, as a rule, when among the participating binary systems there are one, two, or all three systems that are characterized by glass-formation phase diagrams.

This simple rule, without any additional data, allows the prediction of a possibility of glass-formation in any ternary system if the phase diagrams of the binary systems constituent in the ternary system are known.

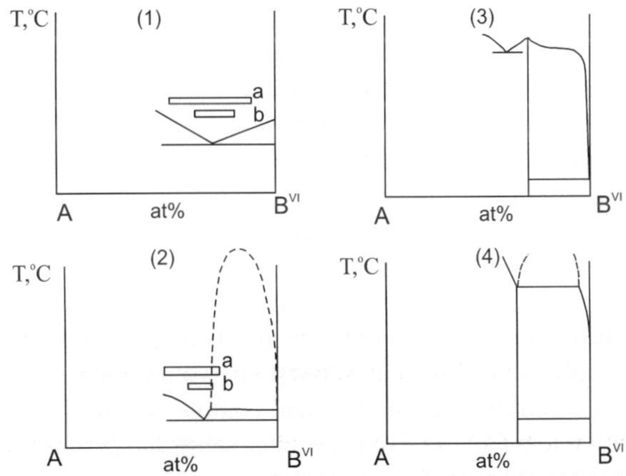

FIG. 2. Types of phase diagrams of binary chalcogenide systems (Minaev, 1982): (1) the glassforming type with the chalcogens-enriched eutectic; (2) the glassforming eutectic type with the phase liquation in the chalcogens-enriched region; (3) the non-glass-forming type with the sharp liquidus rise in the chalcogens-enriched region; (4) the same as 3 but with the phase liquation; (a) the Glass-formation region at the quick quenching of melt; (b) the glass-formation region at the slow cooling of melt.

The glass-formation in ternary systems is connected, as a rule, with phase diagrams of systems characterized by regions with reduced liquidus temperatures, which are usually expressed by the presence of binary and ternary eutectics.

The above mentioned correlates with the liquidus temperature effect (Rawson, 1967), the ‘eutectic’ idea (Tammann, 1903; Lebedev, 1910), as well as with the opinion of Dembovsky and Ilizarov (1978) concerning the relation of glass-formation with the phase diagram appearance.

4.2. Qualitative Criterion of Glass-Formation

Since the liquidus temperature effect of Rawson (1967) has extremes at points corresponding to chemical compounds (excluding chemical compounds melting with peritectic reactions) and eutectic points, it is natural to expect that chemical compounds possess the least GFA (and eutectics, the most), both in binary- and multi-component chalcogenide systems.

When phase diagrams can be obtained, it is easy to find out points where the process of glass-formation is more probable or less probable. Phase diagrams of many binary systems are known. Phase diagrams of ternary systems with chalcogen elements are significantly less well known, and the search for glass-formation regions in these systems is rather difficult.

A further consideration of this concept based on the liquidus temperature effect leads one to say that after eutectic points, the most probable locations for glass-formation in multi-component systems are in curves of phase diagrams connecting points of binary and ternary, ternary and tetradic, etc., eutectic compositions. For ternary systems, it will be in monovariant curves connecting binary and ternary eutectics (Minaev, 1980a). This thesis has something in common with the thesis first established by Kumanin and Mukhin (1947) and then developed by Mukhin and Gutkina (1960): the crystallization ability reaches a minimum in regions of the cooperative crystallization of the given compound with compounds of different chemical composition.

When seeking the approximate determination of glass-forming region locations, Minaev (1977) and Minaev et al. (1978) proposed to replace the use of curves with so-called lines of dilution (DL) of binary, ternary, etc., eutectics by the third, the fourth, etc., components. The dilution line of binary eutectic (DLBE) in a ternary system is the line connecting the eutectic point of the binary system with the vertex of the concentration triangle, corresponding to 100% of the third component content.

The more exact analog of the line going from the binary to the ternary eutectic is the line of dilution of the binary eutectic, not by the third component of total system, but by a particular eutectic subsystem. Such a subsystem would limit telluride systems, for example, by tellurium and the nearest chemical compounds of binary telluride systems. For example, in the system Ge–As–Te such subsystems are systems Te–As_2Te_3–GeTe and As–As_2Te_3–$GeAs_2$ (Fig. 3).

As seen in Figure 3, both glass-formation regions are located along DLBE of the common system e_3(Te–GeTe)–As, e_1(Te–As_2Te_3)–Ge, and e_2(As–As_2Te_3)–Ge as well as along DLBE of particular systems, for example, e_3(Te–GeTe)–As_2Te_3 and e_1(Te–As_2Te_3)–GeTe.

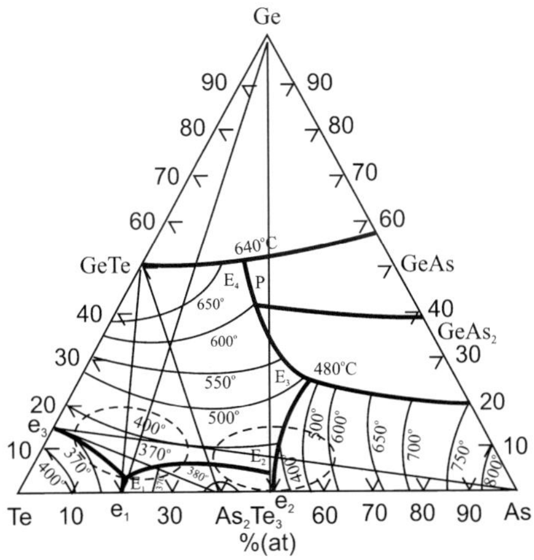

FIG. 3. The projection of the liquidus surface on the concentration triangle, glass-formation regions (the dotted line) and dilution lines of binary eutectics (thin straight lines) in the system Ga–As–Te (Minaev, 1991). e_1, e_2, e_3—binary eutectics; E_1, E_2, E_3, E_4—ternary eutectics.

Naturally, the largest probability of the glass-formation as well as the greatest glass-formation ability will be typical for cross points of DLBE located sufficiently close (in average 2–3 at.% for chalcogenide systems) to ternary eutectics.

Calculations carried out as well as the conformity of DLBE and glass-formation regions of all known ternary systems indicate the propriety of replacement of monovariant curves by dilution lines of binary eutectics aiming to determine coordinates to search locations of real regions of the possible glass-formation.

Thus, one can formulate the qualitative criterion characterizing location of glass-formation areas: glass-formation areas of ternary chalcogenide systems are usually located near lines of dilution of binary eutectics by the third component.

The application of the qualitative criterion of glass-formation (with additional verification based on the quantitative criterion discussed in the next paragraph) has allowed the prediction of glass-formation in several hundreds of ternary chalcogenide systems and, in particular, in several scores of ternary telluride systems based on elements of I^A, I^B, II^B, III^A, IV^A, V^A, VII^A subgroups of the periodic table. The partial experimental verification of this prediction, carried out by Minaev (1980b, 1983), has shown that synthesis of materials at 1000 °C in rotary evacuated (10^{-4} mm Hg) quartz ampoules during 12 h (weight of 10 g) and quenching them in cold water (cooling rate of 10–20 °C s^{-1}) gives the possibility to obtain glasses in 24 new systems: Cu–Si–Te, Cu–Ge–Te, Ga–Si–Te, Ga–Ge–Te, Ga–Pb–Te, Ga–As–Te, In–Si–Te, In–Ge–Te, In–As–Te, Tl–Si–Te, Tl–Ge–Te, Si–Ge–Te, Si–Sn–Te, Si–Pb–Te, Si–Sb–Te, Ge–Pb–Te, Ge–Sb–Te, Sn–As–Te, Pb–As–Te, Ge–Pb–Te, Al–Si–Te, Al–Ge–Te, Al–Pb–Te, and Al–As–Te. Then, a glass-formation region was revealed in the Ga–Tl–Te (Minaev et al., 1968).

Glass-formation regions of above 15 systems are shown in Figure 4.

The application of the qualitative criterion of glass-formation can be expanded for all glass-forming systems in general: halogenide, metallic, oxides, etc. This criterion is likely to be correct for multi-component systems as well.

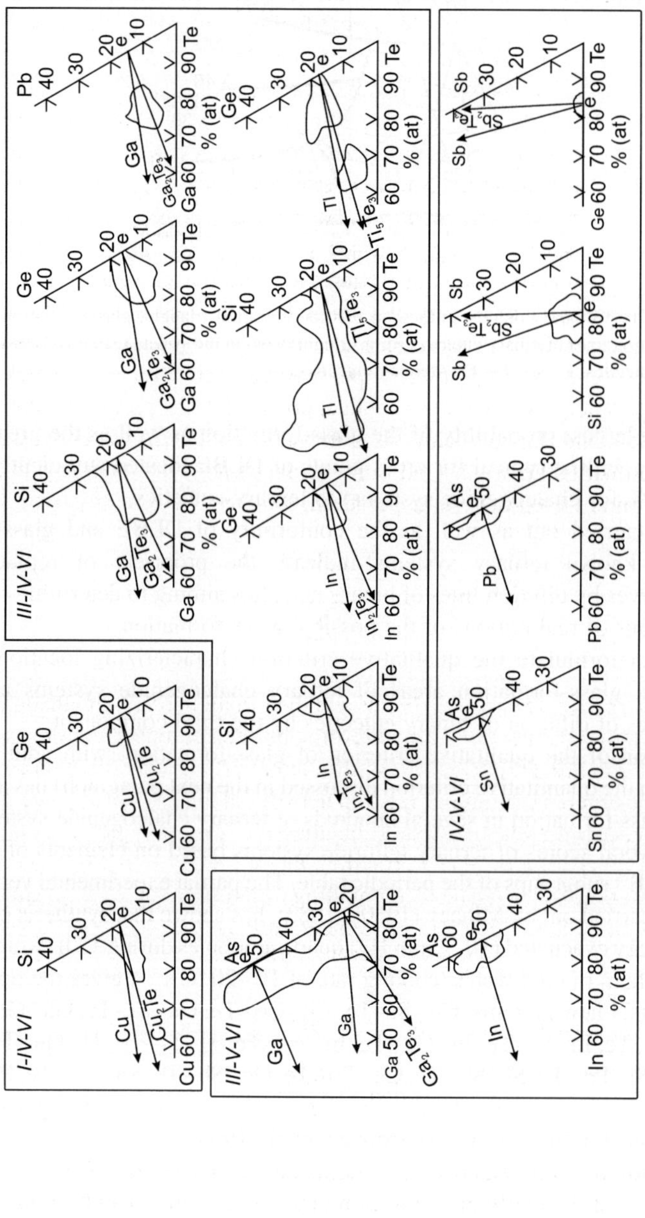

FIG. 4. Glass-formation regions of ternary telluride systems (Minaev, 1991). Arrows denote dilution lines of binary eutectics (DLBE); the element's symbol means the affiliation of the DLBE with the common system; the chemical compound's symbol means the affiliation of the DLBE with the particular system.

4.3. Quantitative Criterion of Glass-Formation

Rawson (1956, 1967) proposed to use as a glass-formation criterion for oxides, the ratio of the energy (the strength) of bonds ($E_{\mathrm{Me-O}}$) to the melting temperature expressed in Kelvin degrees. Taking into account that most of the general features of the glass-formation process are the same in all glass-forming compositions, it has been decided by Minaev (1977, 1978, 1980b) to modify the Rawson's criterion with the aim of using it for multi-component glasses, chalcogenide glasses in particular.

The Rawson's criterion concerns simple chemical compositions—oxides. It is a quotient of the oxide's bond energy and its melting temperature. To extend the approach of Sun and Rawson to multi-component compositions, Minaev has introduced the following 'corrections' in the Sun–Rawson's criterion. Instead of oxide's melting temperature (the denominator in the Sun–Rawson equation), the liquidus temperature has been taken for multi-component alloys, i.e., actually Rawson's idea of the liquidus temperature effect, which he used for qualitative evaluation of glass-formation in complex systems, was used for calculations. Instead of the energy of a single bond, used as a numerator in the Sun–Rawson's criterion for individual oxides, the energy of chemical or, more exactly, covalence–ion binding (CIB) of substance per one averaged atom has been taken, i.e., the sum of products of energies of certain chemical bonds (E_i), the portion of atoms bounded by such bond (M_i), and the half-value of their valence CN (K_i) (actually, each atom is chemically bound with other atoms and, since each chemical bond belongs to two atoms, in order to determine the energy value per atom, it must be divided by two)

$$E_{\mathrm{CIB}} = \frac{\sum_i E_i M_i (K_i/2)}{\sum_i M_i} \tag{8}$$

The presence of concrete chemical bonds and their quantitative ratio are determined by the manner of atomic connection, structure of substance, which therefore is one of the main basis of the modified criterion. Glass structure, in the general case, is characterized by the chemically ordered continuous random network of atoms consolidated by chemical bonds in accordance with valence CNs K dictating the 'chemical order' in the network.

As the modified criterion of glass-formation, the Sun–Rawson–Minaev criterion, the value equal to the ratio of covalence–ion binding of atoms in the multi-component alloy to its melting temperature (the liquidus temperature T_{liq}) in Kelvin degrees has been taken: $E_{\mathrm{CIB}}/T_{\mathrm{liq}}$. This ratio determines, according to Minaev (1978, 1980a), the GFA of a substance.

Thus, the glass-formation ability can be expressed by the formula

$$\mathrm{GFA} = \frac{E_{\mathrm{CIB}}}{T_{\mathrm{liq}}} \tag{9}$$

E_{CIB} can be calculated based on values of chemical bond energies presented by Pauling (1970) or calculated in accordance with the Pauling's formula using his data on interatomic bond energies inside each element as well as in accordance with updated data collected in the Batsanov's monograph (2000).

Pauling's formula

$$E_{A-B} = 1/2(E_{A-A} + E_{B-B}) + 100(X_A - X_B)^2 - 6.5(X_A - X_B)^4 \tag{10}$$

where E_{A-B}, E_{A-A}, and E_{B-B} are the energies of bonds between atoms A and B, A and A, and B and B, respectively; X_A and X_B are the electronegativities of A and B atoms (also according to Pauling).

Usage of a common calculation method of the energy of heterogeneous chemical bonds for all glass-forming systems (according to Pauling's formula) is due to the fact that use of various experimental and calculation methods gives different values for the same bonds E, kJ mol^{-1}:

System	Myuller (1965)	Ioh and Kokorina (1961)	Pauling (1970)
As–S	255.4	202.6	224.4
As–Se	217.7	159.1	174.8
Ge–S	—	230.3	259.0
Ge–Se	234.5	180.0	205.7

The notion 'criterion', in Minaev's opinion (1978, 1980a), is used here for comparison of the internal essential quality of substance—its glass-formation ability—with relation to other substances possessing this quality to a greater or lesser extent. Actually, one compares the correlation between the chemical bond (binding) and the melting (liquidus) temperature, which is, according to Rawson (1956, 1960), the measure of existing thermal energy necessary for rupturing of chemical bonds taking place in the process of atomic re-grouping during crystallization of material's melt.

Both the Sun–Rawson criterion and its modification by Minaev are constructed based on the consolidation of two approaches to the glass-formation problem. The first one is structural–chemical (the CN, the chemical bond), the second is energetic (the bond energy, the thermal energy of substance at crystallization).

In characterizing the concrete glass-formation in a certain system, i.e., the size of the glass-formation region dependent on concrete alloy's cooling conditions, neither in the Sun–Rawson criterion nor in its modification—the Sun–Rawson–Minaev criterion (the SRM criterion), is there any factor characterizing the third approach, kinetic, to describe the alloy's cooling rate on which the size of the concrete glass-formation range is dependent. Both the Sun–Rawson criterion and the SRM criterion remain constant in value for each certain composition. Both at cooling of this alloy at a rate higher than the critical one (V_{cr}), when glass is formed, and at cooling of it at a rate lower than the critical one, when the melt is crystallized, the value E_{CIB}/T_{liq} remains constant, i.e., both criteria are not that of the concrete glass formation.

What are the Sun–Rawson and the SRM criteria in this case? From Minaev's point of view (Minaev, 1978, 1980a, 1991) these criteria are the measure of the glass-formation ability as the physical–chemical essence of a substance is independent of conditions of the concrete glass-formation or crystallization. Glass-formation ability does not depend on the cooling rate or on the intensity of other external factors (pressure, electromagnetic radiation, etc.). It is the property that is inherent (or not) to a substance and is determined by its physical–chemical nature. As an illustrative example, GFA can be compared with the importance of soil fertility in farmland as a factor in overall crop production. Fertile farm soil provides a rich harvest under warm weather and sufficient rains (compare: a large region of glass-formation at the super-cooling quenching), a moderate harvest under rather unfavorable weather (compare: lesser glass-formation region at V_{cr}), and finally, the failure of crops under drought (compare: at $V < V_{cr}$ glass is not formed, the alloy is crystallized!). And all these outcomes occur with the same soil fertility (the same glass-formation ability).

The criterion SRM simultaneously reflects both structural–chemical and energetic approaches to the glass-formation problem and GFA, and clearly shows that in some cases insufficiency of application of only one of them, for example 'the effect of liquidus temperature' (Rawson, 1967) and 'the eutectic law' (Cornet, 1976). It will be shown later in this chapter (Section 4.6) that in Ga–Te and As–Te, some compositions with higher liquidus temperatures are distinguished with greater glass-formation ability than those of neighboring compositions that have lower T_{liq} and eutectic alloys.

For similar cases, the following 'rule of a gentle sloping liquidus' can be formulated, which is a significant addition and correction to the liquidus temperature effect and the eutectic law as well as to Kumanin–Mukhin's rule.

In systems in which the glass-formation region is located near the eutectic or includes the eutectic and expands in the direction to the chemical composition connected with the eutectic by the gentle sloping liquidus curve, the GFA can increase (and the crystallization ability can decrease) at motion from the eutectic to this chemical composition if the covalent–ion binding of alloys increases in this direction to a greater extent than the thermal energy of the system, of which the rate of increase is determined by the steepness of the liquidus curve.

The criterion SRM (GFA) can also be used as a criterion of the concrete glass-formation when taking into account an additional factor that reflects concrete conditions of glass-formation. The most commonly used factor is the cooling rate reflecting the kinetic approach to the glass-formation problem. To compare glass-formation in telluride systems, it is convenient to use the cooling rate of $\approx 180\ °C\ s^{-1}$. At this cooling rate (similar to thin-walled quartz ampoules plunged into water), it is possible to reveal glass-formation in many systems and obtain substances in quantities sufficient for measurements and practical applications.

Based on calculations of GFA of alloys of more than 30 binary and some ternary systems, it has been established (Minaev, 1980a) that at the cooling rate of $\approx 180\ °C\ s^{-1}$ glass-formation takes place, as a rule, at GFA higher than $0.270 \pm 0.010\ kJ\ mol^{-1}\ K^{-1}$. The variations of the value are likely a result of inaccurate experimental measurements of initial values of chemical bonds and liquidus temperatures.

4.4. Glass-Formation of Chalcogens. Glass-Formation in Binary Chalcogen Systems

Glassy sulfur can be obtained by quenching of melt in liquid air at temperatures higher than 160 °C. At lower temperatures, down to the melting interval (113–115 °C), the melt consists practically wholly of molecules S_8. The glass-transition in sulfur occurs at negative temperatures: $T_g = -27\ °C$ (Rawson, 1967).

The selenium melt easily forms glass when cooled to room temperature. The temperature of the beginning of the softening interval is 30.5–31.5 °C according to Rawson's data (Rawson, 1967) and 40 °C according to Borisova's data (1972).

Based on the fact that at solidification of the tellurium melt, the anomalous volume alteration is observed, Toepler (1894) supposed that tellurium forms glass at cooling. Frerichs (1953) observed that tellurium melt does not form glass at cooling. Suhrman and Berndt (1940) have obtained amorphous material by condensing tellurium vapor on a surface cooled by liquid air. The crystallization takes place by the heating of tellurium to 25–30 °C.

According to Donald and Davies (1978) and Davies and Hall (1974), glassy tellurium has been formed by the method of 'shooting' to a cooled copper substrate. Tellurium was cooled from 560 °C at the rate of 10^{10} K s^{-1}. From the drop with mass of 100 mg, flakes of a porous film with the size of 10 × 20 mm and thickness about 20 μm were obtained. The following are temperatures of softening or crystallization of chalcogens (the last one is given for tellurium) obtained by the above method, K: S—240 K, Se—304 K, Te—304 K. The absence of T_g for Te does not allow to state with confidence that glassy tellurium was obtained. Tellurium here is likely in ultra-dispersive state (Minaev and Shchelokov, 1987). T_g could be so close to T_{cr} that it was difficult to be fixed by the method of scanning calorimetry, however.

We failed to find any information on the glass-formation or the amorphization of radioactive polonium.

The comparison of GFA of several elements of VI^A group was given by Borisova (1976). Among chalcogens, the element in which glass-formation most readily occurs both in elemental state and in composition with other elements is not sulfur, following oxygen, but selenium. The author has explained such non-monotonic behavior of GFA at movement from the top to the bottom in the VI group by the presence of the secondary periodicity.

In comparing the data above, the conclusion can be made that GFA of VI^A group elements firstly increases with increase in atomic numbers of elements from sulfur to selenium and then decreases: sulfur < selenium > tellurium > polonium.

Thus, the clear inversion in the main regularity of glass-formation is present: when GFA should decrease with increase in atomic mass, it increases instead at movement from sulfur to selenium.

The main feature of the conclusion concerning this non-monotonous character of GFA is the non-demonstrative, but implied condition—that the glass-formation is considered at normal room temperature. This circumstance does not likely objectively evaluate even the qualitative ratio of GFA of sulfur and selenium. In reality, glass is formed at cooling of the melt (at a certain minimal rate for each substance) below the glass-transition temperature T_g. The glass-formation state should be identified at this temperature as well.

If the sulfur melt is rapidly cooled to a temperature below the glass-transition temperature (−27 °C), sulfur easily forms glass and the inversion 'selenium–sulfur' becomes imaginary and is incorrect (Minaev, 1987b).

To ensure objectivity of the comparison of GFA of VI group elements, it is necessary to carry out the synthesis and the identification of glasses, taking into account the glass-formation nature of each of these elements with maximally favorable conditions for the glass-formation. The quenching of melts must be carried out from temperatures where each of the melts of considered elements contains the optimal concentration of glass-forming associates. Based on the data concerning melt viscosities, for the best glass-formation the quenching of sulfur must be carried out from temperatures higher than 160 °C, tellurium—from the temperature maximally close to the melting temperature because in tellurium the value of the activation energy of the viscous flow is one order of magnitude higher near the melting temperature than that at higher temperatures. It indicates the conservation of the chain structure of tellurium, although tellurium chains are shorter than those of Se or S (Glazor et al., 1967).

Based on the fact that glassy alloys Al_7Te_{93}, Ga_5Te_{95}, and In_9Te_{91}, obtained by Vengrenovich et al. (1987) at the cooling rate of $\sim 10^6$ K s^{-1}, have GFA of 0.256 ± 0.1, 0.244 ± 0.1, and 0.242 ± 0.1 kJ mol^{-1} K^{-1} (according to calculations of Minaev (1980c)), it can be assumed that for the critical cooling rate of 10^6 K s^{-1} the glass-formation ability equals ~0.25 ± 0.1 kJ $mol^{-1}K^{-1}$, which is just the approximate practical criterion of glass-formation at given conditions. Taking into account that the practical glass-formation criterion at the cooling rate of $\sim 10^2$ K s^{-1} is 0.27 ± 0.1 kJ mol^{-1} K^{-1}, it can be expected that at the cooling rate of 10^{10} K s^{-1} (data of Donald and Davies (1978)), tellurium with the glass-formation ability of 0.231 kJ mol^{-1} K^{-1} (Minaev, 1980c), will form glass. So, critical cooling rates of 10^2, 10^6, 10^{10} K s^{-1} correspond to GFA of ~0.27, ~0.25, and ~0.23 kJ mol^{-1} K^{-1}. This exponential dependence of the critical cooling rate on the glass-formation ability is, of course, an approximate one and it is only assumed because it is based on isolated data for cooling rates of 10^6 and 10^{10} K s^{-1}.

Calculations of the glass-formation ability of chalcogens carried out in accordance with the SRM criterion (Minaev, 1981a) have shown that at energies of homogenous bonds of sulfur, selenium, and tellurium of 266 ± 12, 184 ± 12, and 168 ± 12 kJ mol^{-1} K^{-1} and melting temperatures of 119.3, 217, and 449.8 °C, respectively, glass-formation abilities are 0.678 ± 0.1, 0.375 ± 0.1, and 0.231 ± 0.1 kJ mol^{-1} K^{-1}.

Calculated data show regular decrease in glass-formation ability of chalcogens with increase in their atomic numbers.

In the system S–Se, glassy alloys were obtained with the sulfur content up to 42 at.% by Suvorova (1974).

In the system Se–Te (Suvorova, Borisova and Orlova, 1974), alloys with 0–20 at.% Te are in the vitreous state. The quenching of narrow ampoules with melt from 820 °C in the cooled mixture with temperature of −20 °C gives the possibility to obtain glassy alloys containing up to 35 at.% Te (Das, Bever, Uhlman and Moss, 1972).

In the system S–Te, the glass-formation was predicted by Minaev (1987b) and then experimentally obtained with his participation by Valeev et al. (1987) in the range of compositions from pure sulfur to 29% at.% Te. The quenching of narrow wall ampoules with weights of 2 g was carried out in liquid nitrogen. Glasses obtained were with the glass-transition temperature in the region from −26 (for pure sulfur) to −2 °C.

4.5. Glass-Formation in Binary Chalcogenide Systems

The purposeful search for new chalcogenide semiconductor glasses is possible only on the basis of investigations of glass-formation regularities in chalcogenide systems that cannot be revealed without the study of available data on the glass-formation in multicomponent and simplest (binary) systems. However, such analysis cannot give maximum information without a comparison of the features of glass-formation and the structures of phase diagrams of corresponding systems. Further, locations of glass-formation regions, GFA and its relation with peculiarities of phase diagram structures of binary chalcogenide systems will be considered and data will be presented on the prediction of glass-formation regions in systems where glasses have not yet been revealed.

Glassy chalcogenides were first obtained in systems $A^{VA}-B^{VI}$ (vitreous alloys of the As–S system were synthesized by Schultz-Sellak (1870)), then in systems $A^{IVA}-B^{VI}$, $A^{IIIA}-B^{VI}$, $A^{IA}-B^{VI}$, $A^{VIIA}-B^{VI}$. In the same order, the glass-formation in these systems will be considered based on works of Minaev (1979, 1980c, 1981a, 1985a,b, 1989).

Locations and sizes of glass-formation regions are directly connected with cooling rates (quenching) of the melt after synthesis. Relative GFAs of various binary systems have been usually evaluated by sizes of glass-formation regions obtained in most possible similar conditions of synthesis and cooling. GFA of particular alloys—by calculations of Minaev (1980a) using the formula of the SRM criterion: GFA $= E_{CIB}/T_{liq}$ (see above).

4.5.1. *Systems $A^{VA}-B^{VI}$*

Figure 5 (Minaev, 1979, 1991) shows glass-formation regions (horizontal bold bands) and individual glass-forming alloys (rhombs) in $A^{VA}-B^{VI}$ systems superposed with phase diagrams of corresponding systems. The presence of multiple bands indicates the presence of different data on the glass-formation obtained in different conditions (quenching rates from 1–2 to 10^6 K s^{-1}), which are referred to in the listed works of Minaev (1979, 1991) and others.

As mentioned before, for chalcogenide alloys the regularity exists: glass-formation regions are decreased when atomic numbers in each subgroup increase (Goryunova and Kolomiets, 1958, 1960). The consideration of $A^{VA}-B^{VI}$ systems from this point of view shows the following.

In all cases (sulfur, selenium, tellurium), the glass-formation areas of chalcogenide alloys with antimony and bismuth are smaller, or even absent, when compared to the corresponding alloys with lighter arsenic and phosphorus. This regularity is also observed in alloys of selenium with phosphorus and arsenic, and in arsenic alloys with selenium and tellurium. Against this regularity, the following occurrence can be observed: glass-formation regions of phosphorous with sulfur are less than those of arsenic with sulfur, or phosphorous with selenium. Moreover, the glass-formation region of arsenic with sulfur is less than that of arsenic with selenium.

As a rule, the glass-formation region of phosphorus and arsenic with sulfur is less than that with selenium, both from the arsenic-enriched side and the sulfur-enriched side. The latter decrease of the glass-formation region is likely connected with what is described in Section 4.4 as the 'imaginary' inversion selenium–sulfur, and is caused by the fact that sulfur-enriched alloys were cooled to temperatures that were higher than their glass-transition temperatures, below room temperature, and as low as −27 °C for sulfur. The quenching of alloys at temperatures lower than T_g will apparently increase

---

FIG. 5. Phase diagrams and glass-formation regions (bold lines, rhombs) in systems $A^{VA}-B^{VI}$ (Minaev, 1991). In the system P–Te: (a) red phosphorus; (b) white phosphorus.

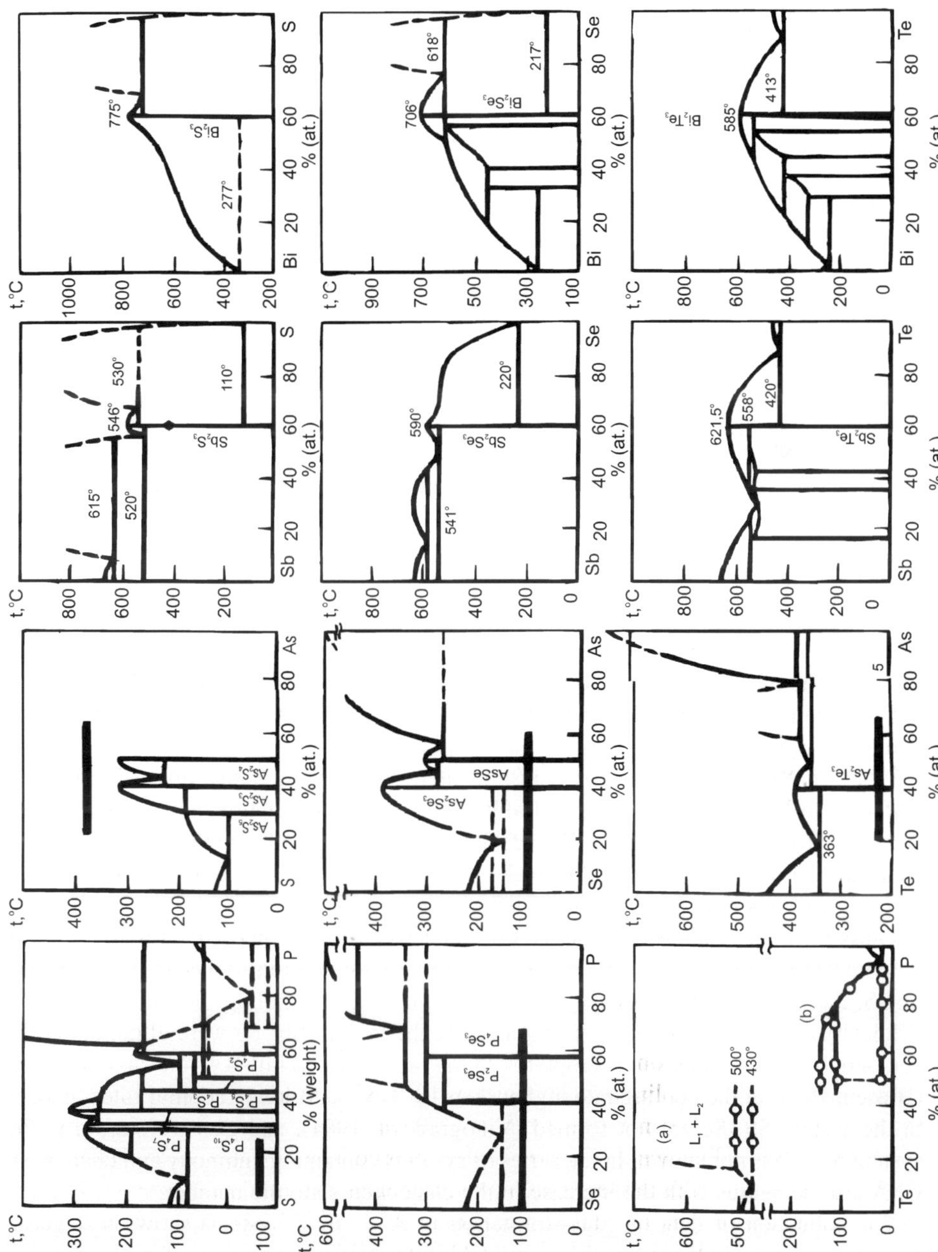

t,°C
% (weight)
% (at.)
P_4S_7
P_4S_{10}
P_4S_3
P_4S_5
P_4S_2
As_2S_5
As_2S_3
As_2S_4
615°
520°
546°
530°
110°
Sb_2S_3
775°
277°
Bi_2S_3
P_2Se_3
P_4Se_3
As_2Se_3
AsSe
590°
541°
220°
Sb_2Se_3
706°
618°
217°
Bi_2Se_3
(a)
$L_1 + L_2$
500°
430°
(b)
363°
As_2Te_3
621,5°
558°
420°
Sb_2Te_3
585°
413°
Bi_2Te_3

glass-formation regions of alloys of phosphorus and arsenic with sulfur, including an increase of as much as 100% for 'arsenic–sulfur' alloys.

Such an explanation is not applicable to data that indicate lesser glass-formation regions from the side enriched with V group elements. Therefore, we must state that in systems A^{VA}–chalcogen, there is an inversion in the regular decrease of the glass-formation ability corresponding with atomic number increase, both for the V^A group (the inversion 'arsenic–phosphorus' for the pair of systems P–S and As–S and the pair of systems P–Te and As–Te), and for chalcogen atomic number increase (the inversion selenium–sulfur for the pair of systems P–S and P–Se and the pair of systems As–S and As–Se).

Phosphorous and sulfur belong to elements of the 4th period of the periodic table, and arsenic and selenium to the 5th period. Therefore, it can be said that in the considered regularity (a decrease in the glass-formation region with increase in the atomic numbers of elements (Goryunova and Kolomiets, 1958, 1960)), there is inversion observed at movement from elements of the 3rd period to elements of the 4th period, the inversion '4–3'. This inversion of the glass-formation ability is connected with 'the secondary periodicity' of properties of elements of the periodic table. The phenomenon of the secondary periodicity will be considered in more detail in connection with the glass-formation in binary and multi-component systems.

The comparison of the calculated GFA and the phase diagram structure in Sb–S glasses shows that besides Sb_2S_3 (GFA $= 0.339$ kJ mol^{-1} K^{-1}) obtained by Melekh and Maslova (1976), homogenous glasses can also be obtained by the same method of synthesis and quenching in the whole range of compositions confined by regions of the melt segregation (Figure 5 for the Sb–S system). Calculations in accordance with the SRM criterion show that the composition $Sb_{37}S_{63}$ has greater GFA than the composition $Sb_{40}S_{60}$. The same can be said about another composition neighboring another region of the segregation—$Sb_{43}S_{57}$. In 1979, Minaev predicted glass-formation in the region of composition $Sb_{37-43}S_{63-57}$. Furthermore, glasses can be formed beyond those borders, but they can be inhomogeneous because of melt segregation. Dalba, Fornasini, Giunta, Burattini and Tomas (1987) have partially confirmed that prediction, having synthesized the glassy composition $Sb_{38}S_{62}$. The prediction was confirmed completely when Shtets, Bletskan, Turianitsa, Bodnar and Rubish (1989) obtained glasses in the region $Sb_{35}S_{62}$–$Sb_{45}S_{55}$, and inhomogeneous glass $Sb_{50}S_{50}$ as well.

Glass-formation regions in systems Sb–S and Sb–Se superimposed at Figure 5 cannot be compared because they were obtained in different conditions. Glasses in the system Sb–S, the Sb_2S_3 glass in particular, were obtained at the cooling rate of ~ 200 K s^{-1} (Melekh and Maslova, 1976), while glasses in the system Sb–Se were obtained by the splat-cooling method as small droplets (25 mg) sputtered in liquid or on a copper plate (Brasen, 1974) at the cooling rate higher than 10^5 K s^{-1}. At lower cooling rates, glasses in the system Sb–Se are not formed (Vinogradova, 1984), and glass-formation in the system Sb–Te is not known. In the series of systems containing antimony and chalcogen, GFA also decreases with the increase in the chalcogen's atomic number.

The collection of data on glass-formation in A^{VA}–B^{VIA} systems shows that glass-formation regions in them are characterized by the prevailing content of chalcogen.

Values of calculated glass-formation abilities of glasses obtained at cooling rates lower than 200 K s^{-1} are in all cases higher than 0.270 kJ mol^{-1} K^{-1} and amount to 760 kJ mol^{-1} K^{-1}. Values of GFA significantly decrease with the increase in atomic mass of elements in each group of the periodic table. For example, in the row S–Se–Te for eutectic alloys with arsenic, the glass-formation ability decreases from 0.716 to 0.281 kJ mol^{-1} K^{-1}; i.e., by 2.5 times. Corresponding to the increase in the atomic mass of chalcogen, limit values of the glass-formation ability (the difference of maximum and minimum values characterizing glass-formation regions) also decrease. For the same arsenic glasses, this value decreases from 0.254 for sulfur to 0.149 for selenium and 0.018 kJ mol^{-1} K^{-1} for tellurium, evidencing lesser possibilities for glass formation.

The SRM criterion is far from the ideal criterion. It works well in the sphere of systems of the same type, such as telluride systems. For systems of different types, such as sulfide, selenide, and telluride, glass-formation region borders are different in GFA value decrease ratios corresponding to increases in chalcogen's atomic number. To compare concrete glass-formation abilities in different types of systems at maximum similar conditions of glass-formation (the same cooling rate), the corresponding coefficient can be introduced, which will reflect peculiarities of the atomic interaction in different types of systems that are unaccounted by the criterion. These peculiarities as well as additional refinements of the criterion were analyzed by Minaev (1991).

4.5.2. Systems A^{IVA}–B^{VI}

A comparison of dimensions of glass-formation regions (Fig. 6) and calculated GFA of A^{IVA}–B^{VI} systems demonstrates the common tendency of alloys of these systems to show decreases in GFA with increase in the atomic mass of elements, corresponding to the generally accepted thesis of Goryunova and Kolomiets (1958, 1960) regarding the loss of glass-formation with increase in the atomic numbers of elements and the metallization of chemical bonds.

This common tendency correlates with energies of chemical bonds A^{IVA}–B^{VI} (kJ mol^{-1} K^{-1}), which decrease with the increase in atomic numbers of elements both in the row Si–Ge–Sn–Pb and in the row S–Se–Te, demonstrating their increasing role of metallization (calculations of energy of bonds were carried out by Minaev (1991) using Pauling's formula (1970) and his data on energies of bonds A^{IV}–A^{IV} and A^{VI}–A^{VI} presented in Section 4.3)

	Si	Ge	Sn	Pb
S	274.0	259.0	251.9	218.4
Se	221.0	205.7	198.6	166.8
Te	186.4	171.5	164.4	138.2

The same can be said about chemical bonds 'chalcogen–chalcogen': $E_{S-S} = 266$, $E_{Se-Se} = 184$, $E_{Te-Te} = 168$ kJ mol^{-1} K^{-1}.

The following facts suggest the tendency of GFA decrease:

- binary sulfide and selenide systems with silicon and germanium are able to form glasses and are not able to form glasses with tin and lead;

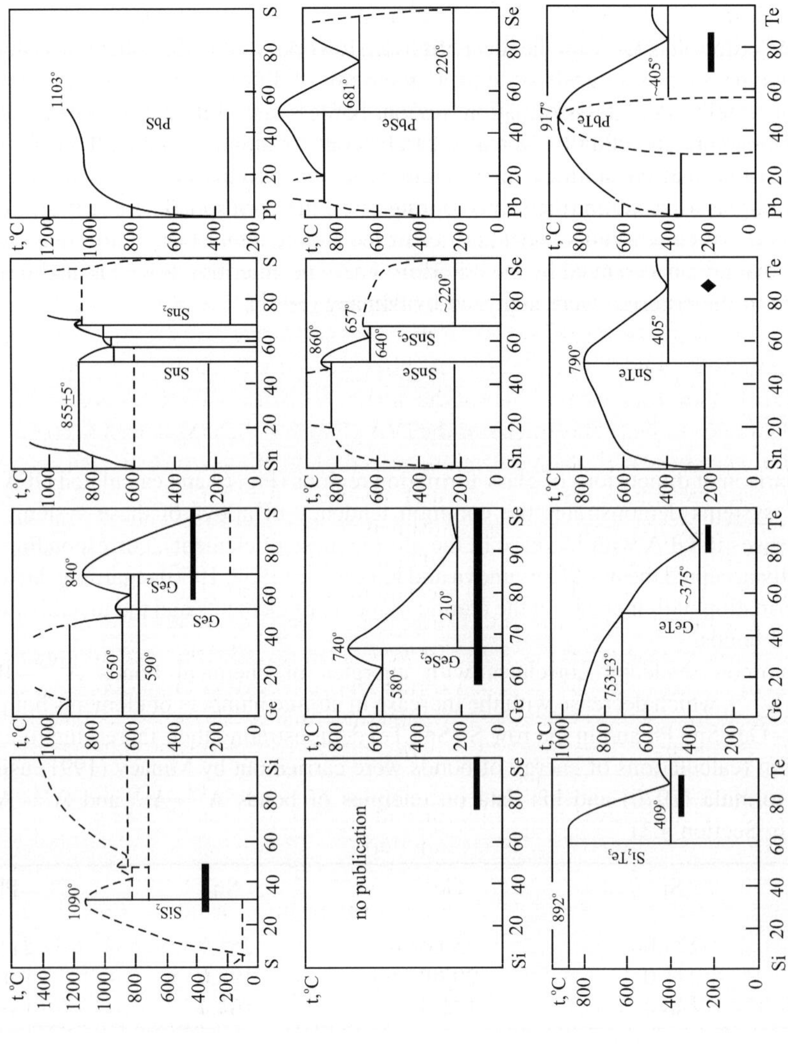

FIG. 6. Phase diagrams and glass-formation regions (bold lines, rhombs) in systems A^{IVA}–B^{VI} (Minaev, 1991).

- in binary telluride systems, glass-formation regions become smaller in the row silicon–germanium–tin;
- glass-formation regions become smaller at transition from binary selenide systems with silicon and germanium to telluride system with the same elements.

Several inversions of the regular GFA decrease with increase in atomic numbers deviate against this tendency, according to Minaev (1981a): glass-formation regions in the system Si–S are smaller at comparable conditions than those in systems Si–Se and Ge–S, and the glass-formation region in the system Ge–Se is larger than those in systems Ge–S and Si–Se. All four inversion pairs of the systems ('Si–Se'–'Si–S', 'Ge–S'–'Si–S', 'Ge–Se'–'Si–Se', 'Ge–Se'–'Ge–S') are of the inversion type '4th period–3rd period' of the periodic table (4–3) that is known from glass-formation in the group of systems A^{VA}–B^{VI} (Minaev, 1979).

The comparison of glass-formation regions in systems Sn–Te and Pb–Te (Fig. 6) show that the glass-formation region in the system with lead is larger than that in the system with tin. For example, in the work of Lasoca and Matyja (1974), eutectic alloys of both systems were obtained in the vitreous form at the cooling rate of 10^5–10^6 K s^{-1}: $Sn_{16}Te_{84}$ and $Pb_{15}Te_{85}$. In the work of Kaczorowski, Dabrowski and Matyja (1977) with the same cooling rate in the system Pb–Te glasses were also synthesized with the lead content from 14.5 to 30 at.%. Therefore, the presence of the inversion of the regular GFA decrease with increase in atomic numbers of elements of the IVA group for alloys with lead in respect to alloys with tin can be stated. Such inversion, according to Minaev (1981a), must appear also in multi-component alloys with tellurium, and in alloys with lighter chalcogens as well. This prediction has been partially confirmed by Minaev (1983) as glass-formation regions have been revealed experimentally in systems Sn–As–Te and Pb–As–Te. In the latter system, the glass-formation region was significantly larger than that in the former, Sn–As–Te. Moreover, the glass-formation region in the system Ga–Pb–Te has been noted, while glasses in the system Ga–Sn–Te were not observed at the same cooling rates.

In this case, we deal with the inversion of the regular alteration of properties with the increase in the atomic number of the element (increasing in the metallization of atomic bonds) for elements of 6th and 5th periods of the periodic table—the inversion '6th period–5th period'.

And there is one more inversion. In sulfide and selenide systems with tin and lead, the glass-formation is not observed, whereas in telluride systems it takes place. This means that the inversion '5th period–4th period' ('5–4')—the inversion 'Te–Se'—is demonstrated. The absence of glass-formation in sulfide and selenide systems is caused by the sharp increase in liquid temperature (by hundreds of degrees), even at the addition of the first portions of tin and lead to sulfur and selenium that sharply decreases the glass-formation ability of the alloy. The problem of the inversion nature of the glass-formation will be considered in following paragraphs in more detail.

4.5.3. *Systems A^{III}–B^{VI}*

Boron, aluminum, gallium, indium, thallium form 15 binary systems with sulfur, selenium, and tellurium. Glass-formation has been noted in only eight of them: B–S, B–Se, Al–Te, Ga–Te, In–Te, Tl–S, Tl–Se, and Tl–Te (Minaev, 1980).

The decreased glass-formation tendency with increase in atomic number of the element of the IIIA group demonstrates itself more completely in systems with tellurium where glass-formation regions at the cooling rate of 180 K s^{-1} are limited by 12–30 at.% Al, 15–25 at.% Ga, the glass-formation with In was not observed (Cornet, 1976).

In the system Tl–Te, the alloy with 30 at.% Tl was obtained in the partially vitreous state. It appears that along with the regular GFA decrease with increase in the atomic number of the A^{IIIA} element, we deal with appearance of the 'Tl–In' inversion—the inversion 6th period–5th period—that we have already observed in systems with IVA elements as the inversion 'Pb–Sn'.

Vengrenovich et al. (1986) used the melt-spinning method for synthesis of A^{IIIA}–Te glasses where the cooling rate is $\approx 10^6$ K s^{-1}. Alloys with 7–28 at.% Al, 5–28 at.% Ga, and 9–28.6 at.% In were obtained by this method. The authors noted that in systems with Al and Ga, the most easily glass-forming alloys are located aside of the eutectic alloy shifted to the compound forming the eutectic with tellurium. This phenomenon will be considered separately in Section 4.6.

Calculations of GFA (Minaev, 1991) of nearest to tellurium glass-forming alloys in systems A^{IIIA}–Te, obtained by the melt-spinning method, gave the following results Al_7Te_{93}—0.256, Ga_5Te_{95}—0.244, In_9Te_{91}—0.242 kJ mol^{-1} K^{-1}.

The data obtained allowed to assume that for the critical cooling rate of 10^6 K s^{-1} the glass-formation ability is equal to 0.25 ± 0.01 kJ mol^{-1} K^{-1}. If this value is compared from one side with the value 0.27 ± 0.01 kJ mol^{-1} K^{-1}, which is typical for the critical cooling rate $\approx 10^2$ K s^{-1} (Minaev, 1980a), and is also compared on the other side with the value 0.23 ± 0.01 kJ mol^{-1} K^{-1}, which is typical for tellurium (according to Minaev's calculations (1980c), and obtained in the vitreous form by Donald and Davies (1978) at the cooling rate of 10^{10} K s^{-1}), the conclusion can be made regarding the exponential increase in the critical cooling rate coinciding with the substance's decreasing ability of glass-formation. This conclusion, made by Minaev (1991), has something in common with the assumption of Dembovsky and Ilizarov (1978) concerning the exponential character of the dependence of the glass-formation ability on the critical cooling rate. We note that Dembovsky, for some reason, considers the glass-formation ability of a substance to be a secondary property with respect to the critical cooling rate. In our opinion, GFA of a substance is unconditionally a basic property of the substance, and the critical cooling rate is just the reflection of the essence of the substance that changes according to the GFA of every concrete alloy.

The SRM criterion that expresses the GFA of a substance and is yet independent of the appearance of the glass-formation in concrete conditions, can be used also as a criterion of concrete glass-formation that reflects certain conditions of glass synthesis. But to two of the factors of the SRM criterion—structural–chemical (E_{cib}) and energetic (thermodynamic (E_{liq}))—one must add the third one, the cooling rate of alloys, which determines the process of the concrete glass-formation, and reflects the kinetic aspect of glass formation.

In systems with sulfur and selenium, the glass-formation regions are observed only for boron and thallium, i.e., there is the inversion again in the main regularity of the glass-formation—the inversion Tl–In (6–5).

In the system B–S, Hagenmuller and Chopin (1962) and Zhukov and Grinberg (1969) have revealed the vitreous compound B_2S_3, in the system B–Se the glass-formation region expands from Se to B_2Se_3 according to data of Boriakova and Grinberg (1969). In both systems, glasses are unstable and easily hydrolyzed in air. Judging by the presented data, in these systems the inversion Se–S (4–3) of the main regularity of glass-formation is observed that is similar to the inversion for some systems of the VA group (Section 4.5.1).

The inversion Se–S (4–3) is also observed, according to data generalized by Minaev (1980c), at the comparison of systems Tl–S and Tl–Se. In the former, the spread of the glass-formation region, according to data of Cervinka and Hruby (1978a,b) is 21.4% (from 50 to 71.4 at.% S), and in the latter, according to data of Cervinka and Hruby (1979), is 33.4% (from 66.6 to 100 at.% Se).

The inversion 'Te–Se' appears in systems with Al, Ga, and In that do not form glasses with sulfur and selenium whereas they form glasses with tellurium (Minaev, 1980c, 1991).

So, in binary systems A^{IIIA}–chalcogen, the main regularity of glass-formation—decrease of the glass-formation region with increase in the atomic numbers of elements—demonstrates itself as follows:

1. At transition from systems B–S and B–Se to corresponding systems with Al where the glass-formation is absent.
2. In the row of telluride systems containing Al, Ga, In (data of Cornet (1976)).
3. At transition from the system Tl–Se to the system Tl–Te.

At the same time, all eight glass-forming systems A^{IIIA}–chalcogen take part in inversion relations with neighboring (with respect to the atomic number of the element of the III group or chalcogen) systems. The inversion 4–3 is observed for pairs of systems 'B–Se'–'B–S', 'Tl–Se'–'Tl–S'; the inversion 6–5 is observed for pairs 'Tl–S'–'In–S', 'Tl–Se'–'In–Se', and 'Tl–Te'–'In–Te' (data of Cornet (1976)); the inversion Te–Se appears in pairs 'Al–Te'–'Al–Se', 'Ga–Te'–'Ga–Se', 'In–Te'–'In–Se' (Minaev, 1980c, 1991). The role of this 'inversion regularity' in systems A^{IIIA}–chalcogen becomes comparable, or even prevailing, with the role of the main regularity of the glass-formation in chalcogenide systems—the decrease in the glass-formation ability with increase in atomic numbers of elements.

The attentive reader perhaps has already noticed the fact that at transition from binary systems of the VA group to the systems of the IVA group, and also at the transition to the IIIA group, the number of inversion pairs in individual systems increases, making the main regularity of the glass-formation—the decrease in the glass-formation ability with increase in atomic numbers of elements (Goryunova and Kolomiets, 1958, 1960)—'more diffused'. This begs the question: how will systems with elements of IIA and IA groups of the periodic table behave with respect to glass-formation. Our point of view will be presented in Sections 4.5.4 and 4.5.6.

4.5.4. *Systems A^{IA}–B^{VI}*

The discovery of common regularities of the glass-formation in binary systems IIIA-, IVA-, VA-, and VIA-subgroups (Goryunova and Kolomiets, 1958, 1960), and their regular inversions (Minaev, 1979, 1980c, 1981, 1991), suggested the possibility of glass-formation in binary chalcogenide systems with other subgroups of the periodic table, and IA–chalcogen systems, in particular. Moreover, it had been already known that in the system $Cs_2S-Sb_2S_3$, the glass-formation region was in the limits of 0–40 and 65–85 mol.% Sb_2S_3 (Salov, Lazarev and Berul, 1977), and that Cs_2S was obtained in the vitreous state. Later, the alloys' tendency toward glass-formation was described by Chuntonov et al. (1982) in work concerning investigations of the equilibrium diagram of the Cs–Te system.

Despite extremely limited experimental data on glass-formation in binary system IA–chalcogen, Minaev (1985) considered the possibility of the systematic glass-formation in the whole set of chalcogenide systems with elements of IA subgroup of the periodic table. In addition to the research stated above, studies had demonstrated the existence of glass-formation in ternary systems, as the alkaline metal based on the system As–Se was presented in works by Borisova (1971), Kokorina, Ioh, Kislitskaya and Melnikov (1976) and Dembovsky and Ilizarov (1978) on glass-formation in systems $Cs_2S-Sb_2S_3$ and $Rb_2S-Sb_2S_3$. Moreover, by the 1980s Minaev (1978, 1980a,b) had already developed the structural–energetic concept of the glass-formation in chalcogenide systems and its basis, the Sun–Rawson criterion of glass-formation as modified by Minaev allowing for the calculation of GFA for multi-component compositions.

Borisova (1971) established the glass-formation region increase in systems 'alkaline metal–As_2Se_3' with increase in atomic numbers of A^{IA} elements. In this compound, one can add without crystallization up to <2 at.% Li, 10 at.% Na, and 30 at.% K. According to Dembovsky's data (1978), the glass-formation region in As–Se-based ternary systems increases in the row Li $<$ Na $<$ K $<$ Rb $<$ Cs. These data contradict the main regularity of the glass-formation in chalcogenide systems with elements A^{IIIA}-, A^{IVA}-, A^{VA}-subgroups of the periodic table established by Goryunova and Kolomiets (1958, 1960). Minaev (1985) calculated energies of chemical bonds, E and glass-formation abilities in binary systems 'alkaline metal–chalcogen' using the Sun–Rawson–Minaev criterion.

Energies of chemical bonds 'metal–metal' and chalcogen–chalcogen and values of the electronegativity from Pauling's monograph (1970) were used in calculations.

The bond energy $A^{IA}-B^{VIA}$, as it can be seen in Table I, increases with the increase in atomic numbers of chalcogens, despite the strength of bonds between alkaline atoms themselves decreases in the same direction, kJ/mol: Li–Li 111, Na–Na 75, K–K 55, Rb–Rb 52, Cs–Cs 45. This fact is explained by that the electronegativity of alkaline metals decreases with increase in atomic numbers (Li 1.0, Na 0.9, K 0.8, Rb 0.8, Cs 0.7), which leads to increase in the difference of electronegativities of pairs alkaline metal–chalcogen, the ion component of the bond and its total energy (Table I). The abnormality in this regularity (the increase in chemical bond energies with increase in atomic numbers of alkaline metals) is observed in connection with the anomalously high increase in the bond energy 'potassium–chalcogen', which is higher than that of rubidium, which is located lower in the periodic table.

Data on liquidus temperatures, used in calculations, were taken from known phase diagrams of $A^{IA}-B^{VIA}$ systems presented in works (Chizhikov and Schastliviy, 1964, 1966; Samsonov and Drosdov, 1972; Kuznetsov, Chuntonov and Yatsenko, 1977). For these diagrams, the tendency of the liquidus temperature to decrease, in the region of

TABLE I

ENERGIES E OF BONDS CHALCOGEN–CHALCOGEN, ALKALINE METAL–CHALCOGEN AND DIFFERENCES OF ELECTRONEGATIVITY X OF ELEMENTS FORMING BONDS (MINAEV, 1991)

Bonds	E, kJ mol^{-1}	X
S–S	266.0	0
Se–Se	184.0	0
Te–Te	168.0	0
Li–S	380.5	1.5
Na–S	383.5	1.6
K–S	396.0	1.7
Rb–S	394.5	1.7
Cs–S	411.5	1.8
Li–Se	319.2	1.4
Na–Se	321.5	1.5
K–Se	332.2	1.6
Rb–Se	331.0	1.8
Cs–Se	350.0	1.7
Li–Te	251.2	1.1
Na–Te	252.0	1.2
K–Te	261.9	1.3
Rb–Te	260.4	1.3
Cs–Te	277.3	1.4

alloys with the predominant content of chalcogen, with increase in atomic numbers of alkaline metals is typical. This tendency is expressed particularly clearly by telluride alloys.

Two tendencies come to light, even before the calculations of GFA are made, using formula (9) in Section 4.3 as well as calculations regarding the regions of glass-formation: the increase in the chemical bond strength and the liquidus temperature decrease in systems $A^{IA}–B^{VIA}$ with the increase in atomic numbers of alkaline metals. These tendencies reflect the main factors facilitating the glass-formation, and one can state a priori that GFA and glass-formation regions in the systems under study will increase with increase in atomic numbers of alkaline metals. Calculations of GFA for systems Na–Te, Rb–Te, and Cs–Te (in systems Li–Te and K–Te the phase diagrams had not been obtained by that time) by Minaev (1985) showed that glass-formation regions must be present at the cooling rate of 180 K s^{-1} in all three systems. In the system Na–Te, the glass-formation region must be located in the limits 41–48 at.% Na, in the system Rb–Te 14–27 at.% Rb and in the point 48 at.% Rb, in the system Cs–Te 12–46 at.% Cs with possible interruption in the point of the chemical compound.

Both the calculated dimensions of the glass-formation region and the calculated maximum GFA of alloys of the systems with tellurium increase when atomic numbers are raised. The dimensions of glass-formation regions in systems Na–Te, Rb–Te, Cs–Te are 7, 13, and 34 at.%, and maximum GFAs are 0.284, 0.327, and 0.356 ± 0.01 kJ mol^{-1} K^{-1}, respectively. Prognostic calculations of GFA of sulfide and selenide systems $A^{IA}–B^{VIA}$ also predict the existence of glass-formation regions in these systems with the tendency of increase in glass-formation regions with increase in atomic numbers A^{IA}. For systems with tellurium, this tendency develops into the regularity (there are no exclusions): the increase in the atomic number leads to increase in GFA without fail.

At present, the author of the work is rather skeptical about the objectivity of a part of the conclusions made in this work (Minaev, 1985) regarding calculations for sulfide and selenide systems. These conclusions seem to be objective in the part concerning the existence of the tendency to increase in GFA and the glass-formation region with increase in atomic numbers of A^{IA} elements, but it is not objective in the part that concerns the concrete quantitative prediction of glass-formation regions.

Actually, in calculations for systems with S, Se, and Te the GFA value of 0.27 ± 0.01 kJ mol^{-1} K^{-1} corresponds to glass-formation borders. This value is the criterion of the concrete glass-formation for telluride systems, but not for sulfide and selenide systems where it is significantly higher. For example, Sb_2S_3 glass was obtained only by quenching in ice water of 1.5–2 g of the melt placed in a flat ampoule that corresponds to the cooling rate of 200–250 K s^{-1} according to Melekh and Maslova (1976). But GFA of the alloy Sb_2S_3 is equal to 0.339 kJ mol^{-1} K^{-1}, i.e., for sulfide glasses the concrete criterion of the glass-formation at this cooling rate seems to be near 0.339 kJ mol^{-1} K^{-1}, and in the vitreous state only alloys with GFA $\geq$ 0.339 kJ mol^{-1} K^{-1} can be obtained. Thus, calculations of concrete glass-formation regions in systems with sulfur and selenium, made by Minaev (1985), are wrong and not presented in this work.

According to Minaev's point of view (1978, 1980a, 1985a,b, 1991) on the inversion nature of the glass-formation, in the case of telluride systems it appears that we deal with a new type of inversion—the thorough, continuous inversion—of the regular decrease in the GFA with increase in atomic numbers of elements typical for chalcogenide systems (Goryunova and Kolomiets, 1958, 1960), the inversion expanding on all elements of the IA subgroup. The tendency for such an inversion is observed in systems with sulfur and selenium as well.

The prognosis of glass-formation in binary selenide systems with alkaline metals, made by Minaev (1985), has received partial experimental confirmation in the later published work of Fedorov, Chuntonov, Kuznetsov, Bolshakova and Yatsenko (1985), where the inclination to glass-formation of the Cs–Se system alloys in the region of the phase diagram adjoining selenium was described.

It seems that no research has been undertaken for the specific purpose of investigating glass-formation in binary systems alkaline metal–chalcogen. Nevertheless, there is information concerning glass-formation in all three chalcogenide (S, Se, and Te) systems with cesium, the element of IA subgroup with the largest atomic number, excluding radioactive francium. It is obvious that revealing the glass-formation region 'accidentally' is easiest in the case when alloys of the system possess increased GFA (when compared to other systems) and broad glass-formation regions.

An indirect confirmation of the correctness of this estimate regarding the projected increase in GFA and glass-formation regions corresponding with increase in atomic numbers of alkaline metals in binary chalcogenide systems is obtained through research detailing a similar increase in glass-formation regions in ternary systems with alkaline metals. As shown earlier, such an increase takes place in systems A^{IA}–As–Se. Moreover, Ribes, Barrau and Souquet (1980) demonstrated that while the glass-formation region in the system Li_2S-GeS_2 extends from pure GeS_2 to 50 mol.% Li_2S, replacement of Li_2S for Na_2S increases the glass-formation region to 60 mol.% Na_2S. In systems $K_2S-Sb_2S_3$, $Rb_2S-Sb_2S_3$, and $Cs_2S-Sb_2S_3$ (data of Berul, Lazarev and Salov (1971) and Salov et al. (1977)) glass-formation regions content 66.7–90, 60–80 and 0–40, and 65–85 (two glass-formation regions in the case of cesium) mol.% Sb_2S_3. The extension of two glass-formation regions with cesium (40 + 20 mol.%) is significantly larger than with rubidium (20 mol.%). But potassium gives the same larger region (23.3 mol.%) than rubidium, which is explained by the non-monotonic character of the regular alterations of properties at transition from 3rd to 4th period, connected with the second periodicity in the periodic table. In this case, the non-monotonic character is manifested, as we have already seen (Table I) in the anomalously large increase in the chemical bond energy potassium–chalcogen when compared to the bond 'sodium–chalcogen' that is higher than the bond energy of located lower rubidium with chalcogen. The result of such an increase is the increase in the total covalent–ion binding of alloys with potassium and the increase in their GFA and, correspondingly, their glass-formation regions.

Thus in the thorough inversion, which is observed in the whole group, there is another inversion restoring pairs of systems with K and Rb to the classical regularity of glass-formation, as demonstrated by Goryunova and Kolomiets (1958, 1960) stating that the glass-formation regions must decrease with increase in atomic numbers of replacing elements of this subgroup.

So, the set of presented experimental data on the glass-formation in binary and ternary chalcogenide systems as well as calculations of glass-formation regions using the Sun–Rawson–Minaev's criterion allow predictions regarding glass-formation in all 15 binary chalcogenide systems with elements of A^{IA}-subgroup of the periodic table. Glass-formation in systems with K, Rb, and Cs can be obtained at rather low cooling rates. The system with sodium and, especially, with lithium will require significantly higher cooling rates.

Glass-formation in these systems is still waiting further research.

4.5.5. Systems $A^{VIIA}-B^{VI}$

In 1970, first mention of glass-formation in the system Se–I appeared, although the author (Oven, 1970) did not name the concrete glass-forming compositions. In 1973, non-crystalline alloys of the Se–I system containing up to 82.5 at.% I were obtained by cooling at the furnace. Vitreous compositions containing 30–70 at.% I were found to be the most stable against crystallization at room temperature, according to Chizhevskaya, Abrikosov and Azizova (1973). Glasses in this system ($Se_{90}I_{10}$ and $Se_{80}I_{20}$) were also obtained in the work of Zamfira, Jecu, Iuta, Vlhovici and Popescu (1982).

Ignatyuk et al. (1980) described glasses in the system Te–I containing 40–55 at.% I.

Nisselson, Sokolova and Soloviev (1980) investigated the system S–Cl. The authors established that for this system, the significant inclination to overcooling with formation of glasses is in the range $\approx 10- \approx 70$ at.% chlorine. Glasses of these system exist only at negative temperatures because the larger part of the liquidus line lies below zero and in the eutectic point goes down to -132 °C.

Vinogradova (1984) supposed the possibility of glass-formation in the system Se–Br, of which alloys, according to Golubkova, Petrov and Kanev (1975), are greatly inclined to overcooling in the concentration range from 40 to 60 at.% Se.

These data provided a foundation to consider the possibility of glass-formation in the whole group of systems 'chalcogen (S, Se, Te)–halogen (F, Cl, Br, I)' by Minaev (1989, 1991) on the basis of the structural–energetic concept of glass-formation and the Sun–Rawson–Minaev criterion initially developed for chalcogenide glasses. Based on calculations of the glass-formation ability in accordance with the SRM criterion, Minaev (1989) predicted the glass-formation in nine binary chalcogenide systems: S–F, S–Br, S–I, Se–F, Se–Cl, Se–Br, Te–F, Te–Cl, and Te–Br. Many of them, like the known glass-forming system S–Cl, must form glasses only at negative temperatures.

The prediction, made in 1989, received a partial confirmation: Lucas and Zhang (1990) synthesized glasses in systems Te–Cl and Te–Br containing 60–67 at.% Cl and 50–70 at.% Br. Ma Hong, Zhang Xiang Hua and Lucas (1991), defined more accurately the Br content in the system Te–Br: 31–41 at.% Br. If these data are compared with those of Ignatyuk et al. (1980) on the Te–I system, one can see that in binary telluride systems the content of the VII group element in glasses gets increased with increase in its atomic number (7 at.% Cl, 10 at.% Br, and 15 at.% I). It means the presence of the inversion with respect to the glass-formation regularity of Goryunova and Kolomiets (1958, 1960) concerning the GFA decrease with increase in atomic numbers.

The set of data collected by Vinogradova (1984) evidences the increase in VII group elements content in glasses with increase in atomic numbers in the row Cl–Br–I in ternary systems with sulfur and selenium: these are systems As–S–(Cl, Br, I), Si–Se–(Br, I), As–Sev(Br, I), i.e., the same inversion is observed in ternary systems as well. Although the data on the glass-formation in binary and ternary chalcogenide systems with elements of VIIA group are by no means complete, one can predict a thorough inversion throughout the whole group regarding the tendency of glass-formation ability to decrease with the increase in atomic numbers of elements.

4.5.6. *Systems A^{IB}–B^{VI} and A^{IIB}–B^{VI}*

The glass-formation in systems A^{IB}–A^{VI} was analyzed by Minaev (1988, 1991) based on comparison of corresponding phase diagrams and the Sun–Rawson–Minaev criterion. Phase diagrams of systems Cu–S, Cu–Se, Ag–S, Ag–Se, and Au–Se were considered by Minaev as the non-glass-forming type of diagrams. Because they are characterized by the phase segregation of the melt, and also because of the sharp increase in the liquidus temperature even with a small addition of the second element to chalcogen, it is very difficult to experimentally obtain homogenous glasses in such systems. These difficulties were overcome in the work of Tsaneva and Bontscheva-Mladenova (1978), who obtained vitreous alloys in the system Ag–Se at concentrations of silver up to 10 at.%. For this, the temperature of the synthesis was increased to 1000 °C, which appears to be out of the temperature range limiting the dome of liquation of melt.

In the work of Perepezko and Smith (1981), glasses in the system Cu–Te were obtained by the droplet-emulsion method. The copper content in the glasses were 19–39 at.%.

Binary telluride systems with copper, silver, and gold have different phase diagrams of the glass-forming eutectic type. This fact, as well as calculations of the SRM criterion, allowed Minaev (1988, 1991) to predict the existence of glasses in Ag–Te and Au–Te, using the same synthesis method as for the system Cu–Te, and to also predict an increase in GFA and glass-formation regions in the row Cu–Ag–Au, indicating the presence of the thorough inversion of the regular decrease of GFA with increase in the atomic numbers of components.

In binary chalcogenide systems with elements of side subgroups of II–VIII groups of the periodic table, glass-formation was not obtained. Glass-formation regions had been predicted by Minaev (1988) in systems Hg–S and Hg–Se at cooling rates of $\sim 10^2$ K s^{-1}, and in the Hg–Te system at cooling rates of $\sim 10^6$ K s^{-1}. Glass-formation regions were also predicted in systems Hg–S and Hg–Se at cooling rates of $\sim 10^2$ K s^{-1}, and in the Hg–Te system at cooling rates of $\sim 10^6$ K s^{-1}.

It is interesting that Dembovsky and Ilizarov (1978) found glass-formation regions in ternary systems A^{IIB}–As–S whose sizes increase in the row Zn–Cd–Hg. This shows the presence of a thorough inversion of the regular decrease in the glass-formation ability with increase in atomic numbers of A^{IIB} elements in at least one group of ternary chalcogenide systems. The presence of such an inversion can be assumed in other chalcogenide systems as well.

4.6. Is the Liquidus Temperature Effect Always Effective?

The increased glass-formation ability in regions with decreased liquidus temperature was known as early as Tammann (1903), Lebedev (1910), and Kumanin and Mukhin (1947). Rawson (1967) named the increased glass-formation ability at decreasing liquidus temperature as the liquidus temperature effect and expressed the point of view that glass-formation is most probable for eutectic compositions. In accordance with the eutectic law of Cornet (1976), the GFA of binary telluride systems with elements of IIIA, IVA and VA subgroups of the periodic table is maximal for compositions located near eutectic ones. The analysis of the glass-formation in these systems emphasizes the importance of the usage of phase diagrams for evaluation of GFA, the role and sometimes insufficiency of the liquidus temperature effect of Rawson (1967) for this purpose putting in the forefront the Sun–Rawson–Minaev's criterion (Minaev, 1978, 1980b) that takes into account both the structural–energetic factor (chemical bonds, the CN, the structure of alloys) and the energetic (thermodynamic) factor (the thermal energy expressed through T_{liq}) of glass formation.

Cornet (1976) presented the glass-formation region in the system Ga–Te as 15–25 at.% Ga. At the same time, it is known that the eutectic in this system is located in the point $Ga_{14}Te_{86}$ and does not form glasses. If one goes along the downward liquidus line from pure tellurium in the direction of the chemical compound $GaTe_3$ (Fig. 7), the glass-formation is not observed even at the lowest eutectic temperature, and the liquidus temperature effect is not observed. But the glass appears later in point $Ga_{15}Te_{85}$, when the liquidus temperature is increased. It means that in this case, the liquidus temperature effect is not effective, and the eutectic law of Cornet is not effective. Cornet made a reservation providently, though, stating that the GFA is maximal for compositions 'near eutectic ones,' but he did not explain the obtained phenomenon—why there is no glass in

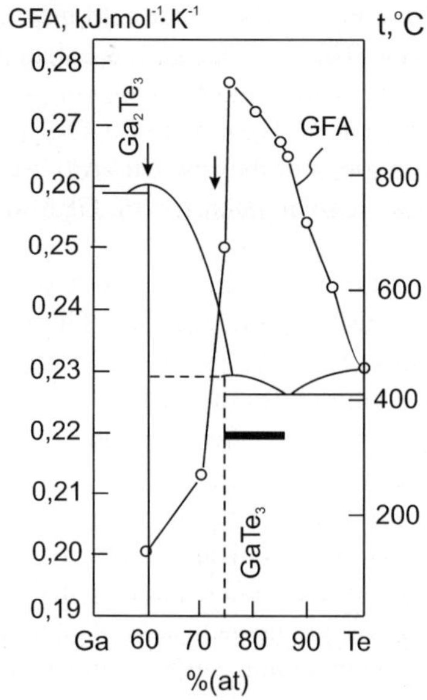

FIG. 7. The phase diagram, the glass-formation region (the bold line), and the glass-formation ability of alloys of the Ga–Te system (Minaev, 1991).

this eutectic. And meanwhile, this 'anti-eutectic' phenomenon, this 'liquidus temperature anti-effect' becomes completely clear when analyzed from the standpoint of the quantitative determination of the glass-formation ability of compositions containing more than one component, from the standpoint of the Sun–Rawson–Minaev criterion (Section 4.3). In accordance with this criterion

$$\mathrm{GFA} = \frac{E_{\mathrm{CIB}}}{T_{\mathrm{liq}}} \tag{9}$$

Figure 7 (Minaev, 1980c) shows the superposition of the GFA values on the phase diagram of the Ga–Te system. In accordance with the liquidus temperature effect, the GFA increases with increase in the gallium content starting from pure tellurium. But this increase does not stop in the eutectic point ($Ga_{14}Te_{86}$), where the GFA is equal to 0.265 kJ mol^{-1} K^{-1}, but increases further despite the increase in the liquidus temperature, against the liquidus temperature effect. At GFA = 0.267 kJ mol^{-1} K^{-1} (the composition $Ga_{15}Te_{85}$), the glass-formation region begins. Cornet stated that the maximal GFA is for the composition $Ga_{20}Te_{80}$. Calculations show that the GFA increases in the direction from the eutectic to this composition, but it continues to increase further and then in the region 24–25 at.% Ga it sharply decreases, seemingly 'having remembered' about the liquidus temperature effect—it is just here the liquidus temperature begins to increase sharply, whereas it had earlier shown a flat slope increase.

The increase in GFA in the range from 0 to 14% Ga (the eutectic) is explained, from one side, by the decrease in the liquidus temperature (decrease of the denominator in the GFA formula), and, from another side, by the increase in the covalent–ion binding of the alloy (the numerator). Actually, the CIB is determined by energies of chemical bonds presented in the alloy. The bond energy Ga–Te is equal to 177.3 kJ mol^{-1} K^{-1} (calculated from the Pauling's expression, Section 4.3), the bond energy Te–Te 168 kJ mol^{-1} K^{-1}. It is clear that with increase in the Ga content (which CN in addition is larger than that of tellurium-3 and 2, respectively), the E_{CIB} increases as well. At the Ga content > 14 at.%, the E_{CIB} continues to grow actively, although the liquidus temperature increases simultaneously. Demonstrating 'flat slope liquidus' action, the temperature increases only insignificantly. As a result, the action of the structural–chemical factor predominates over the thermal factor, and the GFA increases despite the increase (albeit very slow) in the liquidus temperature.

In the work of Vengrenovich et al. (1986), the glass-formation regions 7–28 at.% Al, 5–28 at.% Ga, 9–28.8 at.% In for binary telluride systems by the melt-spinning method (the cooling rate of $\approx 10^6$ K s^{-1}) were described. In systems Al–Te and Ga–Te, the regions of the easiest glass-formation are located outside the eutectic alloy, being shifted to the chemical compounds that form the eutectic with tellurium. It means that in the system Al–Te, which is also characterized by the initial 'flat' and consequent 'sharp' slope, both the action of the liquidus temperature effect and the action eutectic effect are not observed, but the action of the flat slope liquidus rule proposed by Minaev (1980c, 1991) is observed that is one of the components of the structural–energetic concept of the glass-formation in chalcogenide systems proposed by Minaev (1978, 1980b, 1991) as well.

The action of the rule can be predicted for systems As–Te, Si–Te, In–Te, Au–Te, and others.

4.7. Some Energetic and Kinetic Aspects of Glass-Formation and Criteria of Sun–Rawson and Sun–Rawson–Minaev

The Sun–Rawson criterion (1967) and the Sun–Rawson–Minaev criterion (1980a,b), provide a foundation to consider both the structural–chemical and the energetic aspects of glass-formation. The chemical bond of Rawson (1967) and the covalent–ion binding of Minaev (1978, 1980b, 1991) characterize the chemically ordered structural network of atoms of substance, the energy of this structure (E_{str}), i.e., $E_{CIB} = E_{str}$, whereas T_m and T_{liq} in the considered criteria are expressions of the thermal energy E_{therm} necessary for rupturing of chemical bonds, for re-structuring of the given structure, and for transforming it from the liquid structure into the solid-state structure (Rawson, 1967).

Both criteria lack a kinetic component, however, without which they cannot be used to evaluate dimensions of glass-formation regions and even to predict glass-formation itself in given conditions. Both criteria can be used to determine whether the GFA of a given composition is larger or greater than that of another composition under consideration.

Minaev (1978, 1980b,c, 1991) introduced the kinetic factor in his structural–energetic concept of glass-formation as a certain fixed cooling rate, at which the GFA of glass-forming alloys is equal or greater than a certain value as well. Apparently, it implies the well-known characteristic of glass—the critical cooling rate V_{cr}, i.e., the cooling rate at which the alloy of the given composition is still able to form glass. At the lesser cooling rate, a crystal is formed. Thus, in the given system each alloy of the given composition has its own V_{cr}. All alloys of this system with GFA, corresponding to the given V_{cr} and exceeding it, will form glasses at the given V_{cr}. The higher value of V_{cr} expands the glass-formation region, the lower value of V_{cr} narrows it. The above said does not mean that V_{cr} determines the glass-formation ability. On the contrary, GFA of every individual alloy, expressing its genuine physical–chemical essence, requires certain physical–chemical conditions to form glasses, in particular a certain V_{cr} at a certain (1 atm) external pressure acting on the material under cooling. For telluride alloys the latter value is (Minaev, 1980c, 1991) $0.270 \pm 0.010\ \mathrm{kJ\ mol^{-1}\ K^{-1}}$ at the cooling rate of $\approx 180\ \mathrm{K\ s^{-1}}$, $0.250 \pm 0.01\ \mathrm{kJ\ mol^{-1}\ K^{-1}}$ at the cooling rate of $10^6\ \mathrm{K\ s^{-1}}$, and 0. $230 \pm 0.01\ \mathrm{kJ\ mol^{-1}\ K^{-1}}$ at the cooling rate of $10^{10}\ \mathrm{K\ s^{-1}}$. As a result, the Sun–Rawson–Minaev criterion becomes applicable for the prognostic evaluation of the possibility of glass-formation at certain cooling rates. The above mentioned is related with telluride systems whose studies were undertaken by Minaev (1978, 1980a–c, 1991), but requires further development and collection of statistical data on concrete V_{cr}. As for sulfide and selenide systems, which have, as a rule, higher GFA, further research is needed with regard to the possibility of glass-formation at different cooling rates, particularly statistical data concerning critical cooling rates and the comparison with calculated values of the glass-formation ability. Only afterwards will it be possible to give prognostic evaluations of concrete glass-formation regions at certain cooling rates. At present, such evaluations are qualitative or, in the best case, semi-quantitative.

But let us come back to the energetic aspect of the glass-formation problem. The E_{therm} of Rawson (1967), mentioned above, is the product of T_m and the gas constant $R = 8.3143\ \mathrm{kJ\ mol^{-1}\ K^{-1}}$. In the GFA expression of the SRM criterion, Minaev (1991) replaced T_{liq} with $RT_{liq} = E_{therm}$

$$\mathrm{GFA} = \frac{E_{CIB}}{RT_{liq}} = \frac{E_{str}}{E_{therm}} \tag{11}$$

As a result, the ratio of two energies E_{str} (the energy of structure binding) and E_{therm} was obtained. It is the ratio of these energies that determines the glass-formation ability of substances.

For the system As–Te where, as Minaev (1979, 1980c) calculated, $\mathrm{GFA_{Te}} = 0.231 \pm 0.010\ \mathrm{kJ\ mol^{-1}\ K^{-1}}$, $\mathrm{GFA_{As18.8Te81.2}} = 0.274 \pm 0.010\ \mathrm{kJ\ mol^{-1}K^{-1}}$, $\mathrm{GFA_{As20Te20}} = 0.276 \pm 0.010\ \mathrm{kJ\ mol^{-1}\ K^{-1}}$, and $\mathrm{GFA_{As50Te50}} = 0.289 \pm 0.010\ \mathrm{kJ} \times \mathrm{mol^{-1}\ K^{-1}}$, now the GFA will be equal to 27.78, 33.00, 33.20, 33.60, and 34.80 dimensionless units, respectively. The obtained values of GFA are related with compositions (excluding the first—pure tellurium), which form glasses at cooling rates of $\approx 10^2\ \mathrm{K\ s^{-1}}$.

As a matter of fact, the calculated values of GFA show how many times the covalent–ion binding energy is higher than the thermal energy RT_{liq}, which is present in the system during crystallization (melting). As we can see, the glass-formation process in the As–Te system takes place at $E_{CIB} > 30RT_{liq}$.

It is interesting to note that Rawson's 1967 analysis of the kinetic theory of glass-formation as presented by Staveley (1955) and Turnbull and Cohen (1958), notes that the liquid without foreign crystallization centers would not crystallize if the activation energy is higher than 30 RT_m. According to Rawson, the activation energies, which determine origination of nucleus and crystal growth in glass-forming liquids such as SiO_2, GeO_2, B_2O_3, will be apparently the same order of value as the free activation energy of viscous flow (25–30 RT_m) because both crystallization and viscous flow break M–O bonds.

The same can be said about chalcogenide glasses, as the activation energy of viscous flow is apparently a part of the energy of the covalent–ion binding that conforms to the fact of the glass-formation in the As–Te system at $E_{CIB} > 33.2RT_{liq}$.

Rawson (1967) stated that the application of the kinetic theory of glass-formation to more complex systems had not been successful because of difficulties arising at such complications; however, he concluded that the simple criterion of Rawson, as well as more complicated and accurate theses of the kinetic theory of glass-formation, lead to the simple deduction: the glass-formation depends on relative values of strength of bonds (which must be ruptured at the crystallization), and the thermal energy necessary for this rupture. The melt crystallization rate, guided by the kinetic approach to glass-formation, is proportional to $\exp(-B_{M-O}/RT_m)$, where B_{M-O} is the strength of the chemical bond in the B_xO_y oxide, R, the gas constant. Therefore, Rawson continues, if the ratio B_{M-O}/RT_m is large, the crystallization rate will be small and stable glass will be formed as the result.

The criterion of Sun–Rawson–Minaev (1980b, 1982, 1991) has developed this idea further: the glass-formation ability of complex substances depends on the ratio of energies of chemical (for most glasses—covalent-ion) binding of the structure of the given substance and the thermal energy of the system, which is necessary to destroy this binding: $GFA = E_{CIB}/RT_{liq}$.

The comparison of GFA of glasses in the A^{IIIA}–B^{VI} system: 0.270 ± 0.010, 250 ± 0.010, and 0.230 ± 0.010 kJ mol^{-1} K^{-1}, formed at different cooling rates of melt ($\cong 10^2$, 10^6, and 10^{10} K s^{-1}) shows the exponential dependence of the critical cooling rate on the glass-formation ability of alloys that directly corresponds to the above-mentioned Rawson's thesis on the exponential dependence of the crystallization rate on the ratio B_{M-O}/RT_m.

Thus, Rawson's thesis (1967), stated in the general form, has been developed further in the structural–energetic concept of Minaev (1980b, 1991) and brought to concrete quantitative evaluations allowing the prediction of new glass-formation regions at certain cooling rates.

This result owes its appearance to the simultaneous consideration of the glass-formation process from structural–chemical, energetic (thermodynamic), and kinetic positions presented in a rather simplified form.

4.8. Periodic Law and Glass-Formation in Chalcogenide Systems

The analysis of glass-formation in chalcogenide systems, divided according to the principle of participation of elements of 1, 2, 3, 4, 5, 6, and 7 groups of the periodic table, been considered in the previous paragraph, allows to generalize some principal theses concerning the problem of glass-formation and its relation with the geography of individual elements on the map of the periodic table.

According to data of Minaev (1982), elements of all groups of the periodic table take part in the glass-formation in ternary chalcogenide systems, excluding the main subgroup of the eighth group (inert gases), and all periods, excluding the first and the seventh. Almost the same can be said also about binary systems excluding from glass-forming groups the second and the eighth ones. Such frequent occurrence of elements, forming chalcogenide glasses, gives the possibility of consideration of glass-formation as the periodical property of elements constituting glass-forming alloys. This approach is necessary for predicting new glass-forming systems and for seeking glasses with previously unknown combinations of physical–chemical properties.

The number of elements constituting binary chalcogen and chalcogenide systems and the number of binary systems, based on them, distributed in groups of the periodic table are as follows (in brackets—the predicted number of elements and systems) (Minaev, 1987a,b)

Groups	I	II	III	IV	V	VI	VII	Total
The number of elements	2(+6)	−(+1)	5	4	3	3	2(+3)	19(+10)
The number of systems	3(+17)	−(+3)	8	8	7	2(+1)	3(+12)	32(+33)

The comparison of known phase diagrams, glass-formation regions, and properties of glasses in binary and plotted on their bases ternary and multi-component systems reveals the direct genetic relationship between them. The prediction of 33 new binary glass-forming chalcogenide systems allows to predict, in accordance with the revealed genetic relationship, hundreds of new ternary and scores of tetrad glass-forming systems.

As seen from the data presented, moving from I to VII groups of the periodic table, the number of glass-forming systems becomes minimal in II group, approximately equal in III–V groups and decreases in VI–VII groups. Taking into account predicted systems in VII group, the number of systems increases significantly.

The distribution on periods gives the following result:

Period number	1	2	3	4	5	6
The number of elements	−(+1)	1(+2)	5(+1)	5(+2)	5(+2)	3(+2)
The number of systems	−(+3)	2(+6)	8(+6)	10(+7)	7(+5)	5(+6)

As it is seen, the number of glass-forming systems increases with the movement from the second period to the fourth one (the maximum) and than decreases with further movement to the lower part of the periodic table. The regularity remains unchanged after predicted systems are added, including the systems H–S, H–Se and H–Te (Minaev, 1991).

The analysis of the real GFA and GFA predicted with the help of the SRM criterion, shows that in the framework of individual periods of the periodic table the tendency of the glass-formation in binary chalcogenide systems decreases with the movement from systems with elements of the first group to systems with elements of the second group, and then it increases with increase in the atomic number of the non-chalcogen element in the third, fourth, and fifth groups, respectively.

The analysis of the glass-formation, carried out by Minaev (1980a, 1982, 1987a,b, 1991) of all existing chalcogenide systems has confirmed the existence of the following main features observed earlier in individual systems. Direct relation exists between structures of phase diagrams and the glass-formation ability of alloys: the minimum tendency to glass-formation becomes apparent usually for alloys corresponding to the chemical compound composition (excluding peritectical alloys); the glass-formation ability, as a rule, increases with decrease in the liquidus temperature, and this is typical mainly for chalcogen-enriched alloys, and is often maximal for chalcogen-enriched eutectic alloys.

In systems $A^{IIIA}-B^{VI}$, $A^{IVA}-B^{VI}$, and $A^{VA}-B^{VI}$ the general tendency, as indicated by Goryunova and Kolomiets (1958, 1960), which is a decrease in the glass-formation ability of alloys with an increase in atomic numbers of elements in groups of the periodic table becomes apparent, and this correlates with increased metallization and reduced chemical bond energy.

In some binary chalcogenide systems, Minaev (1980a, 1991) and Minaev, Kuznetsov and Fedorov (1982) have either described (systems with Al, Ga, In, Sn, Pb, Cu) or predicted (systems with Ag, Au, Hg, Bi) the phenomenon of the inversion of the regular decrease in the glass-formation ability with increase in chalcogen atomic numbers for alloys with tellurium comparing with alloys with selenium and sulfur (the Te–Se inversion) in accordance with the Sun–Rawson–Minaev criterion.

The cause for such an inversion (Te–Se) is the existence of phase diagrams of the non-glass-formation type for sulfur and selenium with above-mentioned elements that is characterized by a sharp rise in the liquidus temperature with addition to sulfur and selenium even first doses of these elements.

For telluride systems with elements of IA and IB subgroups of the periodic table, the existence of the thorough inversion in the regular decrease of the glass-formation ability with increase in atomic numbers of elements has been predicted (Minaev, 1985a,b, 1991), i.e., the glass-formation ability increases with increase in atomic numbers of elements of the first group. The existence of the tendency to such inversion has been predicted with several exclusions for alloys of alkaline metals with sulfur and selenium.

In binary systems A^{VIIA}–tellurium (where A^{VIIA} is Cl, Br, I), there is the thorough inversion of the GFA decrease at increasing atomic numbers—the glass-formation regions get bigger in the row of systems with chlorine (7 at.%), bromine (10 at.%), and iodine (15 at.%).

According to Minaev (1987a,b, 1991), there are also other types of the inversion in binary systems. Three types of them exist with elements of the fourth period giving larger glass-formation regions with respect to their analogs in groups of the third period. These types of the inversion 'Ge–Si' with S and Se, 'As–P' with S and Te, 'Se–S' with P can be considered as one general class of inversions '4th–3rd period' (4–3).

In the inversion class 5–4, experiments give only one type, 'Te–Se', the prediction adds the inversion 'Ag–Cu'.

The inversion class 6–5 is presented by two experimentally revealed inversion types: 'Pb–Sn', 'Tl–In' and by three predicted types: 'Au–Ag', 'Hg–Cd', and 'Bi–Sb'.

Some aspects of this analysis are presented by Minaev (1987a,b, 1991) (without taking into account cooling rates of melts) in generalized table (Table II) of the glass-formation in binary systems.

As shown, the largest population in the periodic table of elements initiating the glass-formation in binary chalcogenide systems is restricted by the square where p-elements of III–VI groups of 3–6 periods are located.

The increased population of s- and p-elements-initiators of I and VII groups of 1–6 periods has been predicted. From d-elements, only copper at high cooling rates (more than 100 K s^{-1}) gives glasses (with tellurium), and five binary glass-forming systems

TABLE II

THE PERIODICITY OF THE GLASS FORMATION IN BINARY CHALCOGENIDE AND CHALCOGEN SYSTEMS. IN BRACKETS—BORDERS OF GLASS-FORMATION REGIONS (IN AT.% OF NON-CHALCOGEN ELEMENTS), UNDERLINED—PREDICTED SYSTEMS (MINAEV, 1991)

Periods	Groups						
	I	II	III	IV	V	VI	VII
1							*H–S* *H–Se* *H–Te*
2	*Li–S* *Li–Se* *Li–Te*		B–S (B40S60) B–Se (0–40)				*F–S* *F–Se* *F–Te*
3	*Na–S* *Na–Se* *Na–Te*		Al–Te (12–30)	Si–S (31.2–50) Si–Se (0.1–20) Si–Te (10–22)	P–S (5–25) P–Se (0–52)	S–Se (0–100) *S–Te*	CI–S (10–70) *Cl–Se* *Cl–Te* (60–67)
4	*K–S* *Cu–Se* *K–Se* Cu–Te *K–Te*		Ga–Te (15–25)	Ge–S (10–47.6) Ge–Se (0–40) Ge–Te (12–22)	As–S (0–45) As–Se (0–60) As–Te (20–38)	Se–S (0–100) Se–Te (65–100)Se	*Br–S* *Br–Se* *Br–Te* (31–41)
5	*Rb–S* *Rb–Se* Ag–Se *Rb–Te* *Ag–Te*		In–Te (9–28.6)	Sn–Te ($Sn_{16}Te_{84}$)	Sb–S ($Sb_{40}S_{60}$) Sb–Se	*Te–S* Te–Se (65–100)Se	*I–S* I–Se I–Te (40–55)
6	Cs–S Cs–Se Cs–Te Au–Te	*Hg–S* *Hg–Se* *Hg–Te*	TI–S (28.6–50.0) TI–Se (0–33.3) TI–Te ($TI_{30}Te_{70}$)	Pb–Te (14.5–30)			

with d-elements have been predicted: Ag–Te, Au–Te, Hg–S, Hg–Se, and Hg–Te. In these cases, high cooling rates are necessary as well.

The listed types of inversions in the regular decrease of the glass-formation ability are, as mentioned before, one of the forms of the secondary periodicity manifestation. As long ago as Mendeleev (1864, 1947), complications of the periodicity in binary systems have been described as going from properties of elements to properties of individual substances and compounds. As is shown above, such complications of periodicity of glass-formation is also demonstrated in binary chalcogenide glass-forming systems. In the work of Biron (1915), the logic line of Mendeleev's consideration about properties' alterations of elements of the same group in the periodic table was developed and the attention was attracted to the absence in some cases of a monotonic character of alterations of one or another property while going in a group from one element to another in the direction of increase in their atomic numbers. Such a non-monotonic character was called the secondary periodicity.

Shchukarev and Vasilkova (1953) revealed the non-monotonic character of alteration of sums of ionization potentials (in electron-volts) of elements of IIIA and IVA groups with increase in atomic numbers (eV): (1) B—69.97; Al—53.74; Ga—57.02; In—52.37; Tl—56.27; (2) C—147.17; Si—102.62; Ge—103.24; Sn—93.27; Pb—96.71.

In both rows, the tendency of decrease in the sum of ionization potentials is observed, but the sum for gallium, thallium, germanium, and lead is larger than for preceding aluminum, indium, silicon, and tin, respectively. This situation is similar to that described above for glass-formation ability of systems with thallium and indium and lead and tin where the inversion is manifested most clearly.

Schukarev (1954) considers the secondary periodicity to be founded on properties of electronic shells of atoms. The main roles here are played by s-electrons and less important roles by p-electrons. At that the 'plunging' of elliptical s-orbits under shells of 10 d-electrons is significant. Further, the authors relate appearance of the secondary periodicity with d- and f-strengthening of the 'diving' electron bond and the compression of electron shells. The increase in the ionization potential, for example for thallium and lead, is apparently related with this.

Such an explanation is even more acceptable in the case of glass-formation because it is related with strengthening of chemical bonds, one of the main causes of increase in the glass-formation ability in the framework of the structural–energetic concept of Minaev (1980b, 1991).

Shchukarev and Vasilkova (1953) have made the conclusion that the secondary periodicity is related to the structure of the system of elements itself where periods, beginning from the second, are reiterated[3] by pairs, at that the first pair (2nd and 3rd periods) does not have d-electrons, the second pair (4th and 5th periods) contains d-electrons, and the third pair (6th and 7th periods) contains both d- and f-electrons.

This conclusion directly corresponds to the fact of revealing the inversion class 4th–3rd period and 6th–5th period reflecting the alteration of properties, the glass-formation in particular, at turning from one pair of periods of the periodic table to another pair.

[3] In our opinion, instead of 'are reiterated' it would be more exact to write 'follow each other from the standpoint of the regular alteration of properties'.

In the first case, the glass-formation ability increases for elements with complete d-shells (germanium, arsenic, selenium); in the second case, for elements with complete d- and f-shells (gallium, lead, bismuth, etc.).

It is interesting to note that in the systems $A^{VA}-B^{VI}$, there are four pairs of binary systems demonstrating the inversion in the main regularity of the glass-formation, revealed by Goryunova and Kolomiets (1958, 1960)—the decrease in the glass-formation ability with increase in the atomic number. These are pairs according to Minaev (1979, 1980c, 1981a,b): 'P–Se' and 'P–S,' 'As–S' and 'P–Se,' 'As–Se' and 'As–S,' 'As–Te' and 'P–Te.' In the group of systems $A^{IVA}-B^{VI}$, there are six pairs of inverted systems ('Si–Se' and 'Si–S,' 'Ge–S' and 'Si–S,' 'Ge–Se' and 'Si–S,' 'Sn–Te' and 'Sn–Se' 'Pb–Te' and 'Pb–Se,' 'Pb–Te' and 'Sn–Te'). In the group of systems $A^{IIIA}-B^{VI}$ there are seven pairs of inverted systems: 'B–Se' and 'B–S,' 'Al–Te' and 'Al–Se,' 'Ga–Te' and 'Ga–Se,' 'In–Te' and 'In–Se,' 'Tl–S' and 'In–S,' 'Tl–Se' and 'In–Se,' 'Tl–Se' and 'Tl–S.'

At the same time, the experimental data of binary and ternary chalcogenide systems with alkaline metals, presented in Section 4.5.5 as well as calculations, based on the SRM criterion, have allowed Minaev (1985a,b) to predict in the group of systems $A^{IA}-B^{VI}$ the common tendency for this groups to the thorough inversion in the regular decrease of GFA with increasing of the atomic numbers.

In the row of binary chalcogenide systems, containing consequentially elements of V^A, IV^A, III^A, and I^A subgroups of the periodic table, one can observe the decrease in the degree of appearance of the tendency to the regular decrease in GFA with increase in the atomic number of elements in the group down to the point of the predicable inverted regularity (thorough along the whole group)—the inversion for systems $A^{IA}-B^{VI}$ and A^{IB}–Te.

Both classes of the inversion 4–3 and 6–5, revealed in binary systems are also manifested in ternary chalcogenide systems in accordance with the thesis concerning genetic relations between phase equilibrium diagram structures, the glass-formation and glass properties in multi-component and binary systems. The same can be said about the inversion 'Te–Se'.

The examples of the '4–3' inversion are ternary systems Si–P–Te and Si–As–Te, Ge–P–Te, and Ge–As–Te considered by Minaev (1987b, 1991) based on Hilton et al.'s (1966) and Borisova's (1972) data; here glass-formation regions with arsenic are larger than corresponding regions with phosphorus. The inversion 6–5 is manifested, judging by data of Savage and Nielsen (1964) in several systems, for example in Ge–Bi–Se and Ge–Sb–Se ('Bi–Sb'). In the first system, glasses without a crystal phase were obtained up to 3 at.% Bi, in the second system addition of 2 at.% Sb gives partially crystallized glasses.

Facts that are now interpreted as one of variants of the 6–5 inversion (the inversion 'Pb–Sn') were revealed in ternary chalcogenide glasses by Goryunova and Kolomiets long ago. They showed that lead enters in glass-forming compositions in greater amount than tin in systems Sn(Pb)–As–S and Sn(Pb)–As–Se. The same inversion, judging by data of Feltz, Achlenzig, Arnold and Foigt (1974a,b), is observed in systems Ge–Sn–S and Ge–Pb–S. In the first system up to 47.5 mol.% SnS can be added without crystallization, in the second up to 57 mol.% PbS. The same inversion is also revealed in systems As–Sn(Pb)–Te, Si–Sn(Pb)–Te, and Ge–Sn(Pb)–Te according to experimental data of Minaev (1983) and Minaev et al. (1984). The inversion Pb–Sn is also manifested, according to Pazin,

Morozov and Borisova (1979), in tetrad component systems Sn(Pb)–Ge–As–Se—in vitreous alloys of these systems one can introduce up to 20 at.% P and only up to 15 at.% Sn.

So, the inversion Pb–Sn, typical for binary systems, expands also on ternary and tetrad chalcogenide systems. It appears Minaev (1980b, 1991) has supposed that the inversion phenomenon will manifest itself also in multi-component systems.

There is no doubt that other types of inversions will behave the same way, i.e., they will appear in multi-component systems.

It is interesting that the inversion Pb–Sn appears to exceed the bounds of chalcogenide systems as such and expands to oxide glasses as Minaev (1991) considers. For example, in the system $SnO-SiO_2$, only the composition with 50 mol.% SnO was obtained in the glassy state whereas in the system $PbO-SiO_2$ the upper border of the glass-formation amounts to 75 mol.% PbO (Mazurin, Streltsina and Shvayko-Shvaykovskaya, 1975–1979).

In the same works, eight more binary oxide systems are presented where one of components is PbO (other components are B_2O_3, Al_2O_3, GeO_2, and others), and there are no single systems with tin oxides.

The thorough inversion predicted for systems with alkaline metals appears in ternary systems based on As–Se. According to data of Dembovsky and Ilizarov (1978), in these systems the glass-formation region increases in the row: lithium < sodium < potassium < rubidium < cesium. The increase in the glass-formation region in systems alkaline metal–As_2Se_3 in the row lithium < sodium < potassium was stated by Borisova (1971). The glass-formation region in the system $Cs_2S_3-Sb_2S_3$, according to data of Kokorina et al. (1970), is significantly larger than that in the system $Rb_2S-Sb_2S_3$.

It is interesting that the latter inversion manifests itself also in some binary and ternary oxide systems as is shown in works of Minaev and Timoshenkov (2002).

The thorough inversion, according to Dembovsky and Ilizarov (1978), is also stated in the group of ternary systems As–S–II^B, where II^B—Zn, Cd, Hg, and is forecasted in other ternary systems with elements of the IIB subgroup.

In the conclusion, we will consider glass-formation in ternary sulfide systems with rare earth elements and Ga. The most systematically studied are the systems (Cervelle, Jaulmes, Laurelle and Loireau-Lorach, 1980): $Ln_2S_3-Ga_2S_3$, where Ln—La, Ce, Pr, Nd, Sm, Eu, Gd, Tb, Dy, Ho, Er, and the system $Y_2S_3-Ga_2S_3$. Glass-formation regions of these systems become smaller with increase in atomic numbers of lanthanides. The exception is europium, which does not form glass in this particular system. It is interesting that europium violates regularities, typical for other lanthanoids, in other properties as well (T_m, T_{boil}, the atomic radius, the density, and others). Yttrium—the element of 5th period located in the III group above lanthanides—imparts less glass-formation ability to alloys of this system than lanthanum and following it are Ce, Pr, Nd with greater atomic mass located in 6th period. Thus, the described above inversion 5–6 in the regular decrease of GFA with increase in atomic numbers of elements is manifested here as well.

5. Conclusion

The structural–energetic concept of glass-formation (with elements of kinetic approach) in chalcogenide systems has been presented in this chapter.

The bases of the concept proposed are the following theses:

- the glass-formation in binary chalcogenide systems is typical mainly for chalcogenide-enriched compositions and determined by glass-forming types of phase diagrams characterized by low-temperature eutectics;
- the classification of ternary systems has been proposed whose feature is the combination of types of phase diagrams of binary systems; a ternary system is glass-forming if between participating binary systems there is at least one system characterized by a glass-forming phase diagram;
- in accordance with the qualitative criterion of the glass-formation in ternary chalcogenide systems, glass-formation regions are usually located near lines of dilution of binary eutectics by the third component (element) of the common ternary system or/and near lines of dilution of binary eutectics by the chemical compound that is the component of a particular ternary system—the member of the common ternary system;
- in accordance with the quantitative criterion of glass-formation applied to chalcogenide systems with any number of components (the criterion of Sun–Rawson–Minaev), the glass-formation ability of a substance is the ratio of the chemical, as a rule covalence–ionic, binding of one mole of atoms of this substance and its liquidus temperature at the normal pressure; the glass-formation ability is the property inherent to (in greater or lesser extent) a substance, it is determined by its chemical nature and does not depend on external impacts on the glass-formation process, the cooling rate in particular;
- as the criterion of the actual (concrete experimental) glass-formation, the average value of the glass-formation ability of boundary compositions (glass–crystal) of glass-formation regions of the same type of glasses, obtained at a certain cooling rate, is taken; in telluride compositions, glasses are formed in the case of cooling rates $V_c \approx 10^2$ K s^{-1} at GFA $\geq 0.270 \pm 0.010$ kJ mol^{-1} K^{-1}, in the case of cooling rates $V_c \approx 10^6$ K s^{-1} at GFA $\geq 0.250 \pm 0.010$ kJ mol^{-1} K^{-1}; a criterion of the concrete experimental glass-formation cannot be formulated without taking into account the kinetic factor expressed in this case through cooling rates of melts;
- in systems with two chalcogens, the glass-formation was known for systems S–Se and Se–Te; the usage of the Sun–Rawson–Minaev criterion has allowed to predict and experimentally reveal the glass-formation region (0–29 at.% Te) in the system S–Te where glasses exist at negative temperatures;
- elements of all groups of the periodic table, excluding the second and the eighth, take part in the glass-formation in binary chalcogenide systems; in ternary systems some elements of the second and eighth groups take part in the glass-formation as well;
- in binary and ternary chalcogenide systems with elements of the third, fourth, fifth, and sixth groups of the periodic table, the common tendency to regular decrease in the glass-formation ability of alloys with increase in atomic numbers of elements in the group is manifested that correlates with increase in the metallization and decrease in the energy of chemical bonds;
- in binary systems several inversions in the regular decrease of GFA with increase in atomic numbers of elements in the group has been revealed:

(a) the inversion Te–Se: the presence of glass-formation regions for Al, Ga, In, Sn, Pb, Cu with Te and absence of such with S and Se having lesser atomic numbers than Te;
(b) the inversion 4th–3rd period (4–3): 'Ge–Si' with S and Se, i.e., the glass-formation region of Ge with S or Se is larger than that of Si with S or Se, respectively; 'As–P' with S and Te, 'Se–S' with B, P, and As;
(c) the inversion 6th–5th period (6–5): 'Pb–Sn' with Te, 'Tl–In' with S and Se;
(d) the thorough—in the whole group—inversion in decreasing of GFA with increase in atomic numbers of A^{VIIA} elements (Cl, Br, I) in the A^{VIIA}–tellurium system;

- in the row of binary chalcogenide systems containing consequentially elements of V^A—P, As, Sb, Bi, IV^A—Si, Ge, Sn, Pb, III^A—B, Al, Ga, In, Tl, and I^A—Li, Na, K, Rb, Cs subgroups of the periodic table, decrease in the degree of manifestation of the tendency to the regular decrease in the glass-formation ability with increase in the atomic numbers of elements in the group to the point of the predictable opposite (inverse) regularity (the thorough—along the whole group—inversion) for systems I^A–chalcogen is observed;
- the thorough inversion in the regular decrease of GFA with increase in atomic numbers has been also predicted for some binary and ternary chalcogenide systems containing elements of IB (Cu, Ag, Au) and IIB (Zn, Cd, Hg) subgroups of the periodic table;
- regularities of the glass-formation (including the inversion) revealed in binary chalcogenide systems have been also experimentally revealed in ternary chalcogenide systems;
- inversions in the regular decrease of GFA with increase in atomic numbers of elements of corresponding groups are one of forms of the manifestation of the secondary periodicity of properties of elements of the periodic table;

Practically each of the proposed theses can play, to some extent, a prognostic role for revealing the glass-formation in systems where it was not known before. This role becomes significantly more important at simultaneous accounting of all prognostic aspects of the structural–energetic concept of glass-formation that have been used to predict some binary and ternary glass-forming chalcogenide systems, the prognosis which was confirmed for binary systems Cs–Se, S–Te, Te–Cl, Te–Br, and some ternary telluride systems Cu–Si–Te, Ga–Si–Te, Ga–Pb–Te, Si–Sb–Te, Ge–Sb–Te, and others.

The structural–energetic concept of glass-formation in chalcogenide systems can be used, to substantial extent, for revealing of peculiarities of glass-formation and predicting of it in oxide, halide, and other systems.

References

Batsanov, S.S. (2000) *Structural Chemistry. Book of Facts (in Russian)*, Dialog-MGU Publishers, Moscow, pp. 292.

Berul, S.I., Lazarev, V.B. and Salov, A.V. (1971) Thermo-graphic investigations of $Me_2X–Sb_2X_3$ phase diagrams, *J. Non-Org. Chem. (in Russian)*, **16**, 3363–3366.

Biron, E.V. (1915) Phenomena of secondary periodicity, *J. Russ. Phys.–Chem. Soc.*, **47**, 964–988.

Boolchand, P., Bresser, W.J., Georgiev, D.G., Wang, Y. and Wells, J. (2001) Evidence for intermediate phase in chalcogenide glasses. In *Phase Transitions and Self-Organization in Electronic and Molecular Networks* (Eds, Phillips, J.C. and Thorpe, F.) Kluwer Academic/Plenum Publishers, New York, Dordrecht, pp. 65–84.

Boolchand, P. and Thorpe, M.F. (1994) Glass-forming tendency, percolation of rigidity, and one-fold-coordinated atoms in covalent networks, *Phys. Rev.*, **B50**, 10366–10368.

Boriakova, V.A., Grinberg, Ja.H., Zhukov, E.G., et al., (1969) Phase diagrams of B_2Se_3–Se system, *Proc. AS USSR. Non-Org. Mater.*, **5**, 477–480.

Borisova, Z.U. (1971) Glass-formation in chalcogenide systems and Periodic Table of elements, *Proc. AS USSR. Non-Org. Mater.*, **7**, 1720–1724.

Borisova, Z.U. (1972) *Chemistry of Glassy Semiconductors*, LGU Publishers, Leningrad, pp. 248.

Borisova, Z.U. (1976) Glass-formation in chalcogenide systems and the Periodic Table. *Structure and Properties of Non-crystalline Semiconductors*, Nauka Publishers, Leningrad, pp. 6–12.

Brasen, D. (1974) Glass-formation and switching effect in Sb–Se system, *J. Non-Cryst. Solids*, **15**, 395–398.

Cervelle, B.D., Jaulmes, S., Laurelle, P. and Loireau-Lorach, A.M. (1980) Variation with composition of refraction indexes of lanthanum sulfide and gallium sulfide glasses and indexes of some related glasses, *Mater. Res. Bull.*, **15**, 159–164.

Cervinka, L. and Hruby, A. (1978a) The structure of amorphous TlS_x system, *J. Non-Cryst. Solids*, **30**, 191–209.

Cervinka, L. and Hruby, A. (1978b) The structure of amorphous TlS_x and $TlSe_x$ systems. In *Amorphous Semiconductors-78*. Proceedings of International Conference Pardubice, Prague, pp. 211–214.

Cervinka, L. and Hruby, A. (1979) The structure of amorphous $TlSe_x$ system, *J. Non-Cryst. Solids*, **34**, 275–295.

Chizhevskaya, S.N., Abrikosov, N.Kh. and Azizova, B.B. (1973) Investigations of interaction of selenium with iodine in solid and liquid states, *Proc. AS USSR. Non-Org. Mater.*, **9**, 218–221.

Chizhikov, D.M. and Schastliviy, V.P. (1964) *Selenium and Selenides*, Nauka Publishers, Moscow.

Chizhikov, D.M. and Schastliviy, V.P. (1966) *Tellurium and Tellurides*, Nauka Publishers, Moscow.

Chuntonov, K.A., Kuznetsov, A.N., Fedorov, V.M. and Yatsenko, S.P. (1982) Synthesis of alloys and phase diagram of Cs–Te system, *Proc. AS USSR. Non-Org. Mater.*, **18**, 1108–1112.

Cornet, J. (1976) The eutectic law for binary Te-based systems: a correlation between glass-formation and eutectic composition. In *Structure and Properties of Non-crystalline Semiconductors*. Proceedings of the Sixth International Conference on Amorphous and Liquid Semiconductors, Leningrad, USSR, November 18–24, 1975 (Ed., Kolomiets, B.T.) Nauka Publishers, Leningrad, pp. 72–77.

Dalba, G., Fornasini, P., Giunta, G., Burattini, E. and Tomas, A. (1987) Local structure in non-crystalline Sb–S semiconductors, *J. Non-Cryst. Solids*, **97–98**(part I), 411–414.

Das, G.C., Bever, M.B., Uhlman, D.R. and Moss, S.C. (1972) Relaxation phenomena in amorphous selenium–tellurium alloys, *J. Non-Cryst. Solids*, **7**, 251–270.

Davies, H.A. and Hall, J.B. (1974) Some aspects of splat-quenching in inert atmosphere and of the formation of non-crystalline phases in Al–1.3 at.% Cu, germanium and tellurium, *J. Mater. Sci.*, **9**, 707–717.

Dembovsky, S.A. (1977) Empiric Theory of Glass-formation in Chalcogenide Systems, *J. Non-Org. Chem.*, **22**, 3187–3199.

Dembovsky, S.A. (1978) Some aspects of glass-formation in chalcogenide systems, *Phys. Chem. Glass*, **4**, 522–528.

Dembovsky, S.A. and Ilizarov, L.M. (1978) Application of empiric theory of glass-formation to calculations of region borders in chalcogenide systems and coordination of melts, *Proc. AS USSR. Non-Org. Mater.*, **14**, 1997–2006.

Donald, L.W. and Davies, H.A. (1978) Prediction of glass-forming ability for metallic systems, *J. Non-Cryst. Solids*, **30**, 77–85.

Fedorov, V.M., Chuntonov, K.A., Kuznetsov, A.N., Bolshakova, G.A. and Yatsenko, S.P. (1985) Phase diagram of the cesium–selenium system, *Proc. AS USSR. Non-Org. Mater.*, **21**, 1960–1961.

Feltz, A., Achlenzig, E. and Arnold, D. (1974a) Glasbildung im System SnS–GeS–GeS_2 und die Mossbauer-Spektren der Glaser, *Ztschr. Anorg. und Allg. Chem.*, **403**, 243–250.

Feltz, A. and Foigt, B. (1974b) Bleithiogermanat (II, IV)-Glaser and ihre Eigenschaften, *Ztschr. Anorg. und Allg. Chem.*, **403**, 61–67.

Frankenheim, M.L. (1835) *Die Lehre von der Cohäsion umfassend die Elastizitat der Gase, die Elastizitet und Cohärenz der Flüssigen und festen Körper, und die Kristalkunde*, A. Schulz und Komp., Breslau.

Frerichs, R. (1953) New optical glasses with good transparency in the infrared, *J. Optic Soc. Am.*, **43**, 1153–1157.

Funtikov, V.A. (1987) Stable electronic configurations and properties of glassy semiconductors. In *New Ideas and Physics of Glass* (Ed., Matveev, G.M.) Moscow Mendeleev's Chemical–Technological Institute, Moscow, pp. 141–148.

Glazov, V.M., Chizhevskaya, O.N. and Glagoleva, N.N. (1967) *Liquid Semiconductors*, Nauka Publishers, Moscow.

Goldschmidt, V.M. (1926) Geochemische Verteilungsetze der Elemente. VIII. Untersuchungen über Bau und Eigenschaften von Kristallen. *I. Math. Naturvidenskapelig Klassen*, Skrifter Norske Videnskaps Akad., Oslo.

Golubkova, G.V., Petrov, E.S. and Kanev, A.N. (1975) Investigations of the Se–Br system, *Proc. Siberian Branch AS USSR. Ser. Chem. Sc.*(5), 39–44.

Goryunova, N.A. and Kolomiets, B.T. (1958) Glassy semiconductors. IV. On the problem of regularities of glass-formation, *J. Tech. Phys.*, **28**, 1922–1932.

Goryunova, N.A. and Kolomiets, B.T. (1960) Glassy semiconductors. IX. Glass-formation in compound chalcogenides based on arsenic sulfide and selenide, *Solid State Phys.*, **2**, 280–283.

Hagenmuller, P. and Chopin, P. (1962) Sun le systeme bore-soufre, *C.r. Acad. Sci.*, **255**, 2259–2260.

Hilton, A.R., Jones, C.E. and Brau, M. (1966) Non-oxide IVA–VA–VIA chalcogenide glasses, *Phys. Chem. Glasses*, **7**, 105–126.

Hruby, A. (1972) Evaluation of glass-forming tendency by means of DTA, *Czech. J. Phys.*, **B22**, 1187–1193.

Ignatyuk, V.A., Stavnistyi, N.N. and Minaev, E.N. (1980) Glassformation and parameters of the short-range order in the I-Te system, In *Structure, Physical-Chemical Properties and Application of Non-Crystalline Semiconductors*. AS USSR, AS MSSR, Kishinev, pp. 117–118.

Ioh, L.G. and Kokorina, V.F. (1961) Optical glasses transparent in IR spectrum range $\lambda = 11 \div 15$ microns (Parts I and II), *Optic.–Mech. Ind., Leningrad, USSR*(4), 39–41 (6, pp. 48–53).

Kaczorowski, M., Dabrowski, B. and Matyja, H. (1977) Crystallization of amorphous tellurium–lead alloys obtained by splat-cooling, *Mater. Sci. Eng.*, **29**, 189–192.

Kobeko, P.P. (1952) *Amorphous Substances*, AS USSR Publishers, Moscow.

Kokorina, V.F. (1971) Influence of chemical bond on glass-formation and glass properties. *Vitreous State*, Nauka Publishers, Leningrad, pp. 87–92.

Kokorina, V.F., Ioh, L.G., Kislitskaya, E.A. and Melnikov, V.V. (1976) New investigations of glass-formation and properties of non-oxide semiconductor glasses. In *Structure and Properties of Non-Crystalline Semiconductors* (Ed., Kolomiets, B.T.) Nauka Publishers, Leningrad, pp. 39–43.

Kolomiets, B.T. and Goryunova, N.A. (1955a) Electrical properties and structure of some ternary semiconductor systems. *Theoretical Problems and Investigations of Semiconductors and Processes of Semiconductor Metallurgy*, AS USSR Publishers, Moscow, pp. 38–58.

Kolomiets, B.T. and Goryunova, N.A. (1955b) Properties and structure of ternary semiconductor systems, *J. Technol. Phys.*, **25**, 984–994.

Kumanin, K.P. and Mukhin, E.Ya. (1947) On control of crystallization ability of industrial glasses, *Optic.–Mech. Ind.*, **1**, 21–25 (Leningrad, USSR (in Russian)).

Kuznetsov, A.N., Chuntonov, K.A. and Yatsenko, S.P. (1977) Rubidium–tellurium phase diagram. *Theses of Reports of All-Union Conference on Chemistry and Technology of Rare Alkaline Elements*, Nauka Publishers, Moscow, pp. 122–123.

Lasoca, M. and Matyja, H. (1974) Studies of structural relaxation of some chalcogenide glasses using D.S.C. measurements, *J. Non-Cryst. Solids*, **14**, 41–47.

Lebedev, A.A. (1910) *Proc. St. Peterburg Polytechnic Institute*, **13**, 613.

Lebedev, A.A. (1921) On polymorphism and annealing of glass, *Proc. State Optic. Inst., Sanct-Peterburg*, **2**(10), 1–21.

Lebedev, A.A. (1924) On annealing of optical glass, *Proc. State Optic. Inst., Leningrad*, **3**(4), 1–24.

Lucas, J. and Zhang, X.H. (1990) The tellurium halide glasses, *J. Non-Cryst. Solids*, **125**, 1–16.

Ma Hong Li, Zhang, X.H. and Lucas, J. (1991) "High T_g" tellurium selenium halide glasses. *J. Non-Cryst. Solids*, **135**, 49–54.

Mazurin, O.V., Streltsina, M.V. and Shvayko-Shvaykovskaya, T.P. (1973, 1975, 1977, 1979) *Properties of Glasses and Glass-forming Melts. Handbook*, Vol. 1–6, Nauka Publishers, Leningrad.

Melekh, B.T. and Maslova, Z.V. (1976) Glassy antimony sulfide and some of its properties, *Phys. Chem. Glass*, **2**, 189–190.

Mendeleev, D.I. (1864) *Glass Production*, St. Petersburg.

Mendeleev, D.I. (1947) *Principles of Chemistry*, 13th ed., Goshimizdat Publishers, Moscow.

Minaev, V.S. (1977) On the problem of prediction of glass-formation regions in ternary chalcogenide systems and some "violations" of glass-formation regularities. *Deposit Fund No. DE-2213*, Publication of Central Science and Information Institute Electronics, Moscow.

Minaev, V.S. (1978) Criterion of glass-formation in chalcogenide systems. *Proceedings of International Conference on Amorphous Semiconductors—78*, AS ChSSR Publishers, Pardubice, Prague, pp. 71–74.

Minaev, V.S. (1979) Regions of existence of glassy semiconductor materials in systems $A^{VA}–B^{VIA}$ and their relation with phase diagrams. Electronics Engineering Series: Materials, Vol. 1, CSII Electronics, Moscow, pp. 52–60.

Minaev, V.S. (1980a) Prediction and seeking for new chalcogenide glassy semiconductors, Electronics Industry, 8(92)–9(93), CSII Electronics, Moscow, pp. 67–71.

Minaev, V.S. (1980b) Structural–energy concept of glass-formation in chalcogenide systems, Electronics Engineering Series: Materials, Vol. 9 (145), CSII Electronics, Moscow, pp. 39–48.

Minaev, V.S. (1980c) Peculiarities of glass-formation in systems $A^{III}A–B^{VI}A$ and their relation with phase diagrams structure, Electronics Engineering Series: Materials, Vol. 9 (145), CSII Electronics, Moscow, pp. 29–38.

Minaev, V.S. (1981a) Glass-formation in the main subgroup of the sixth group of the Periodic Table, Electronics Engineering Series: Materials, Vol. 11 (160), CSII Electronics, Moscow.

Minaev, V.S. (1981b) Glass-formation ability of chalcogenides of IVA-group of the Periodic Table and its relation with phase diagrams, Electronics Engineering Series: Materials, Vol. 8 (157), CSII Electronics, Moscow, pp. 45–51.

Minaev, V.S. (1982a) Classification of glass-formation diagrams of ternary chalcogenide systems, Electronics Engineering Series: Materials, Vol. 7 (168), CSII Electronics, Moscow, pp. 47–53.

Minaev, V.S. (1983) New glasses and some peculiarities of glass-formation in ternary telluride systems, *Phys. Chem. Glass*, **9**, 432–436.

Minaev, V.S. (1985a) Classification of varieties of solid state of substances based on differences of short-range orders. *Abstracts of All-Union Conference "Glassy Semiconductors"*, AS USSR, Leningrad, pp. 137–138.

Minaev, V.S. (1985b) Glass-formation in chalcogenide systems $A^{IA}–B^{VIA}$. Experimental data and prognosis, Electronics Engineering Series: Materials, Vol. 9 (208), CSII Electronics, Moscow, pp. 44–50.

Minaev, V.S. (1987a) Periodic law and glass-formation in chalcogenide systems. Collected Reports of All-Union Seminar on New Ideas in Physics of Glass, Vol. 2, pp. 125–132.

Minaev, V.S. (1987b) Structural–energy concept of glass-formation in chalcogenide systems. *Abstracts of Seminar on Structure and Nature of Metal and Non-Metal Glasses*, Udmurtian State University, Izhevsk, USSR, pp. 68.

Minaev, V.S. (1988) Possibilities of glass-formation in binary chalcogenide systems $A^{IB}–B^{VI}$ and $A^{IIB}–B^{VI}$. *Deposit Fund, No. P4674*, Publication of Central Science and Information Institute Electronics, Moscow.

Minaev, V.S. (1989) Glass-formation in systems $A^{VIIA}–B^{VI}$. Experimental data and prognosis. *Deposit Fund, No. P4710*, Publication of Central Science and Information Institute Electronics, Moscow.

Minaev, V.S. (1991) *Glass-Forming Semiconductor Alloys*, Metallurgy Publishers, Moscow.

Minaev, V.S., Glazov, V.M. and Aliev, I.G. (1984) Formation and properties of chalcogenide glasses in systems $A^{IV}–B^{IV}–Te$ (A^{IV}, B^{IV}—Si, Ge, Sn, Pb). *Proceedings of International Conference on Amorphous Semiconductors—84*, Vol. 1, AS BNR, Bulgaria, Sofia-Gabrovo, pp. 204–206.

Minaev, V.S., Khan, V.P., Mladentseva, E.V., Sotnikova, M.N., Chuzhanov, F.V. and Shchelokov, A.N. (1988) Glass-formation in the Gallium–Thallium–Tellurium System, *Proc. AS USSR. Non-Org. Mater.*, **24**, 697–699.

Minaev, V.S., Kuznetsov, Yu.N. and Fedorov, V.A. (1982) Classification of glass-formation diagrams in ternary chalcogenide systems. *Proceedings of International Conference on Amorphous Semiconductors—82*, AS SRR, Romania, Bucharest, pp. 100–102.

Minaev, V.S., Kuznetsov, Yu.N., Petrova, V.Z. and Boguslavskiy, A.N. (1978) Some features of glass-formation in ternary chalcogenide systems. *Proceedings of International Conference on Amorphous Semiconductors—78*, AS ChSSR Publishers, Pardubice, Prague, pp. 79–82.

Minaev, V.S. and Shchelokov, A.N. (1987) Short-range order and classification of varieties of solid state of substance, *Proc. AS USSR. Non-Org. Mater.*, **23**, 1021–1026.

Minaev, V.S. and Timoshenkov, S.P. (2002) On the general regularity of glass-formation in binary chalcogenide ($A^{I}–B^{VI}$) and oxide ($A^{I}2–B_xO_y$) systems where A^{I}—Li, Na, K, Rb, Cs. Proceedings of 6th ESG Conference Glass Odyssey, Montpellier, France, CD-ROM.

Mukhin, E.Ya. and Gutkina, N.G. (1960) *Crystallization of Glasses and Methods of Its Prevention*, Gosoboronizdat, Moscow.

Myuller, R.L. (1940) Structure of solid glasses on electro-conductivity data, *Proc. AS USSR. Ser. Phys.*, **4**, 607–615.

Myuller, R.L. (1960) Chemical peculiarities of polymeric glassforming substances and nature of glass-formation. *Vitreous State*, AS USSR Publishers, Moscow, pp. 61–71.

Myuller, R.L. (1965) Chemistry of solid state and vitreous state. *Chemistry of Solid State*, Leningrad State University Publishers, Leningrad, pp. 9–18.

Myuller, R.L., Baydakov, L.A. and Borisova, Z.U. (1962) Electro-conductivity of As–Se–Ge system in vitreous state, *Bull. Leningrad State Univ.*, **17**(10, Part 2), 94–102.

Nisselson, L.A., Sokolova, T.D. and Soloviev, S.I. (1980) Investigation of S–Cl system, *J. Non-Org. Chem.*, **25**, 520–526.

Oven, A.E. (1970) Semiconducting glasses, *Contemp. Phys.*, **11**, 227–286.

Ovshinsky, S.R. (1976) Lone pair relations and the origin of excited states in amorphous chalcogenides. In *Structure and Excitations of Amorphous Solids*, Proc. AIP Conf. No. 31 (Eds, Lukovsky, G. and Galeener, F.L.) American Institute of Physics, New York, pp. 31.

Pauling, L. (1970) *General Chemistry*, W.H. Freeman and Company, San-Francisco.

Pazin, A.V., Morozov, V.A. and Borisova, Z.U. (1979) Glass-formation region and some properties of lead selenide–germanium selenide–arsenic selenide (PbSe–GeSe–AsSe) glasses, *Bull. Leningrad State Univ.* (22, Part 4), 84–89.

Perepezko, J.H. and Smith, J.S. (1981) Glass-formation and crystallization of highly undercooled Te–Cu alloys, *J. Non-Cryst. Solids*, **44**, 65–83.

Phillips, J.C. (1979) Topology of covalent non-crystalline solids I: Short-range order in chalcogenide alloys, *J. Non-Cryst. Solids*, **34**, 153–181.

Poray-Koshits, E.A., Besborodov, M.A., Bobkova, N.M., Brekhovskikh, S.M., Ermolenko, N.I. and Mazo, E.E. (1959) Glass structure. *Diagrams of Vitreous Systems*, Belarusian State University Publishers, Minsk, pp. 13–35, Chapter I.

Rawson, H. (1956) The relationship between liquidus temperature, bond strength and glass-formation. *Proceedings of IV International Congress on Glass*, Impremenie Chaix, Paris, pp. 62–69.

Rawson, H. (1967) *Inorganic Glass-Forming Systems*, Academic Press, London.

Ribes, M., Barrau, B. and Souquet, J.L. (1980) Sulfide glasses: glass-forming regions, structure and ionic conduction of glasses in $Na_2S–XS_2$ ($X = Si, Ge$), $Na_2S–P_2S_5$ and $Li_2S–GeS$ systems, *J. Non-Cryst. Solids*, **38–39**, 271–276.

Salov, A.V., Lazarev, V.B. and Berul, S.I. (1977) *J. Non-Org. Chem.*, **22**, 1437.

Samsonov, G.V. and Drosdov, S.V. (1972) *Sulfides*, Metallurgy Publishers, Moscow.

Savage, J.A. and Nielsen, S. (1964) Preparation of glasses transmitting in the infrared between 8 and 15 μ, *Phys. Chem. Glasses*, **5**, 82–86.

Schultz-Sellak, C. (1870) Diatermanisie einen Reihe von Stoffen für Wärme sehr geringer Brechbarkeit, *Ann. Phys. Chemic*, **139**, 182–187.

Shchukarev, S.A. (1954) The Periodic Law of D.I. Mendeleev as the Main Principle of Modern Chemistry, *J. Gen. Chem., (in Russian)*, **24**, 581–592.

Shchukarev, S.A. and Vasilkova, I.V. (1953) The phenomenon of secondary periodicity on the example of magnesium compounds, *Bull. Leningrad State Univ.* (2), 115–120.

Shtets, P.P., Bletskan, D.I., Turianitsa, I.D., Bodnar, M.P. and Rubish, V.M. (1989) Production, structure and photoelectric properties in the Sb–S system, *Proc. AS USSR. Non-Org. Mater.*, **26**, 933–937.

Smekal, A.G. (1951) The structure of glass, *J. Soc. Glass Technol.*, **35**, 411–420.

Sosman, R.B. (1927) *The Properties of Silica*, American Chem. Society Monograph Series, New York.

Stanworth, J.E. (1952) Tellurite glasses, *J. Soc. Glass Technol.*, **36**, 217T–241T.

Staveley, L.A.K. (1955) Homogeneous nucleation in supercooled liquids. *The Vitreous State*, Glass Delegacy of the University of Sheffield, pp. 85–90.

Suhrman, R. and Berndt, W. (1940) Uber die irreversiblen anderungen des elektrishen wilderstandes und hichtreflexions-vermogens von bei tiefen temperaturen kondensierten antimon- arsen-, tellur-, eisen- und sibler-schiehten, *Z. Physik*, **115**, 17–46.

Sun, K.H. (1947) Fundamental condition of glass-formation, *J. Am. Ceram. Soc.*, **30**, 277–281.

Suvorova, L.N. (1974) Investigations of glass-formation and crystallization of glass in the S–Se–Te system. Abstract of thesis for Doctor of Chemistry degree, Leningrad State University.

Suvorova, L.N., Borisova, Z.U. and Orlova, G.M. (1974) Glasses in the S–Se–Te system, *Proc. AS USSR. Non-Org. Mater.*, **10**, 441–445.

Tammann, G. (1903) *Kristalliziren und Schmelzen. Leipzig.*

Tammann, G. (1935) *Vitreous State*, ONTI, Moscow.

Tananaev, I.V. (1972) Scientific fundamentals of non-organic materials production for new technologies, *Bull. AS USSR*, **2**, 21–29.

Tarasov, V.V. and Savitskaya, Ya.S. (1953) *Proc. AS USSR*, **88**, 1019–1022.

Thorpe, M.F. (1983) Continuous deformation in random networks, *J. Non-Cryst. Solids*, **57**, 335–370.

Toepler, G.M. (1894) Bestimmung der volumanderung beim schmelzen fur eine ansahl von elementen, *Ann. Phys. Chem.*, **53**, 343–378.

Tsaneva, K. and Bontscheva-Mladenova, Z. (1978) Untersuchung der glasartigen phasen im system Se–S–Ag, *Monatsh. Chem.*, **109**, 911–918.

Turnbull, D. and Cohen, M. (1958) *J. Chem. Phys.*, **29**, 1049.

Turnbull, D. and Cohen, M. (1959) *J. Chem. Phys.*, **31**, 1164–1169.

Uhlman, D.R. (1977) Glass-formation, *J. Non-Cryst. Solids*, **25**, 43–85.

Vallev, N.Kh, Oblasov, A.K., Gvozdyrev, A.V. and Minaev, V.S. (1987) The nature of low-temperature glasses in the S-Te system. *Abstracts of Seminar on Structure and Nature of Metal and Non-Metal Glasses*, Udmurtian State University, Izhevsk, USSR, pp. 51.

Vengrenovich, R.D., Lopatnyuk, I.A., Mikhalchenko, V.P. and Kasian, I.M. (1986) Production and crystallization of amorphous Te-based alloys. *Proc. II All-Union Conf. Material Science and Technology of Chalcogenide and Oxygen-Containing Semiconductors*. Chernovtsy, USSR, **1**, 150.

Vinogradova, G.Z. (1984) Glassformation and phase equilibriums in chalcogenide systems. *Binary and Ternary Systems*. Nauka Publishers, Moscow.

Winter, A. (1955) The glassformers and the periodic system of elements, *Verres Refract.*, **9**, 147–156.

Yatsenko, S.P., Kuznetsov, A.N. and Chuntonov, K.A. (1977) Phase equilibrium in the Se–Rb system, *Proc. AS USSR. Non-Org. Mater.*, **13**, 738–739.

Zachariasen, W.H. (1932) The Atomic Arrangement of Glass, *J. Am. Chem. Soc.*, **54**, 3841–3851.

Zamfira, S., Jecu, D., Iuta, I., Vlhovici N. and Popescu, M. (1982) Preparation and properties of glasses in the system Se–I. Proceedings of International Conference on Amorphous Semiconductors—82, Bucharest, Romania, pp. 141–143.

Zhukov, E.G. and Grinberg, Ia.H. (1969) Production of B_2S_3 single crystals. Proc. AS USSR. *Non-Organic Materials*, **5**, 1646–1647.

CHAPTER 2

ATOMIC STRUCTURE AND STRUCTURAL MODIFICATION OF GLASS

A. Popov

MOSCOW POWER ENGINEERING INSTITUTE (TECHNICAL UNIVERSITY), 14 KRASNOKAZARMENNAYA ST., MOSCOW, 111250, RUSSIA

1. Structural Characteristics of Solid

The necessity to understand the structure is determined by the fact that it defines all the major properties of both crystalline and non-crystalline substances. The distinctive sign of crystals is a long-range order of arrangement of atoms or translation symmetry. As a rule, non-crystalline substances are determined as materials that do not have long-range order of arrangement of atoms. The negative character of this definition is not only contrary to the common rules—to define from the general to the particular—but also has very little useful data. At the XIX International Glass Congress (Edinburgh, 2001) Cormack, Du and Zeitler (2001) pointed out that 'commercial as well as research activity is taking place in the absence of anything like a complete understanding of the atomic structure of glass'. In this connection, it is necessary to answer the following question: which characteristics of the structure are necessary and sufficient for defining non-crystalline state of solid?

Let us start from the simplest case from the standpoint of a structure—ideal single crystal. In order to describe its complete structure it is enough to know the structure of an elementary cell or a short-range order of the arrangement of atoms. It is necessary to add a defective subsystem for a whole definition of any real single crystal. For describing the structure of polycrystals in addition to short-range order and defects, one should take into consideration the morphology of material i.e., crystal size distribution, crystal texture, formation of spherulites and so on.

As for non-crystalline solids four levels of structural characteristics should be considered for describing their structure:

- short-range order of atomic arrangement;
- medium-range order of atomic arrangement;
- morphology;
- defect subsystem.

ISBN 0-12-752187-9
ISSN 0080-8784

TABLE I
STRUCTURAL CHARACTERISTICS OF SOLIDS

Structure subsystems	Solid states			
	Ideal single crystal	Real single crystal	Polycrystal	Non-crystalline solid
Short-range order	+	+	+	+
Medium-range order	−	−	−	+
Morphology	−	−	+	+
Defect subsystem	−	+	+	+

The above-mentioned characteristics of solids are summarized in Table I, where it is shown that the amount of characteristics necessary for describing the structure of the substance increases with the growth of its complexity. While one characteristic for describing the structure of an ideal single crystal is sufficient, it is necessary to use four characteristics to describe the structure of non-crystalline solids. It is evident from this that a more complete definition of non-crystalline solids in comparison with the definition based on the absence of long-range order of atomic arrangement can be formulated. Non-crystalline solids are the materials that require the use of the parameters of short- and medium-range orders of atomic arrangement, morphology and defect subsystem to fully describe their structure. At the same time, as for non-crystalline substances the very terms of short-range order and particularly medium-range order of atomic arrangement are currently under discussion. So let us examine these terms in detail.

2. Short-Range and Medium-Range Orders

It is known that in the absence of long-range order of atomic arrangement in both non-crystalline solids and fluids, a certain so-called ‘local order’ remains.

Unlike crystals where order of atomic arrangement at any level is pre-defined by translation symmetry, understanding of the local order in disordered systems requires specification: what is dimension of fields of local order in atomic arrangement? Which parameters are required and sufficient in order to provide its full description?

On studying elements of order in non-crystalline materials, one can choose short-range order of atomic arrangement, determined by the chemical nature of atoms that forms the given substance (valency, bond length, bond angle). As a rule, it is assumed that the field of short-range order includes the atoms that are the nearest to the atom chosen as a central one and that form the first co-ordinate sphere (atoms 1 and 3 in respect to atom 2 in Fig. 1a, atoms 1, 3, 4, 5 in respect to atom 2 in Fig. 1b) (Aivasov, Budogyn, Vikhrov and Popov, 1995). The parameters of short-range order are: the number of the nearest neighbor atoms (first coordination number), their type, the distance between them and the central atom (radius of the first coordination sphere, r_1) (Fig. 1), their angle position in respect to the central atom defined by bond angles (valency angles φ). The given definition limits the short-range order to the first coordination sphere. However, the above-mentioned parameters of short-range order define not only the first coordination

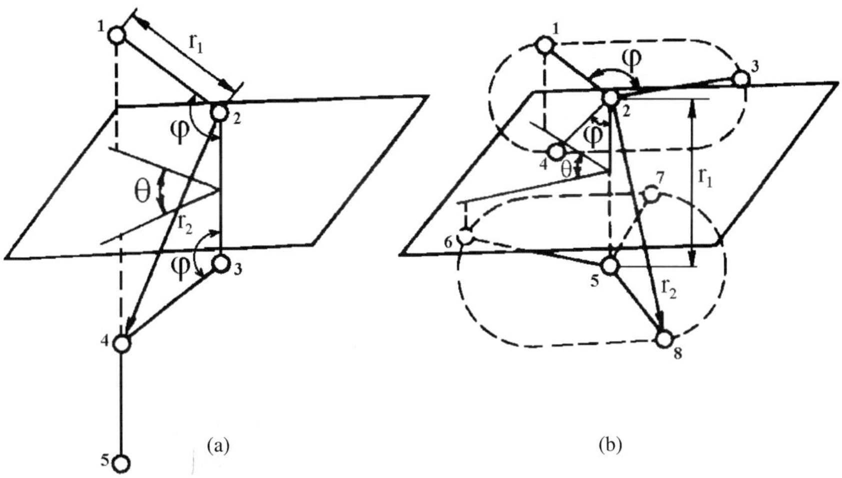

FIG. 1. Characteristics of atoms' relative disposition in the case of linear (a) and tetrahedral (b) structures: r_1, r_2 are the first and second coordination sphere radii, φ the bond angle, θ the dihedral angle.

sphere, but at least in part the second one as well. Thus, the radius of the second coordination sphere r_2 (Fig. 1) is determined by the radius of the first coordination sphere and valency angles:

$$r_2 = 2r_1 \sin(\varphi/2) \tag{1}$$

This contradiction can be resolved if we pass from the geometric parameters of short-range order to the power parameters of interaction among atoms (Popov and Vasil'eva, 1990). We can include in the field of short-range order those atoms, respective position of which is defined by the strongest interactions. For semiconductors with predominance of covalent type of chemical bonds the strongest interactions are defined by the parameters of covalent relations (bond length—energy of interaction v_s and bond angle—energy of interaction v_b, Fig. 2). Thus, the field of short-range order includes atoms of the first coordination sphere as well as those atoms of the second coordination sphere the position of which which respect to the chosen central atom is determined by the covalent interaction.

The introduction of the notion of short-range order does not allow to describe in full the local order in atomic arrangement observed in disordered systems because it does not answer the questions like how the fields of short-range order are connected with each other and also does not explain considerable length of the ordered fields in non-crystalline materials. Experimental proofs of quite long ordered fields triggered introduction of the notion of medium-range order in atomic arrangement in non-crystalline materials. One can consider different microcrystalline and cluster models of structure of these substances as the first attempt to explain the presence of medium-range order though the above-mentioned models arose historically before the introduction of the term 'medium-range order' in atomic arrangement.

At present, medium-range order is linked, as a rule, with the distribution of dihedral angles (Fig. 1, angle θ). But the concrete definitions of this term have so far remained

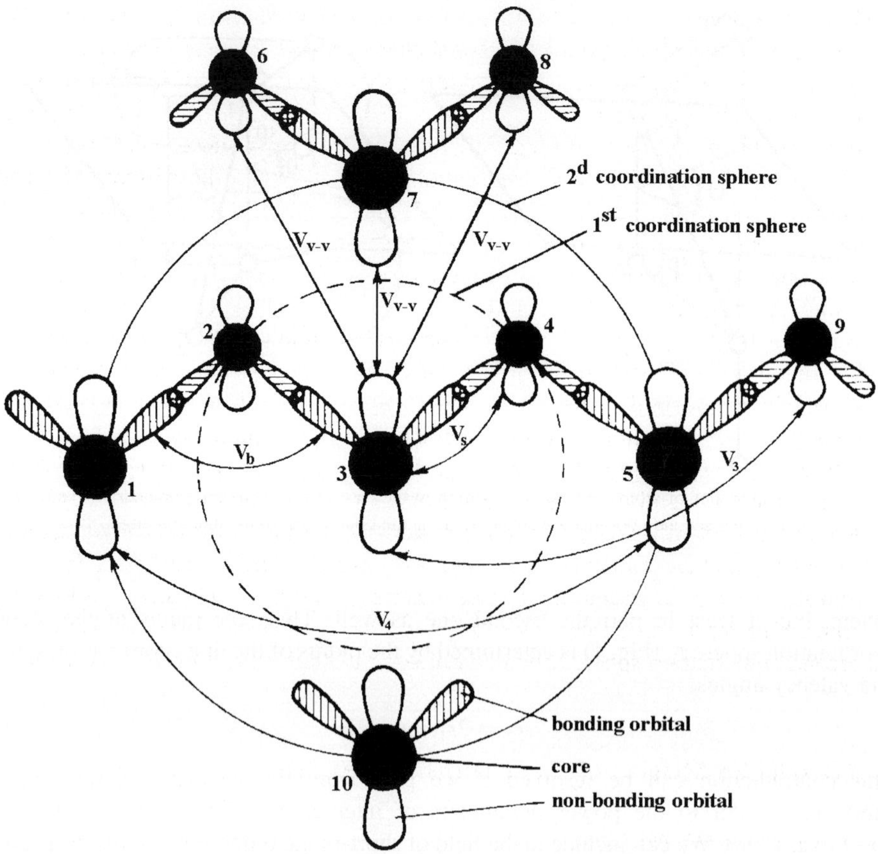

FIG. 2. Interatomic interaction in the case of linear polymer (see text).

open to discussion. Hence, in works by Lucovsky (1987) medium order is defined as the regular distribution of dihedral angles for a distance of about 10 atoms. It should be said that in case of linear polymers, for example chalcogens, it applies only to the atoms belonging to one molecule (chain or ring) and the atoms of other molecules even if they are positioned closer to the atom chosen as the central one, are excluded from the elements of medium-range order. Elliot (1987) divides the field of medium-range order into three levels: the field of local medium-range order (mutual disposition of neighboring structural units), medium field of medium-range order (mutual disposition of clusters) and long field of medium-range order linked to spatial order of different fields of a structural network. Voyles, Zotov, Nakhmanson, Drabold, Gibson, Treacy and Keblinsky (2001) offered paracrystalline atomistic model of amorphous silicon for explaining medium-range order. The inconsistent interpretation of the term medium-range order becomes particularly evident when analyzing materials that have different kinds of chemical bonds, for example, linear polymers. In this case, the above-mentioned characteristics are not sufficient to take into account of even the second and third neighbors—the atoms belonging to other molecules.

More rational approach of defining medium as well as short-range order of atomic arrangement is transitional from geometrical characteristics to energy characteristics of mutual atom interaction. As mentioned earlier, short-range order is determined by the strongest interaction between atoms v_s and v_b (Fig. 2). For example, in case of linear polymer (selenium, sulfur) short-range order includes first the nearest neighbors (atoms 1 and 3, if atom 2 is considered as the central one, Fig. 2) and those atoms of second coordination sphere that are of the same molecule as the central atom (atom 4 as its position in respect to atom 2 is defined by interaction v_s and v_b, Fig. 2). Interaction of atoms of the second order is related to the long-pair electrons of atoms that are in the same or different molecules, Van der Waals's interaction between atoms of neighboring molecules (v_3, v_4, v_{v-v}, Fig. 2). These interactions determine medium-range order of the atomic arrangement. Therefore, medium-range order is formed by atoms that are partially positioned in the second coordination sphere (in case of linear polymer—atoms of neighbor molecules), and the atoms of coordination spheres of higher orders.

So there is no long-range order of atomic arrangement in non-crystalline materials, but there are short and medium-range orders. In the case of semiconductor materials with predominance of covalent kind of chemical interaction, short-range order is determined by the interaction of covalent bonded atoms and includes the first and partially second coordination sphere. Medium-range order is determined by the interaction of long-pair electrons, Van der Waals's interaction and it is formed by atoms partially positioned in the second coordination sphere and atoms of coordination spheres of higher orders.

3. Investigation Methods of Disordered System Structure

3.1. Experimental Methods

The absence of translation symmetry in disordered systems greatly complicates the problems of investigation of their atomic structure as compared with crystals. The direct methods of investigating structure in disordered systems similar to crystals are diffraction of short-wave length radiation on the atoms of researched substance and X-ray spectroscopy.

The core of diffraction methods is the registration of spatial picture of intensity of monochromatic radiation coherently dispersed by investigated object, transitional from it to distribution of intensity in back space and calculation using Fourier transformation of micro distribution of density of the substance. However, if in case of crystals where the data received in this way give total information about the spatial distribution of atoms in the object, for disordered systems they provide only spherical symmetrical atomic radial distribution function (RDF), that is statistic in character and indicates the probability of an atom being positioned at the given distance from the atom chosen as a central one.

Typical RDF of a glassy semiconductor (bulk and film samples of As_2S_3) is shown in Figure 3 (Smorgonskaya and Tsendin, 1996). Disposition and area of the first peak of RDF allows to determine the number of the nearest neighbors of an atom (first coordination number) and distances between them (radius of first coordination sphere r_1). In case of elementary materials and double compounds of stoichiometric composition in

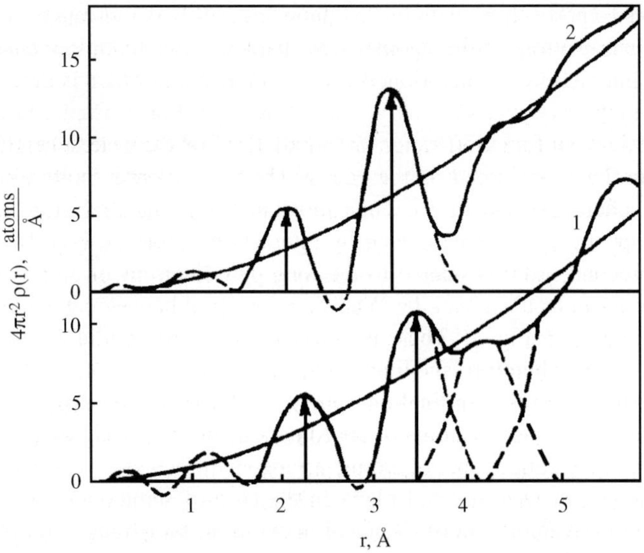

FIG. 3. RDF of bulk (1) and thin film (2) a–As_2S_3 (Smorgonskaya and Tsendin, 1996).

which only different type atoms are connected chemically, disposition of the second peak (r_2) and its halfwidth (δ_2), as a rule give information about average meaning of bond angle φ (Eq. (1)) and about range of its changes $\Delta\varphi$:

$$\delta_2 = r_1 \, \Delta\varphi \cos(\varphi/2) \tag{2}$$

In some cases, where the second coordination peak has a complicated shape it is necessary to carry out preliminary analysis to understand the reasons for changes in the shape of peak before using Eqs. (1) and (2). For example, the second peak RDF of glassy selenium that is an inorganic polymer, consisting of chain and ring molecules, has a shoulder on the side of large r. This is caused by the fact that the contribution to the second peak is made by atoms in the same molecule (two atoms that are situated at the distance r_2 from atom chosen as a central one) and atoms of the neighboring molecules. The decomposition of the second peak into two sub-peaks of the Gaussian form when the area of the first sub-peak corresponds to coordination number 2 allows using Eqs. (1) and (2) for the first sub-peak and estimate with their help average meaning of bond angle and range of its changes.

The values of coordination numbers and radii of coordination spheres of the third and higher order do not give enough direct information about spatial distribution of atoms beyond the first coordination sphere.

As for binary materials of non-stoichiometric composition and multi-component non-crystalline materials, RDF received from the results of diffraction measuring does not provide enough information even for interpretation of short-range order parameters. In this case, the method of X-ray spectral structural analysis based on extended X-ray absorption fine structure (as a rule, k-absorption) is more useful. Analysis of extended

X-ray absorption factor allows calculating parameters of first coordination sphere around an atom absorbing X-ray emission quantum. In the case of multi-component material investigation one can get extended X-ray absorption fine structure separately for atoms of each element by changing the energy of X-ray radiation. This allows calculating parameters of first coordination spheres around atoms of each element. But in this case too the information is limited by parameters of short-range order in atomic arrangement.

Thus, diffraction and X-ray spectral analysis methods allow to define parameters of short-range order in atomic arrangement and give some information (mainly qualitative) about the structure beyond short-range order, but do not provide possibility to reproduce spatial disposition of atoms in non-crystalline materials on the basis of experimental data.

Another group of structural investigation methods of disordered systems is vibration spectroscopy, including spectroscopy of infrared absorption and Raman scattering (RS), as a rule in the frequency range of 400–44 cm^{-1} (Skreshevsky, 1980). In both cases, details of received spectra are connected with vibration of atoms and chemical bonds in structural network of materials and therefore give information about forces, operating within structural units. However, infrared and Raman spectra do not duplicate each other, since their rules of selection for transition between oscillatory levels are different.

In case of crystals, harmonic approach is used for decoding spectra of vibration spectroscopy. In this case, definite bands of absorption (also called group frequencies or characteristic lines) in vibration spectrum correspond with single bonds and groups of atoms inside structural units of different chemical compounds.

Transition from crystal to disordered systems further and significantly complicates vibration spectrum calculation due to the loss of long-range order. This is why the method of comparative analysis of obtained spectra with vibration spectra of crystals analogous in chemical composition is usually used for interpreting experimental spectra of non-crystalline materials. The comparison of frequencies of characteristic lines and their intensity in both spectra allows to draw definite conclusions about the changes of atomic interaction during transition from crystalline state to the disordered state.

Thus, methods of vibration spectroscopy give information about the presence and character of certain bonds and groups of atoms in structural units of investigated substance. At the same time there is no theoretically substantiated analytical correlation that would allow to reliably calculate short inter-atomic distances based on the results of vibration spectroscopy not to mention parameters of medium-range order. This is due to the fact that inter-atomic distances are only one of many parameters that are the part of cinematic factors of interaction of complicated vibration task.

In addition to diffraction methods and vibration spectroscopy, so-called indirect methods can give certain information about the structure of disordered systems and their change under the influence of different factors. Indirect methods are based on structural dependencies of physical and chemical properties of a substance. As such dependencies exist practically for all material properties, success is defined by the choice of the most structurally sensitive property of a substance used to measure indirect properties. In case of glassy substances structurally sensitive characteristics that are included in the first place are density, viscosity, solubility, thermal conductivity, heat capacity, sound velocity, refracting index and their temperature dependencies, results of differential scanning calorimetry. Since electrical conductivity changes as a rule by several orders of magnitude during transition from glassy to crystalline state, its measurement allows to investigate kinetics of such phase transitions.

Experimental investigation methods of non-crystalline material structure shows that in the best case they give information about the short-range order in atomic arrangement, but neither none of them separately nor all of them combined provide comprehensive data about medium-range order and therefore they do not allow to reproduce spatial disposition of atoms in disordered systems. At present a simulation of disordered material structure is used for solving such a task.

3.2. Atomic Structure Simulation

To determine spatial disposition of atoms in disordered systems one can use structure simulation with further comparison of characteristics calculated on the base models with experimentally defined characteristics of the modeled object. There are two methods of creating structural models of non-crystalline materials: physical simulations that are based on physical objects (wire, tubes, balls, etc.) with additional checks of model adequacy and correction of atomic coordinates; and computer simulations, which are structured by the entry of data regarding basic atomic disposition and further transformation until characteristics of the model and simulated object are matched. Both approaches are not free from defects. Therefore, in the case of physical simulation for model construction it is necessary to use the rules of construction based on common conception about the structure of substance (possible mutual disposition of structural units, the meaning of dihedral angles and so on) in addition to the data received from direct experiment (first coordination number, radius of first coordination sphere). Thus, the model is based on both the objective data and subjective perceptions of the author. Some methods of computer simulation are free from the above-mentioned defects. However, insufficient amount of limitations received from direct experiments causes final arrangement of atoms in a model to become only one of the possible configurations that provides an adequate model of the simulated object.

Physical structural model of non-crystalline selenium that has 539 atoms was built by Long, Galison and Alben (1976). They have not introduced any limitations on the dihedral angle value in their model. Tetrahedral cells that have two hard covalent bonds and two flexible bonds representing interaction between chain molecules were chosen for modeling. When designing the model the following limitations were used

(1) presence of large cavities was not permitted;
(2) presence of regions with parallel packing of chain molecules similar to microcrystals was not allowed;
(3) large deviation of covalent bond lengths and angles as well as break up or rolling of chain molecules within model were excluded.

Atomic coordinates of the model were programmed and the calculated model strain resulting from deformation of bond length and angles of covalent bonds as well as intermolecular interaction was minimized. After minimization of strain in the model mean square deviation (MSD) of length of covalent bonds was 0.89% and bond angles were 3.6% (in respect to 105°). Density of modeled selenium was different from the density of

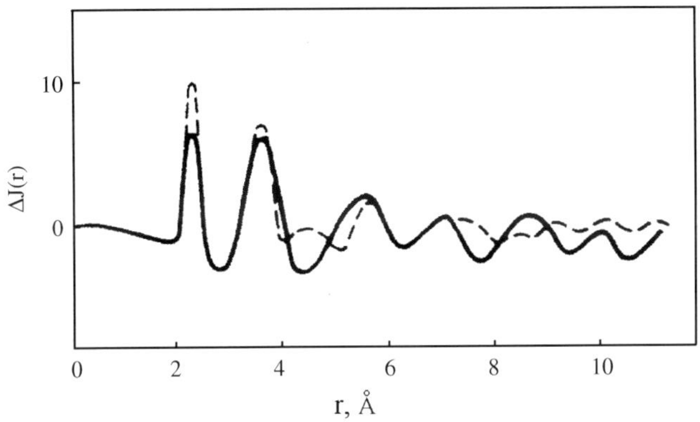

FIG. 4. Theoretical (solid line) and experimental (dashed line) RDF of a–Se (Long et al., 1976).

non-crystalline selenium by less than 3%. RDF of the model and experimental RDF of non-crystalline selenium are shown in Figure 4.

The main advantage of physical models is their descriptive value. Considerable effort is involved in designing large models. The purpose of these macro physical models is mainly for simulating elementary cells of non-crystalline substance when interpreting first peak of RDF. As mentioned in Section 3.1 for multi-component and even non-stoichiometric binary non-crystalline materials RDF received from diffraction measurements does not give enough information for definitive interpretation of elements of short-range order. Constructing physical models of elementary cells, comparing areas and coordinates of first peaks calculated on the basis of the models with similar parameters of experimental RDF allow to determine mutual atomic configuration at least within the first coordination sphere. Coincidence of calculated and experimental parameters of RDF proves correctness of the chosen model to a certain extent. Let us consider the usage of physical models of elementary cells for interpreting short-range order in non-crystalline arsenic chalcogens of non-stoichiometric composition As_x X_{100-x} where X is S, Se, Te (Michalev, 1983; Michalev and Popov, 1987).

Construction of models is carried out by combination of fragments consisting of 3–4 atoms provided that valence of elements is preserved and there are no broken chemical bonds. It is advisable to make the selection of fragments based on the analysis of combination scattering or infrared absorption spectra. Separate fragments (Fig. 5a) are combined in an elementary cell for which correlation of atoms of arsenic and chalcogens that form it is determined by the chemical formula of the studied matter. If it is impossible to fulfill all indicated requirements, the structural network of the substance can be formulated based on elementary cells' composition closest to the chemical formula of the substance. Thus, for example, model of the structure of substance As_{70} Se_{30} (formula As_7Se_3) can be formed on the basis of elementary cells As_6Se_3 and As_6Se_2 with a ratio of 4 : 3. In Fig. 5b and c the shaded areas correspond to elementary cells.

The contribution of every fragment of elementary cell to the square of RDF first peak is determined by the formula

$$F_i = 2K_A K_B m_{A/B} n \tag{3}$$

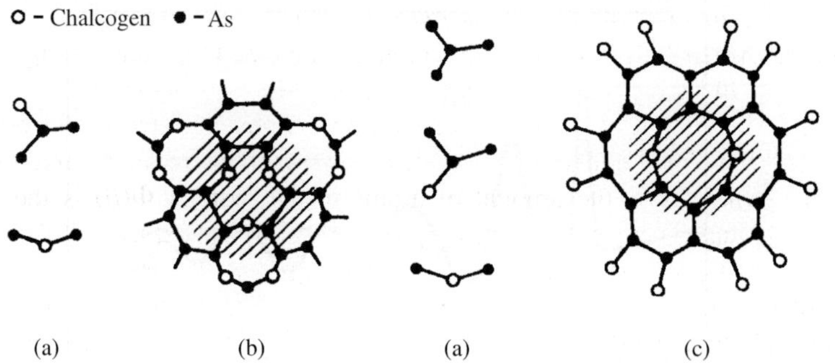

FIG. 5. Fragments of elementary cell (a) and elementary cells of As_6Se_3 (b) and As_6Se_2 (c).

where K_A and K_B are the scattering ability of atoms A and B constituting a fragment; $m_{A/B}$ is the amount of atoms B around atom A, and n number of atoms A.

Calculations are summed up for all the atoms of the elementary cell

$$F_{\sum} = \sum_{l} F_i k_i \tag{4}$$

where k_i is the amount of i fragments in structural cell in accordance with the formula.

Models of elementary cells of glassy arsenic chalcogens in different compositions that give the best conformity with the experimental RDF and that matches with RS spectra are shown in Figure 6. Completed models give a chance to estimate the spatial disposition of atoms in arsenic chalcogens non-stoichiometric compositions and changes caused by changes of chemical composition of the matter. So, for example, when arsenic content is increased by more than 33%, transition of the structure from chain to layer takes place (elementary cells As_2X_3 and others).

In computer simulation of the atomic structure of glassy materials, molecular dynamics and Monte Carlo methods (Morigaki, 1999) are widely used. The method of molecular dynamics is used for studying the kinetic properties of matters (for example,

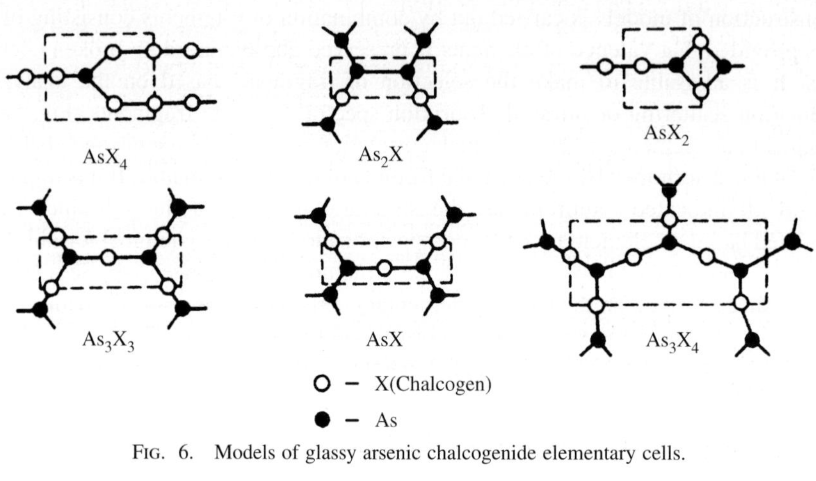

FIG. 6. Models of glassy arsenic chalcogenide elementary cells.

for simulating processes of phase transition: crystallization, melting, glass transition (Adler and Hoover, 1968; Poluchin and Vatolin, 1985)) and for constructing models of atomic structure of glassy materials (Barreto, Alves, Mort and Jackson, 2001; Cormack et al., 2001). The method is based on the assumption that movement of atoms can be described by Newton's equations. Force acting on an atom is a sum of vectors $\sum_j^{N-1} \nabla\Phi(ij)$, where N is the amount of atoms in the system, $\Phi(ij)$ is the pair-wise interaction potential of atoms (assumed that type of potential is defined), ∇, Hamiltonian. Starting coordinates of atoms are defined pseudo-randomly (that is, with additional condition of prohibition to place two or more atoms at one point of space) or by the periodic crystalline network configuration. Initial speed of atoms has random directions and equal absolute meanings chosen so that full kinetic energy of system was true at the given temperature in accordance with the classic formula:

$$T = \frac{1}{3k}\left(\sum_i^N m_i v_i^2\right) \tag{5}$$

where k is the Boltzmann constant, expression in brackets is the average meaning of pulse of the atoms. When initial coordinates and speeds are set, the atoms by turn are set free and the system begins to approach an equilibrium state. The result of simulation is a series of atomic configurations matching different points in time.

Unlike a purely deterministic equation of molecular dynamics, the Monte Carlo method is a numeral calculation in which probability elements are included (Renninger et al., 1974; Kozlov and Krikis, 1978). The characteristic feature of the method is the construction of statically random process—Markuv chain where separate states represent various configurations of the examined system that are obtained by the random removal of its particles. Every new configuration is accepted or rejected. The criterion for a decision is probability of existence of a new configuration estimated by the Boltzmann factor $\exp(-\Phi_{Nj}/k\mathrm{T})$ (Φ_{Nj} is the potential energy of given configuration) or by the similarity of RDF calculated for the given configuration with the experimental RDF. In the first case the algorithm of model creation is based on the following approach. As a rule a starting pseudo-random disposition of atoms is created. An atom is selected randomly or in turn and its random relocation from point i to point j is examined. If in this case total potential energy of the model decreases then transition is considered to be allowable and the previous configuration is replaced with the new one. If total potential energy increases then transition can take place only with probability $p_{ij} = \exp(-\Delta\Phi_N^{ij}/kT)$, where $\Delta\Phi_N^{ij}$ is the change of potential energy. With such algorithm potential energy of sequentially examined configurations has a tendency to attain an equilibrium value.

In another version of structure modeling using Monte Carlo method the probability of any configuration is estimated by the similarity between RDF of the configuration and the experimental RDF. If sequential relocation of the atom enhances the similarity of RDFs then a new configuration is accepted. Otherwise, it is rejected. This approach does not result in models with equilibrium structures and so it must be followed by a relaxation of model's energy.

The main shortcoming of Monte Carlo method is that computer experiment is very time-consuming, which leads to the limitation of the size of the created models.

The gradient radiant method of atomic structure simulation was developed to overcome this shortcoming (Vasil'eva, 1989). Unlike Monte Carlo method based on random search for optimum position of atoms in a model, the gradient method undertakes targeted search for final the disposition of atoms. This allows to significantly reduce the time needed to obtain adequate models of structure, especially in the case of materials where various types of chemical bonds co-exist.

Let us examine the gradient method in simulating the structure of glassy selenium—two-fold coordinated material with covalent bonds inside molecules and Van der Waals intermolecular interaction (Popov, Vasil'eva and Khalturin, 1981; Vasil'eva, Popov and Khalturin, 1982; Vasil'eva and Khalturin, 1986) as an example. Simulation of structure by the gradient method includes several stages. First, the number of atoms in the model is chosen, which allows us to calculate the size of the model while taking into consideration the atomic density of simulated matter. An average density is defined by the expression

$$u = \rho/M\, m_{\mathrm{H}} \tag{6}$$

where ρ is the experimental density of simulated substance; M is the atomic weight and $m_{\mathrm{H}} = 1.65 \times 10^{-24}$g—mass of hydrogen atom. In addition to the size of the model it is necessary to define its form. Cubic or spherical forms are used widely and the choice of a particular form is mainly determined by the method of taking into consideration of the model's finite size.

Another way to take into account the finite size of a model is to use various correction factors in calculating characteristics of the model. So, in calculating, for example, an RDF of a spherical model errors caused by finite size of a model are compensated by the division of RDF by the correction factor defined by expression

$$D(r) = 1 - 1.5\frac{r}{d} + 0.5\left(\frac{r}{d}\right)^3 \tag{7}$$

where d is the model's diameter and r is the radial distance from the center of the model. At the same time, not every characteristic of a model can be corrected satisfactorily with the help of correction factors. This is why only the internal part of the model instead of its whole volume is used for calculation, so that any atom in it has a normal environment that corresponds to an atom in the volume of material. For all of the above reasons, the examined model of non-crystalline selenium took the form of a sphere with a diameter of 2.9 nm and included 420 atoms.

After defining the size and form of the model, the next stage is the creation of initial disposition of atoms. Pseudo-random distribution of atoms in a model is most acceptable as a starting disposition because it excludes the influence of subjective initial assumptions on the results of simulation. During the formation of initial disposition of atoms two limitations should be observed, namely, the number of atoms must equal the pre-defined number of atoms in the model, and the distance between any atoms must not be less than a minimum distance defined by position of the first peak in experimental RDF.

In the third stage of simulating, initial atomic disposition is rearranged in order to obtain given likeness of RDF of model with experimental RDF. Degree of their difference is estimated by the MSD of these functions

$$\mathrm{MSD} = \sum_{i=1}^{N} [\mathrm{RDFE}(I) - \mathrm{RDFM}(I)]^2 \tag{8}$$

where N is the number of points, where RDF is compared, RDFE(I), RDFM(I) are the values of experimental and calculated RDF for the model at point I. The purpose of rearranging atoms in the model is to minimize the value of MSD. When it is necessary to obtain the best possible fit between RDF of the model and experimental RDF at a certain range (for example, in the region of RDF first peak) different weight factors for different ranges of values (I) can be included in Eq. (8)

$$\mathrm{MSD} = K_1 \sum_{I=1}^{N_1} [\mathrm{RDFE}(I) - \mathrm{RDFM}(I)]^2 + K_2 \sum_{I=N+1}^{N} [\mathrm{RDFE}(I) - \mathrm{RDFM}(I)]^2 \tag{9}$$

where K_1 and K_2 are the weight factors.

When rearranging initial disposition of atoms using gradient method (as opposed to Monte Carlo method where new atomic coordinates are defined with the help of generation of random numbers) the new atomic disposition corresponding to the minimum value of MSD is found as follows. Gradient MSD in every direction of coordinate axes is determined for each atom. Based on the obtained values of the gradient, a direction in which MSD decreases with the largest speed is chosen. A new disposition of atom in the model that corresponds to the minimum MSD in the chosen direction is defined in the next scheme (Fig. 7). A cylinder with a radius that equals the minimum distance between atoms is built in the direction of the maximum value of

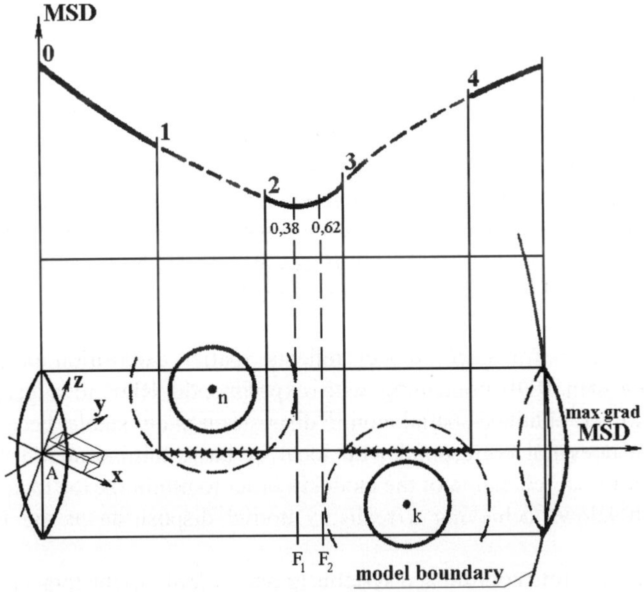

FIG. 7. Procedure of atom position corresponding to MSD minimum value finding.

MSD gradient. Number of atoms, that have their centers covered by the cylinder, are defined. Then spheres with a radius that equal the minimum distance among atoms are built around these atoms. The areas of direct line of maximum gradient MSD covered by these spheres are prohibited for the placement of rearranged atom. After that the values of MSD at the points of overlap between spheres and the line of maximum gradient MSD are calculated sequentially and the allowed area with the disposition of atom equaling minimum MSD is chosen. In Figure 7 it is the part between points 2 and 3. Minimum MSD between points 2 and 3 is defined by method of 'golden section.' If moving an atom in any direction does not lead to decrease of MSD, then this atom is left in its initial position. The procedure is repeated sequentially for each atom until desirable similarity between RDF of the model and the simulated object can be achieved.

When calculating the RDF of the model it is necessary to take into account the influence of heat fluctuation of atoms on the shape of RDF. As a rule, atomic vibration is supposed to be spherically symmetric and is accounted for by the distribution of initial positions of atomic centers of mass in accordance with Gauss law. This necessitates calculation of the average square deviation of atomic centers of mass in response to heat fluctuations. This parameter can be defined theoretically on the basis of Debye formula

$$\sigma = \sqrt{3\hbar^2 T/4\pi mk\theta^2} \tag{10}$$

where k is the Boltzmann constant, $\hbar$ is the Planck's constant, T the temperature, K, m the atomic weight, θ the Debye temperature; or alternatively using half-width of the first peak of experimental RDF (Skreshevsky, 1980)

$$\sigma = \sqrt{0.18L_{1/2}^2} \tag{11}$$

where $L_{1/2} = R_2 - R_1$, R_2 and R_1 are the meaning of 'r' when function $4\pi r^2\rho(r)$ equals half of its maximum value in the region of the first RDF peak.

The fourth stage is the finding of covalent bonded atoms or uniting atoms in molecules that have strong covalent bonds. Undertaking this stage after minimizing MSD of RDFs is justified by the fact that rearrangement of free atoms (i.e., atoms that are not linked in molecules) at the previous stage allows to achieve better similarity of experimental and model RDF faster.

Thus, after the fourth stage, model reflects spatial disposition of atoms united into molecules with RDF coinciding with experimental RDF to a given degree of accuracy. However, obtained spatial atomic disposition is not stable, because energy of such a system does not correspond with local or basic minimum. This is why it is necessary to carry out relaxation of the model in order to minimize the total energy of the system, which allows achieving a realistic atomic disposition at the fifth stage of simulation.

In the case of two-fold coordinated systems (glassy selenium) the total energy of atoms consists of four components:

(1) bond–stretching energy between the given atom and its covalently bonded nearest neighbors

$$V_S = 0.5\alpha(R_{l,i}^2 - d^2)^2 \tag{12}$$

where α is the constant, $R_{l,i}$ the bond length between atoms l and i and d the most probable distance between nearest neighbors, determined by the position of the first peak of the experimental RDF;

(2) bond-bending energy

$$V_B = 0.5\beta[(\mathbf{R}_{l,i}\cdot\mathbf{R}_{l,j}) - c]^2 \tag{13}$$

where β is the constant, $\mathbf{R}_{l,i}$ and $\mathbf{R}_{l,j}$ are vectors, binding atom l with atoms i and j and c, the constant, which is selected in such a way as to make V_B equal zero at the given bond angle φ_0. The meaning of φ_0 is selected as a rule to be equal to a bond angle in the appropriate crystalline material (for selenium $\varphi_0 = 105°$);

(3) energy of the Van der Waals interaction between the given atom and the atoms which have no covalent bond with it

$$V_{l,i} = \begin{cases} -\dfrac{A}{R_{l,i}^6} + \dfrac{B}{R_{l,i}^{12}} & \text{if } R_{l,i} \leq R_c, \\ 0 & \text{if } R_{l,i} \geq R_c, \end{cases} \tag{14}$$

where A and B are the constants defining repulsion and attraction of atoms and selected in such a way that $V_{l,i}$ equals the minimum value possible with the given distance $R_{l,i}^0$ between atoms not directly connected by covalent bonds (in the case of selenium the meaning $R_{l,i}^0$ is determined in accordance with the position of the second peak of the experimental RDF and is 3.7 Å); R_c is the maximum distance where the Van der Waals interaction is taken into consideration (as a rule, 4–5 Å),

(4) bond-twisting energy

$$V_D = \gamma\{[\mathbf{R}_{ij}\mathbf{R}_{jk}][\mathbf{R}_{jk}\mathbf{R}_{kl}] - K\}^2 \tag{15}$$

where γ is a constant, i, j, k, l the indexes of sequentially bonded atoms in molecule, K is the constant, the value of which should provide minimum meaning V_D at the given meaning of dihedral angle $\theta = \theta_0$ and is calculated based on the formula (Fig. 1a):

$$K = (r')^2 \cos\theta_0 = [r_1 \cos(\varphi_0 - 90°)]^2 \cos\theta_0 \tag{16}$$

Value θ_0 is selected as a rule as being equal to the dihedral angle in a certain crystalline modification of the matter (for selenium $\theta_0 = 102°$).

When designing models like these, substantial uncertainty exists in the selection of values for constants α, β and γ. This is why these values are usually selected to provide the required level of similarity between the experimental and model RDF.

Minimization of energy is carried out by sequential movement of each of the atoms in the direction of the quickest decrease of its total energy. This direction is determined by calculating energy gradient of atom, that is, by calculating the force that influences the

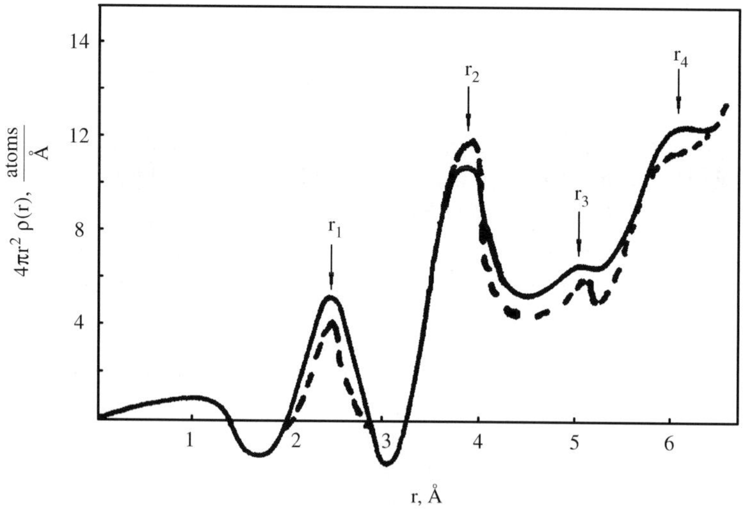

FIG. 8. Comparison of RDF of a–Se for model (dashed line) with experiment (solid line) (Vasil'eva and Khalturin, 1986).

atom. The search for position of an atom that would correspond to its minimal total energy in the selected direction is carried out with the help of the algorithm used at the third stage of simulation.

It should be noted that after minimization of the system's energy is carried out, the similarity between the model and experimental RDF as a rule degrades. In order to enhance the similarity between the experimental and model RDF and to further decrease the total energy of model, the third, fourth and fifth stages of stimulation process are repeated to achieve the desired value of MSD at the minimum of total energy of the model. RDF of model is shown in Figure 8. Figure 9 shows the distribution of bond length and bond angles, as well as dihedral angles in the obtained model of glassy selenium.

4. The Results of Structural Research of Glassy Semiconductors

4.1. Atomic Structure of Glassy Selenium

Selenium is in the VI group of periodic table. The structure of outermost electron shell is $4s^2p^4$. Hybridization of electron orbitals in selenium is small; therefore, as a rule, only p-electrons form chemical bonds. In elementary selenium two p-electrons of each atom form covalent bonds creating molecules in the shape of rings or high polymer chains and other two p-electrons stay in a non-bonding state as lone-pair electrons.

Selenium exists in several allotropic crystalline and non-crystalline forms (Table II) (Baratov and Popov, 1990). Thermodynamic stable form of selenium is trigonal selenium, formed by spiral chain molecules Se_n. All other forms of selenium turn into trigonal modification when exposed to thermal treatment.

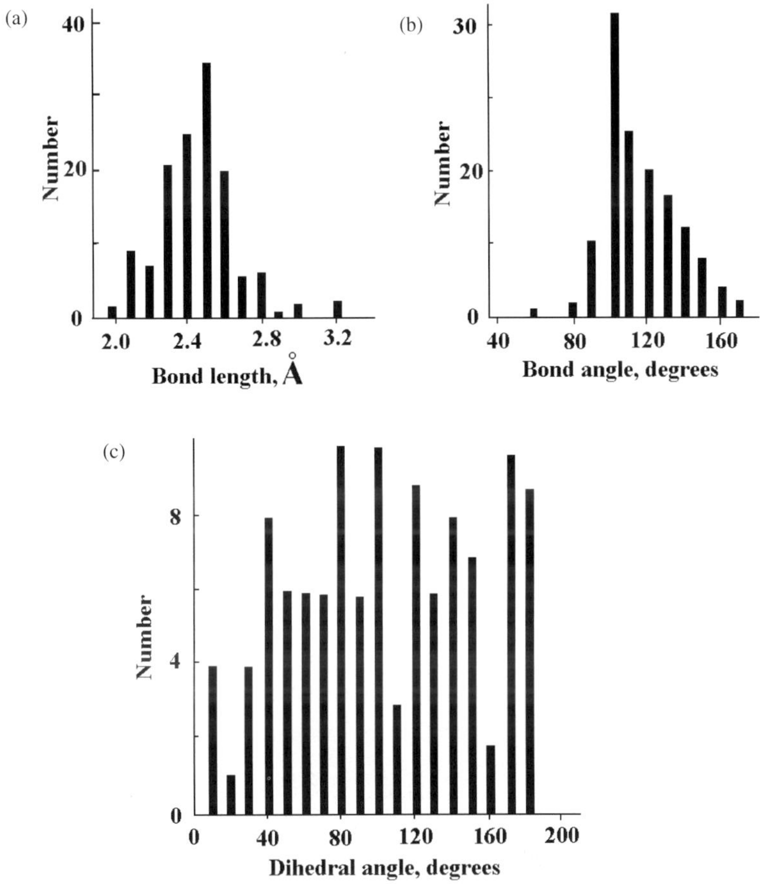

FIG. 9. Distribution of bond lengths (a), bond angles (b) and dihedral angles (c) in the model of a–Se.

Crystalline forms of selenium are studied quite well, but at the same time strict classification of allotropic non-crystalline forms is absent. As seen in Table II there are at least three allotropic forms of solid non-crystalline selenium. Red amorphous selenium is produced by chemical restoration, for example, H_2SeO_3, or by sharp quenching of superheated vapor of selenium. It is unstable even at the temperature of about 300 K. The structural models of red amorphous selenium are rather contradictory. However, the analysis of obtained data allows to suggest that red amorphous selenium consists of ring-shaped molecules Se_6. At 30–40 °C red amorphous selenium is turned into a black amorphous modification. This transition has an irreversible endothermic effect that is probably connected with splitting of ring molecules. Information about black amorphous form of selenium is limited by the fact that there is no long-range order in atomic arrangement in this material.

Glassy selenium is the most wide-spread non-crystalline form of selenium. At the same time information about structure and properties of glassy selenium is greatly different and

TABLE II

ALLOTROPIC FORMS OF SELENIUM

N	Form	Type of molecules	Bond length (nm)	Bond angle (°)	First coordination number	Lattice constants (nm)		
						a	*b*	*c*
Crystalline								
1	Trigonal	Spiral chains Se_n	0.233	103.1	2	0.436	–	0.495
2	α-monoclinic	Rings Se_8	0.232	105.9	2	0.905	0.908	1.160
3	β-monoclinic	Rings Se_8	0.234	105.5	2	1.285	0.807	0.931
4	α-cubic	–	0.297	–	6	0.297	–	–
5	β-cubic	–	0.248	–	4	0.575	–	–
6	Rhombo-hedral	Rings Se_6	0.235	101.1	–	1.136	–	0.442
7	Ortho-rhombic	–	–	–	–	2.632	0.688	0.434
Non-crystalline								
8	Red amorphous	Rings	0.23	–	≈2.4	–	–	–
9	Black amorphous	–	–	–	–	–	–	–
10	Glassy	Chains and/or rings	0.23	105	2.1	–	–	–

sometimes even contradictory. It is caused by the fact that glassy state includes various forms of selenium, that differ from each other in terms of ratio, size, form and mutual packing of structural units and therefore they have different properties.

While examining molecular shapes of crystalline forms of selenium (Table II) one can suggest that the same molecules (rings Se_8, rings Se_6, spiral polymer chains Se_n) are present in the glassy matter too. The dependence of glassy selenium properties on the conditions of its preparation is explained in this case by the change of ratio of different molecules and the change of level of polymerization of molecules. So, films of glassy selenium that were prepared in research by Nabitovich (1970) and Cherkasov and Kreitor (1974), consisted of only ring molecules Se_8 (amorphous analog of monoclinic selenium) or of only chain molecules Se_n (amorphous analog of trigonal selenium) and as a result had widely different properties.

But the absence of long-range order in glassy selenium determines the possibility of wider changes to the molecular structure, rather than a simple mixing of molecular forms of different crystalline modifications. At first it is revealed in the possibility of changing the value or the sign of dihedral angle in molecules and in the possibility of forming defects so characteristic of glassy semiconductors.

The absolute values of dihedral angles for trigonal and monoclinic modifications are close to each other (102 and 101°). The difference is that the sign of dihedral angle in ring molecules of monoclinic form is changed under transition to each following atom so that atoms are placed in *cis*-coupling configuration and are closed into rings (Fig. 10a). The sign of dihedral angle in chain molecules of trigonal form is constant for the whole molecule. In this case, atoms are in *trans*-coupling configuration and form an endless spiral chain (Fig. 10b).

When long-range order is lost, the requirement to have a strictly defined absolute value and sign of dihedral angle is no longer necessary. This is why models of flat zigzag chains (dihedral angle equals zero) (Richter and Breiting, 1971), free rotation chain model (Fig. 11a) in which dihedral angle can have any value (Malaurent and Dixmier, 1977), disordered chains where changes only to sign, but not to value of dihedral angle are permitted (Fig. 11b) (Lucovsky and Galeener, 1980) were suggested for describing structure of selenium. In the latter case a molecule can have elements of both monoclinic forms (rings) and elements of trigonal form (spiral chains).

Other factors that influence glassy selenium structure are quasi-molecular defects and valence alternation pairs (VAP). VAP concentration in selenium can achieve the value comparable with concentration of molecules (5×10^{18}–5×10^{19} cm^{-3}) that must have an effect on the structure of matter due to formation of intermolecular bonds by

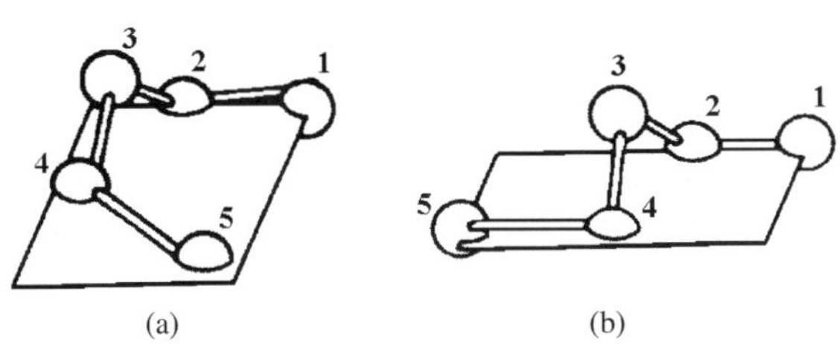

FIG. 10. *Cis*- (a) and *trans*- (b) coupling configurations for molecular bonding in selenium.

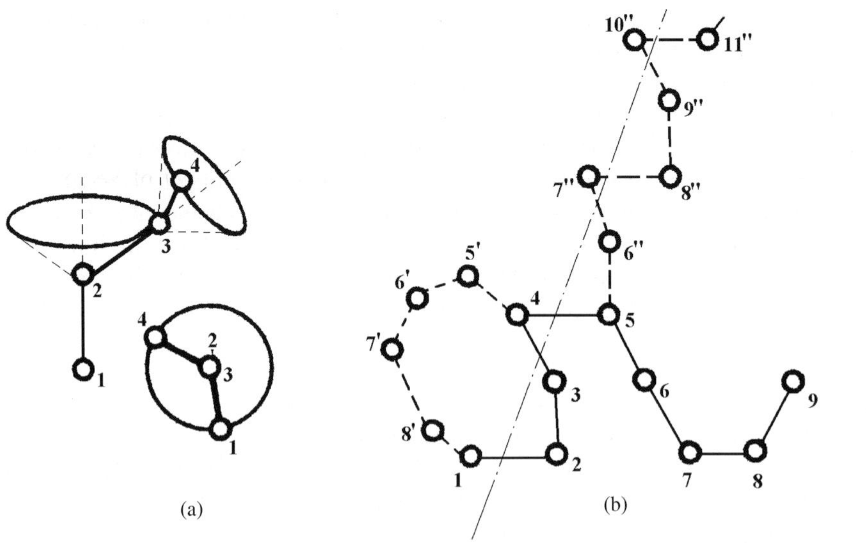

FIG. 11. 'Free rotation chain' (a) and 'disordered chain' (b) models of non-crystalline selenium molecules.

Se_3^+-centers. The introduction of quasi-molecular defects also leads to the formation of intermolecular bonds and increasing selenium atom coordination.

The structure of glassy selenium was widely investigated with the help of diffraction methods and the analysis of extended X-ray absorption of fine structure (EXAFS). The first peak of RDF (that is determined by the nearest neighbors in the same molecule) is well isolated from the rest of this curve. The second peak of RDF has a shoulder on the side of large 'r.' It is caused by the fact that contribution to the second peak is made both by the atoms situated in the molecule (partial coordination number = 2) and the atoms of neighboring molecules. Decomposition of the second peak into two sub-peaks of Gaussian form under the condition that the area of the first sub-peak corresponds to coordination number 2 allows to estimate an average value of bonds angle (100–105°) and the value of deflection from medium meaning (~8°). The number of trigonal form coordination radii (Table II) is absent in the RDF of glassy selenium. An example could be radius 4.36 Å that fits the network constant *c* in trigonal form. The absence of appropriate coordination sphere is the result of loss of periodical disposition of molecules. The disposition of the third peak RDF of glassy selenium (4.7–4.9 Å) is close to the value of network constant *c* of trigonal form (4.95 Å). This coordination radius can prove that there exist at least spiral chain molecule fragments in non-crystalline substances.

X-ray diffraction investigation of liquid selenium at the range of temperatures 230–430 °C by Poltavzev (1984) showed small (within 2–3%) displacement of the first and second maxima of RDF of melted selenium explained by the author as the change of ratio between *cis*- and *trans*-configuration in selenium molecules with the changing temperature.

As diffraction method alone does not allow to determine atomic structure of non-crystalline matter unambiguously, it is necessary to use other methods alongside with diffraction method. Basic details of IR- and RS-spectra of glassy selenium and α-monoclinic and trigonal modification (Baratov and Popov, 1990) are given in Table III. The comparison of infrared absorption spectra shows that peaks of absorption in glassy selenium at 95 and 254 cm^{-1} as well as shoulder at 120 cm^{-1} are in good agreement with bands of fundamental absorption in α-monoclinic form of selenium. This served as the basis for an assumption that a considerable amount of Se_8 molecules is present in glassy selenium. On the other hand deep band of absorption at 135 cm^{-1} in glassy selenium is close to peak 144 cm^{-1} in trigonal selenium and the shoulder at 230 cm^{-1} directly corresponds to the peak in trigonal selenium, which is interpreted as proof of presence of spiral chain molecules. The analysis of RS (Table III) leads to similar conclusions. Glassy selenium spectrum is characterized by a wide peak of complicated form with the maximum at 250 cm^{-1} and a shoulder at 235 cm^{-1}, the position of which covers the peaks at 237 and 250 cm^{-1} on the spectra of trigonal and monoclinic forms. Other peaks of glassy selenium spectrum are close to fundamental modes of crystalline forms as well. Thus, when the results of IR- and RS-spectroscopy are interpreted, the structure of glassy selenium is viewed as a rule, because a mixture of molecular forms of monoclinic and trigonal modifications with predominating amount of ring molecules Se_8. But such interpretation is not in agreement with other experiments. So, for example, high viscosity of selenium at the temperatures higher than glass transition region proves predomination of polymeric molecules rather than monomeric ring molecules in a substance.

In this connection, the work by Lucovsky and Galeener (1980) is very interesting. It shows the possibility to interpret the RS spectra from the position of the model of disordered chain. Let us consider A_1 vibration modes that are Raman active in the cases of *cis*- and *trans*-coupling configurations of selenium atoms in a molecule (Fig. 12). In the case of *cis*-coupling characteristic to ring molecules two A_1 symmetry modes are possible: displacement in the bonding plane with the frequency of 256 cm^{-1} and displacement perpendicular to this plane with the frequency of 113 cm^{-1}. At the same time for *trans*-coupling configuration only displacement in the bonding plane with a frequency of 256 cm^{-1} is possible. Thus, in the molecule that has the form of a disordered chain (containing the fragments of rings and spiral chains, Fig. 11b) all the atoms take part in vibrations with the frequency of 256 cm^{-1} and only part of atoms in the fragments of rings contributes to the mode of 113 cm^{-1}. This explains considerably higher intensity of peak 250 cm^{-1} in RS spectra of glassy selenium.

Differential solubility of glassy selenium in CS_2 or CH_2J_2 is used as an alternative method for estimating the ratio of ring Se_8 and polymer Se_n molecules. As these solvents dissolve the monoclinic form well and do not dissolve the trigonal form of crystalline selenium at all, it is assumed that ring monomers Se_8 are turned into solution. In the early works using this method, an amount of matter turned into solution achieved 20–40%. But later on it was defined more exactly that solution of glassy selenium in CS_2 depends on the degree of illumination of a sample in the process of solution, which is probably caused by the structural changes in a substance under the influence of radiation. When radiation has no influence on the process of solution of glassy selenium in the mentioned solvents, the amount of a substance that turns into solution is as a rule less than 5–10%. This testifies to low concentration of monomer molecules and predominance of polymer ones in glassy selenium that is also confirmed by high viscosity of the material.

TABLE III

THE MAIN FEATURES OF SELENIUM INFRARED AND RAMAN SPECTRA

Selenium		Infrared absorption					Raman scattering			
Glassy	Location (cm^{-1})	95	120	135	230	254	110–115	140	235	250–256
	Type	Peak	Shoulder	Peak	Shoulder	Peak	Peak	Shoulder	Shoulder	Peak
α-Monoclinic	Location (cm^{-1})	92 – 97	120	–	–	254	113	–	–	250
	Type	Doublet	Peak	–	–	Peak	Peak	–	–	Peak
Trigonal	Location (cm^{-1})	–	–	144	230	–	–	143	237	–
	Type	–	–	Peak	Peak	–	–	Peak	Peak	–

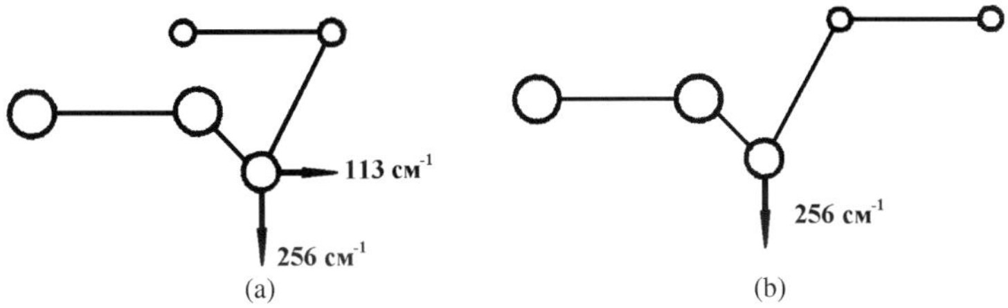

FIG. 12. Atomic displacements of the A_1 symmetry modes in the case of *cis*- (a) and *trans*- (b) coupling configurations (Lucovsky and Galeener, 1980).

The results of viscosimetry are often used to obtain information about the degree of selenium chain molecule polymerization. Generally, the viscosity of selenium should be determined by the following factors: strength of intermolecular and intramolecular interactions, the ratio of the amounts of ring and chain molecules and degree of polymerization of the latter, types and concentration of defects. The level of influence that these factors have on viscosity of selenium will be different under various temperatures. The analysis of temperature dependence of non-crystalline selenium viscosity (Popov, 1980b) showed that at the range of temperatures of 60–80 °C the viscosity of selenium is determined mainly by two factors, namely: by intermolecular interaction, the value of which corresponds to the temperature of measurement and molecular structure of material determined mainly by the regimes in sample production. So, the degree of chain molecule polymerization (the amount of 8-atom monomers in a molecule) is determined by the expression

$$\log\eta = A(T)P^{1/2} \tag{17}$$

where $A(T)$ is the factor dependant on the temperature of measuring, P the degree of polymerization molecules and η is the viscosity and as a rule is within 10^3–10^4.

The examined results of research of glassy selenium structure allow to determine accurately enough the short-range order in atomic arrangement, as well as to confirm the low content of monomeric ring molecules; estimate the degree of polymerization of chain molecules; and finally select most probable models of chain molecular structure—the models of free rotating and disordered chains. At the same time there is not enough data to describe spatial disposition of atoms in the material. As mentioned in Section 3 this task is completed with the help of simulation of non-crystalline material structure.

The typical distribution of bond lengths, bond angles and dihedral angles for one of the models of glassy selenium layers obtained by the method of sublimation in a vacuum at different substrate temperature is shown in Figure 9. (RDF of the model is shown in Figure 8). The distribution of the bond lengths in the model has a clear maximum that corresponds to both form and disposition of the first maximum of the experimental RDF. The spread of bond length values increases with the decrease of substrate temperature within forming layers. The distribution of bond angles is represented by curves with the range of values between 60 and 180° with the maximum in the interval of 100–110°. Although the difference between bond angle average values in the model and the value of 105° characteristic for crystalline form increases slightly with the decrease of the substrate temperature, the distribution of bond angles as a whole in models produced at different substrate temperatures differ only slightly. The values of dihedral angles are at the range of 10–180° and their distribution does not have explicit maxima, which confirms realization of free rotation chain model with arbitrary value of dihedral angles in researched samples.

The analysis of the data proves that the structure of investigated layers of glassy selenium consists of deformed chain molecules with small predomination of *trans*-coupling configuration in comparison with *cis*-coupling configuration. The level of structural ordering increases with the increase in the substrate temperature. The latter is proved not only by statistic characteristics of models, but also by the analysis of their energy characteristics: total energy of system and its four components (energy of distortion of bond lengths, bond angles, dihedral angles and energy of Van der Waals interaction). It is necessary to point out that the main difference in the meaning of total energy is determined by the difference of the Van der Waals component, that is, the difference of energy of intermolecular interaction. It corroborates that the changes of degree of structural ordering with the changes of the preparation conditions of getting patterns are determined mainly by the changes in the mutual packing of molecules, that is, in medium but not in the short-range order in atomic arrangement.

The doping of admixtures also leads to the change of molecular structure of glassy selenium. This is why admixtures in selenium are usually divided into three groups in accordance with their influence on the structure: isoelectron admixtures (oxygen, sulfur, tellurium), branching admixtures (elements of the fourth and fifth groups) and univalent admixtures (hydrogen, alkaline metals, thallium). Let us examine the influence of each mentioned group of admixtures on the structure and properties of glassy selenium.

Similar construction of the outermost electron shell of atoms of the sixth group most likely excludes the formation of new structural units under doping of selenium by isoelectron admixtures. The mixture of tellurium with selenium leads to the formation of mixed-up molecules and the strong decrease in the level of chain molecule polymerization. On RDF curve an additional peak appears at the distance 2.8 Å equal to bond length Te–Te. The decrease in the degree of polymerization leads to a decrease in crystallization activation energy of selenium with admixture of tellurium. The lower potential of tellurium atom ionization promotes the expansion of the photoconductivity spectrum of doped selenium in a long wave field. Resistivity and activation energy of conductivity of selenium monotonic decrease with the growth of tellurium concentration.

With the addition of sulfur in the spectrum of RS of selenium a peak at 355 cm^{-1} appears and the intensity of this peak is growing with the increase in the content of ligand. This peak is connected with the formation of mixed-up rings of sulfur and selenium. Thus, when doped with sulfur the number of mixed ring molecules increases and the level of polymerization of molecules Se_n decreases.

Small concentrations of oxygen have a significant influence on the properties of glassy selenium. So the resistivity of glassy selenium decreases 10^6 times with the addition of 5×10^{-3} at.% oxygen and its photoconductivity grows at an order of magnitude with the addition of 2.5×10^{-2} at.% oxygen (Lacourse, Twaddell and MacKenzie, 1970). The presence of oxygen decreases the intensity of selenium photoluminescence at an order of magnitude and significantly changes the type of frequency dependence of conductivity (Baratov and Popov, 1990).

Similar to other cases of isoelectron admixtures, the formation of new structural units when doping selenium by oxygen is unlikely. Taking into consideration large electronegativity of oxygen in comparison to selenium (3.5 for oxygen and 2.4 for selenium) one can assume interaction of oxygen with charged defects. In this case, four options are possible (Popov, 1978).

The first option is realized when the concentration of oxygen is lesser than amount of chain molecules Se_n and, therefore, the concentration of negatively charged defects Se_1^-. In this case because of higher electronegativity of oxygen the latter quite probably will form negatively charged defects O_1^-. The concentration of Se_1^- will decrease. However, as the initial amount of these defects exceeds the number of oxygen atoms the following expression will be realized

$$N_{Se_3^+} = N_{Se_1^-} + N_{O_1^-} \tag{18}$$

and the addition of oxygen in such quantity should not change the electro conductivity of selenium significantly (in the above-mentioned expression the concentration of free carriers is omitted because Fermi level is near the middle of gap and material, therefore, is near the intrinsic semiconductor).

The second option corresponds with the concentration of oxygen atoms that exceeds the number of chain molecules but by not more than two times:

$$N_{Se_n} < N_O < 2N_{Se_n} \tag{19}$$

In this case under the limitation of selenium molecules with the atoms of oxygen an amount of unsaturated bonds of the latter exceeds the quantity of unsaturated bonds of selenium atoms. So, the transition of the part of oxygen atoms to O_1^- state is possible only by means of catching electrons from selenium atoms inside molecules as a result of greater interaction of unpaired electrons of oxygen atoms with the lone pair electrons of selenium atoms. Such action is equal to the formation of a hole:

$$O_1^0 + e = O_1^- + p \tag{20}$$

In this case Eq. (18) should be written as follows:

$$N_{Se_3^+} + p = N_{O_1^-} \tag{21}$$

which means that the number of negative defects exceeds the number of positive defects. This in turn determines the position of Fermi level nearer to the valence bond edge and increases p-type electro conductivity of selenium.

In the third option the number of oxygen atoms exceeds the quantity of Se_n molecules by more than two times. In this case all (or almost all) chain molecules are limited by oxygen atoms and additional atoms of oxygen are situated inside molecules in the O_2^0 state. Therefore, increasing the number of oxygen atoms should not influence greatly on the electro conductivity of selenium.

The final option corresponds to concentration of more than 0.1 at.% oxygen, that represents the limit of solubility of oxygen in selenium. Further increase of oxygen concentration leads to the formation of the second phase and does not have significant effect on the electroconductivity of material.

The dependence of electroconductivity of non-crystalline selenium on the oxygen concentration must have three different phases. In the first phase with small concentration of oxygen the electroconductivity remains constant; in the second phase there is a sharp growth of electroconductivity within the narrow range of oxygen concentration and at the third phase electroconductivity does not change with the increase of oxygen concentration. The given situation is true if the increase of oxygen concentration does not influence the level of polymerization, that is, the length of chain molecules of selenium. At the same time, there is data available that point out acceleration of selenium crystal growth in the presence of oxygen that testifies to the reduction of polymerization levels of molecules in line with the increase of oxygen concentration. This will lead to some dependence of electroconductivity on oxygen concentration in the third phase of the curve.

Experimental dependence of resistivity of non-crystalline selenium on oxygen concentration is shown in Figure 13 (Lacourse et al., 1970). We can see that there is a qualitative coincidence of the experiment with the model in the second and third phases. With respect to small oxygen concentration, there is no sufficient experimental points in order to come to a conclusion on the presence of the first phase.

So, additional oxygen leads to the formation of acceptor levels in selenium owing to the interaction of oxygen atoms with the structural defects of selenium. The formation of acceptor levels occurs only within a narrow range of oxygen atom concentration.

Among the univalent admixtures the influence of halogens on selenium is better researched. Let us consider the behavior of halogen atoms (X) added to molten selenium (Popov, Geller, Karalunets and Ipatova, 1980). Atoms of halogens have the electron configuration s^2p^5 and can generally be situated in selenium at following states (Fig. 14):

- to terminate the ends of polymer molecules Se_n (Fig. 14a)

$$Se_1^0 + X_0^0 \rightarrow Se_2^0 + X_1^0 \tag{22}$$

- to be situated between selenium molecules; in this case halogen atom has one unpaired p-electron and can either interact with lone pair electrons of the nearest

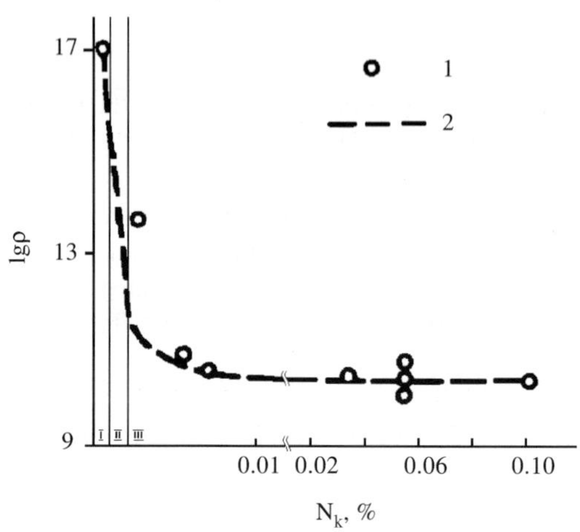

FIG. 13. Dependence of non-crystalline selenium resistivity on oxygen concentration (1—experiment, Lacourse et al., 1970; 2—model).

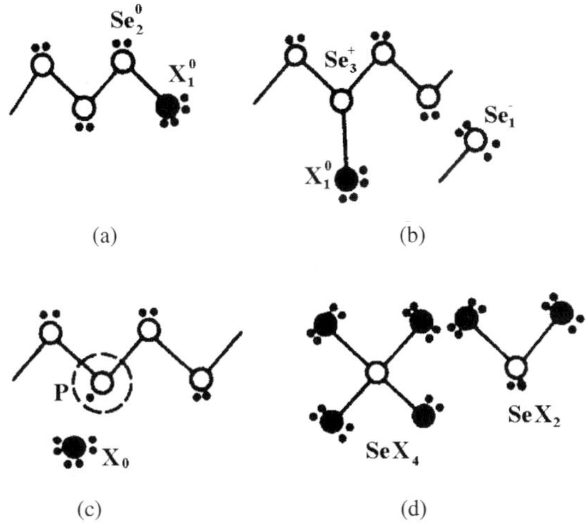

FIG. 14. Various bonding configurations of halogen atom (X) in non-crystalline selenium.

selenium atom and form covalent bond or capture an electron because of greater electronegativity of halogen. In the former case the valence coordination of the selenium atom increases and Se_3^+-center is formed (Fig. 14b):

$$Se_2^0 + X_0^0 \rightarrow Se_3^0 + X_1^0 \rightarrow Se_3^+ + X_1^0 + e \tag{23}$$

The released electron can form Se_1^--center with selenium atom at the end of polymer molecule:

$$e + Se_1^0 \rightarrow Se_1^- \tag{24}$$

The latter case leads to the creation of negatively charged ion of halogen and a hole (Fig. 14c)

$$\begin{gathered} Se_2^0 - e \rightarrow (Se_2 + p) \\ X_0^0 + e \rightarrow X_0^- \end{gathered} \tag{25}$$

– to form chemical compounds with selenium under the following reactions (Fig. 14d):

$$\begin{gathered} 2X + Se \rightarrow SeX_2 \\ 4X + Se \rightarrow SeX_4 \\ 2SeX_2 \rightarrow SeX_4 + 2Se \end{gathered} \tag{26}$$

The probability of this process must rise with increasing temperature.

Summing up the data of different kinds of halogen atom states in the molten selenium, the neutrality condition can be written as follows

$$N_{\sum Se_3^+} + p = N_{Se_3^+Se} + N_{Se_3^+X} + p = N_{Se_1^-} + N_{x_0^-} + e \tag{27}$$

where $N_{\sum Se_3^+}$ is the Se_3^+-centers total concentration, $N_{Se_3^+Se}$ is the concentration of Se_3^+-centers, bonded with three atoms of selenium; $N_{Se_3^+X}$ the concentration of Se_3^+-centers bonded with two atoms of selenium and an atom of halogen; $N_{Se_1^-}$ is the concentration of Se_1^--centers; $N_{X_0^-}$ is the concentration of halogen ions and e and p are the concentration of electrons and holes accordingly.

The ratio between the concentrations of different kinds of defects corresponding to the equilibrium at the temperature of the melt remains the same in specimens of amorphous selenium in case of quenching of the melt.

The above-mentioned kinds of interaction of halogen atoms with selenium will influence on the molecular structure and properties of the latter in a different way. So, the termination of chain-ends with atoms of halogens (reaction (22)) will tend to decrease VAP concentration that should lead to a shift of the reaction

$$2Se_2^0 \leftrightarrow Se_1^- + Se_3^+ \tag{28}$$

to the right till the restoration of an equilibrium concentration of these centers at the given temperature of melt. This, in its turn, will lead to a decrease in degree of polymerization of chain molecules and must bring to decreasing viscosity and activation energy for crystallization of matter under other conditions. At the same time, the above-mentioned process should not influence greatly on electrical properties of selenium in particular on its resistivity.

Reaction (23) will reduce the concentration of those Se_3^+-centers which bond the molecules of selenium between themselves ($N_{Se_3^+Se}$). At the same time, the total concentration of $N_{\sum Se_3^+}$ will not change significantly. The reduction in the amount of bridges between molecules must also lead to decreasing viscosity of matter even at a constant level of polymerization. On the other hand, the examined changes at the first approximation would not have a strong influence on electroconductivity of selenium.

The creation of negative ions of halogens (reaction (25)) should not have any significant influence on viscosity and activation energy of crystallization, but it will increase the concentration of holes and electroconductivity of material. Reaction (26) leads to a reduction of the number of halogen atoms participating in the above processes while concentration of halogen in a sample as a whole does not change.

The results of experimental investigation of influence of halogen (bromine) admixture on the properties and structure of selenium prove the correctness of the described model (Popov et al., 1980).

An addition of arsenic to selenium decreases concentration of ring molecules that is proved by the results of Raman spectra investigation. Besides, arsenic atoms bond separate molecules of selenium with covalent bonds that leads to rising glass transition temperature and activation energy for crystallization of selenium in line with increasing arsenic concentration. Admixture of germanium affects selenium similarly (Mott and Davis, 1979).

4.2. Atomic Structure of Chalcogenide Glasses

Arsenic chalcogenides at a considerable range of component ratio are disposed to the formation of a non-crystalline phase under different methods of production. Hence, during cooling of the melt glass formation regions include for system As–S compositions with the content of arsenic from 5 to 45 and from 51 to 65 at.% and for the system arsenic–selenium–from pure selenium to composition with 60 at.% of arsenic, for system arsenic–tellurium compositions from 46 to 58 at.% of arsenic (Vinogradova, 1984). Preparation of arsenic chalcogenides films by different vapor deposition methods remarkably broadens the mentioned ranges of components ratio, which provide the formation of non-crystalline substances. In the case of system Sb–S only compositions near to stoichiometric one (Sb_2S_3) have been prepared in glassy state (Vinogradova, 1984; Rubish, Kuzenko, Poltavzev, Turyniza, Michalev and Popov, 1993).

The structure of non-crystalline arsenic chalcogenides is explained on the basis of conceptions of continuous network. In terms of chemical order, i.e., correlation between heteropolar (arsenic–chalcogen) and homopolar (arsenic–arsenic, chalcogen–chalcogen) chemical bonds, two extreme cases are possible: completely random network, in which chemical bond distribution is purely static, and chemically ordered network where heteropolar bonds realize anywhere if it is permitted by the chemical composition and by the requirement of continuity of the network. In real materials, there is some intermediate position determined by the difference of energy value of various bonds, the proportions of components and to a large extent by the conditions of material preparation. So, for example, according to data of different researchers concentration of homopolar bonds in films of As_2Se_3 change from 10 to 35%.

Most investigations of glassy arsenic chalcogenides structure were carried out on materials of stoichiometric composition As_2X_3 (where X is a chalcogen: sulfur, selenium, and tellurium). Particular feature of such compounds is a laminated atomic disposition and the predominance of covalent atomic bonds in layers that is confirmed by isolation of the first peaks of RDF. The basic structural unit is pyramidal blocks $AsX_{3/2}$.

There are not so many investigations devoted to the structure of non-stoichiometric compositions of non-crystalline arsenic chalcogenides. The pyramidal blocks in glasses rich in selenium are assumed to be connected to each other with additional atoms of selenium that increase their free mutual orientation. Free orientation of pyramidal blocks decreases and horizontal length of layers increases in line with the growth of arsenic content. Electron diffraction and X-ray diffraction investigations of non-crystalline layers of arsenic chalcogenides at a wide range of compositions showed (Michalev, 1983; Rubish et al., 1993) that there is a peak at $S = 1.2...1.4\ Å^{-1}$ testifying to the presence of the medium-range order in atomic disposition of these materials on the curves of scattered intensity of electrons $I_n(s)$. Intensity of this peak is minimum for the compositions rich in chalcogen (samples $As_{7.5}S_{92.5}$, $A_{19}Te_{81}$) and increases along with growing arsenic content for all investigated systems. This provides evidence of ordering of material structure with the increase of structural network rigidity. The growth of atomic number of chalcogen (from sulfur to tellurium) under other equal conditions leads to the reduction of the first peak intensity on the curves $I_n(s)$. This can be explained by the changes of structural network rigidity as a result of increase of chemical bond metallization. Practically, for all the compositions on RDF there is a small peak in the region from 4.5 to 5.3 Å, which is equal to the most probable distances between layers in the materials.

Comparison of RDFs of arsenic chalcogenides allows to note the following tendencies, observed when either the kind of chalcogen or ratio of components is changed. In system As–Te under the change of component ratio from compositions rich in tellurium to stoichiometric compound As_2Te_3 ,and further to compounds rich in arsenic the first peak of RDF moves to the left of a value close to doubled covalent radius of tellurium (2.8 Å), to the value that equals the sum of covalent radii of tellurium and arsenic (2.58 Å) for compositions nearest to stoichiometric one, and then to the value of doubled covalent radius of arsenic (2.42 Å). Such change of position of the first peak of RDF can be explained by the gradual replacement of chemical bonds between atoms of tellurium in samples rich in this element for heteropolar bonds As–Te, typical for stoichiometric compound As_2Te_3 with their further replacement with the bonds between arsenic atoms by increasing the content of the latter to more than 40%. The distinctive feature of RDF of As–Te samples of compositions close to stoichiometric ones is a split second peak that defines distances between the nearest unbounded atoms of arsenic and chalcogen. These distances are determined by the length of bond As–Te and bond angles As–X–As, X–As–X. The positions of maxima of the first and second peaks of RDF of sample As_2Te_3 allow to determine the value of bond angles: the most probable value of bond angles As–Te–As is 84° and for Te–As–Te is 99°.

Some displacement of the first maximum to the left under reduction of arsenic content is typical for the RDF of As–S system that is caused by the growth of a number of chemical bonds between atoms of sulfur. In the system As–Se the position of the first maximum practically does not change while changing of components ratio that is explained by a small difference of arsenic and selenium covalent radii.

A direct interpretation of area of RDF peaks to determine coordination numbers within binary compositions is possible only for samples of stoichiometric compositions (if we suppose that there are only heteropolar bonds) for the first peak. In order to correctly interpret the results of diffraction investigation of non-stoichiometric binary compounds, one needs additional information about interaction of atoms in a substance within an elementary cell due to presence of homopolar bonds in non-stoichiometric compounds. Such information is obtained from spectra of RS that allows identifying structural units of investigated substance.

The RS spectra of glassy arsenic selenides are shown in Figure 15 (Rubish et al., 1993). (The arsenic curve is taken from a work by Greaves, Elliott and Davis (1979).) RS spectrum of glassy arsenic consists of peaks at 200, 240, 285 cm^{-1} and a system of peaks in the region of 110–160 cm^{-1}. Good prepared volume specimens of glassy As_2Se_3 have as a rule only one peak at 227 cm^{-1} that proves the presence of one kind of structural unit $AsSe_{3/2}$ which in turn confirms correct interpretation of RDF of this compound based on the assumption about the existence of heteropolar bonds only.

Comparison of RS spectra of specimens of system As–Se with different components ratio shows that the peak of stoichiometric compound As_2Se_3 at 227 cm^{-1} is in spectra of all specimens to $As_{65}Se_{35}$. However, intensity of this peak decreases under the increase of arsenic content and only the shoulder is observed in spectra of specimens $As_{65}Se_{35}$ under the given frequency shift. A small excess of arsenic content over stoichiometric composition (sample $As_{45}Se_{55}$) leads to the appearance of a peak at 160 cm^{-1} and a shoulder in the region 240 cm^{-1}. Emergence of these details in the spectrum is linked to the origin of bonds As–As in the glass network. However, the absence of a considerable part of peaks characteristic to RS spectrum of non-crystalline arsenic in the spectrum of $As_{45}Se_{55}$ testifies to the absence of structural arsenic units (atom As linked with three atoms of As) in this sample. Persistence of a large intensity peak at 227 cm^{-1} along with the emergence of the above-mentioned details allows to assume that pyramidal blocks $AsSe_{3/2}$ remain as predominant structural units in a sample of composition $As_{45}Se_{55}$ and

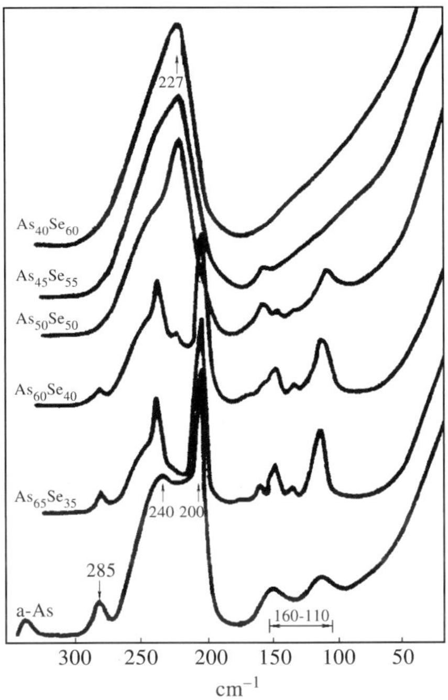

FIG. 15. Raman spectra of glassy arsenic chalcogenides (a–As after Greaves et al., 1979).

that new structural units appear alongside in which one atom of selenium is replaced by one atom of arsenic—$As(Se_2As)_{1/2}$. Rise in the arsenic content to 50% makes a shoulder at 240 cm^{-1} in more expressed RS spectrum and leads to a series of peaks in the zone of 100–160 cm^{-1} that could be a reflection of the fact that the amount of structural units $As(Se_2As)_{1/2}$ and possibility of the presence of structural units consisting only one atom of selenium $As(As_2Se)_{1/2}$ increases. At the same time, the absence of a peak at 200 cm^{-1} in SR spectra typical to non-crystalline arsenic testifies to the absence of arsenic structural units in this matter. SR spectra of samples with 60 and 65% arsenic are characterized by greater reduction of intensity of the peak at 227 cm^{-1} typical to As_2Se_3 (that confirms a small concentration of structural units $AsSe_{3/2}$), rise of intensity of peaks at 240 cm^{-1} and at a range of 110–160 cm^{-1}. Besides, peaks at 200 and 285 cm^{-1} become apparent and thus all peaks typical to non-crystalline arsenic are present in these spectra. The obtained data testifies the predominance of structural units $As(Se_2As)_{1/2}$ and As $(As_2Se)_{1/2}$ in these substances as well as to the existence of arsenic structural units in them.

Analysis of RS results in substances of systems As–S and Sb–S at a range of compositions from $Sb_{36}S_{64}$ to $Sb_{43}S_{57}$ (Rubish et al., 1993) shows that changes of spectra together with change of components ratio are analogous to the considered ones for system As–Se.

Additional information about structure of arsenic chalcogenide glasses can be obtained by indirect methods by measuring various physical and chemical properties of the matter. Qualitative results of alteration of heteropolar and homopolar bonds ratio in the materials

TABLE IV

GLASS FORMING REGION IN SILICON (GERMANIUM)–CHALCOGEN SYSTEMS

Chalcogen	Silicon (at. %)	Germanium (at. %)
S	31–50	10–45
Se	0–20	0–40
Te	10–22	10–25

of systems As–Se, As–S can be obtained using the method of differential solution (Baratov, Popov, Bekicheva and Michalev, 1983). It is known that in the process of dissolution of arsenic sulfides and selenides alkaline solvents have more impact on heteropolar bonds (As–Se, As–S) than on homopolar ones (As–As, Se–Se, S–S). Investigation of As–Se materials solubility in 10% KOH showed (Michalev, 1983) that rate of dissolution falls significantly and the quantity of insoluble residuum rises under the increase of arsenic content over stoichiometric composition ratio. It conforms to the conclusion made on the basis of analysis of RS spectra that the amount of As–As bonds increases in this case.

The chalcogenides of the group IV elements have smaller regions of glass formation (Table IV) and demand harder regimes of melt quenching in comparison with arsenic chalcogenides. In their glassy state, silicon and germanium have a valency of four. The main features of structural change with change of composition are similar to those for arsenic chalcogenides observed above. Tetrahedrons $GeX_{4/2}$, $SiX_{4/2}$ that consist of a central germanium atom (silicon) bonding with four atoms of chalcogens are the structural units of composition GeX_2, SiX_2. In compositions enriched by chalcogens with respect to the indicated compositions, the number of tetrahedrons is proportional to the atomic concentration of the group IV elements and surplus atoms of chalcogen are united in chains linking tetrahedrons among themselves. Under higher concentration of chalcogen, separate molecules of the latter appear and form solid solution with structural network of material. So, RS spectra of glasses $Ge_{20}Se_{80}$ testify the existence of ring molecules Se_8 in material. In the materials enriched by germanium a peak connected with structural units $Ge(X_3Ge)_{1/2}$ (Mott and Davis, 1979) appears on the RS spectra.

5. Structural Modification of Non-Crystalline Semiconductors

5.1. LEVELS OF STRUCTURAL MODIFICATION

In the initial investigations concerning non-crystalline semiconductors Kolomietz (1960) established that the most essential feature of these materials is their weak sensitivity to impurities. In spite of some solutions being found later (for example, chemical modification of chalcogenide glassy semiconductor films (Ovshinsky, 1977), the control of conductivity in transition metal ion glasses by adding oxidizing agent (Hogarth and Popov, 1983), doping of hydrogenated amorphous silicon (Spear and Le Comber, 1977), the problem of control over the properties of non-crystalline semiconductors and the problem of reproducible synthesis of non-crystalline semiconductors with prescribed properties still remain pressing.

As an alternative to the control over semiconductor properties by doping, the method of the structural modification of non-crystalline semiconductor properties has been proposed (Popov, 1980a; Popov and Shemetova, 1982; Popov, Michalev and Shemetova, 1983) that consists of controlling the properties by changing the structure of a material without changing its chemical composition. The physical basis of the method is the fact that the electron energy term of non-crystalline materials has several minima corresponding to various metastable states of the system (Anderson, Halperin and Varma, 1972; Bal'makov, 1975). At first the method was developed for chalcogenide glassy semiconductors and used the changes of atomic structure at the level of medium-range order.

Along with this as shown in Section 1 there are four levels of structural characteristics necessary for common description of structure of non-crystalline solids, namely: short-range order in atomic disposition, medium-range order in atomic disposition, morphology and subsystem of defects. Therefore, structural changes are possible not only at the level of medium-range order, but also at the other levels of structural characteristics mentioned above. In that way one can conclude that on the whole there exist four levels of structural modification differed by various changes of material structure namely: level of short-range order (level 1), level of medium-range order (level 2), level of morphology (level 3) and level of defect subsystem (level 4) (Table V; Popov, Vorontsov and Popov, 2001).

The structural changes at the level of short-range order lead to variations of all basic properties of material. For example, polymorphic crystalline modifications of carbon (diamond, graphite and carbine) possess fundamentally different physico-chemical properties because of the different hybridizations of electron orbitals and different atomic structures at the short-range order level. Amorphous carbon films incorporate structural units of different allotropic modifications, with the relative content of these units determined by film growth modes and varying widely for the same preparation method. Correspondingly, the coordination of atoms varies (between 2 and 4) together with other parameters of the first coordination sphere (Vasil'eva and Popov, 1995). When films of amorphous hydrogenated carbon (α-C:H) are obtained by rf-ion-plasma sputtering in an argon–hydrogen atmosphere, merely changing the substrate temperature and discharging power may give films (Popov, Ligachev, Vasil'eva and Stuokach, 1995), in which the optical gap varies by two orders of magnitude (between 0.02 eV for graphite-like films and 1.85 eV for films with predominance of the diamond-like phase); and the dark conductivity by more than 10 orders of magnitude (between 7 and $2 \times 10^{-10}\,\Omega^{-1}\,\mathrm{cm}^{-1}$). The dependencies of resistivity ρ (Figure 16a) and the width of optical gap E_g (Fig. 16b) on substrate temperature and the charge power (p) are shown in Figure 16.

Changing the medium-range order without changing the short-range order mainly affects the macroscopic properties of material (viscosity, micro-hardness, Young modulus and photo-contraction of films as shown in Table V). In Figure 17 dependence of density of glassy GeS_2 on the value of applied pressure (Miyauchi, Qiu, Shojiya and Kawamura, 2001) is shown and in Figure 18 the dependence of microhardness changes in different chalcogenide glasses as a result of changes of their thermal history on the criterion of efficiency of structural modification (CESM) is shown (Section 5.3)

TABLE V
LEVELS OF STRUCTURAL MODIFICATION

Level	Structural changes	Method of treatment	Characterization of sensitive properties	Groups of sensitive properties	Examples of sensitive properties
1	Short-range order	Various methods and modes of preparation	All properties	All properties	All properties
2	Medium-range order	External factor treatment during preparation or thermal treatments	Properties associated with rearrangement of structural units	Mechanical properties, phase transitions	Viscosity, hardness, Young modulus, photo-contraction of films, temperature and activation energy of crystallization
3	Morphology	Changes in the preparation and treatment modes	Properties dependent on microheterogeneities	Electrical, optical	AC conductivity
4	Defect subsystem	Changes in preparation modes, treatments affecting the defect subsystem	Properties dependent on the distribution of the density of localized states and on the Fermi level position	Electrical, photo-electric	Field-dependent conductivity

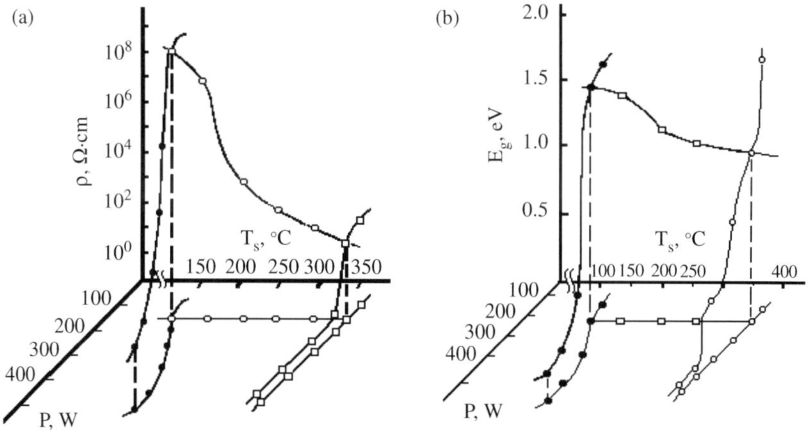

FIG. 16. Resistivity ρ (a) and optical gap E_g (b) of a–C : H (rf-sputtering) films in relation to their preparation modes. (T_s is the substrate temperature and P is the rf discharge power.)

(Popov et al., 1983). At the same time, the properties governed by the electronic structure (electronic spectrum) of the material (conductivity, photoconductivity), that depend mainly on the short-range order, change relatively weakly.

Morphology changes affect the properties sensitive to microinhomogeneties (Ligachev, Popov and Stuokach, 1996a) whereas changes in the defective subsystem modify the spectrum of localized states in the mobility gap, shift the Fermi level, and modify the properties related to the electron subsystem (Table V). Energy distribution of the density of localized states, $N(E)$ in the mobility gap of a-Si:H before (solid line) and after (dashed line) exposure to UV radiation (dose 10^{19} cm^{-2}) that was found using (1) constant photoconductivity method; (2) modeling of the temperature dependence of conductivity; and (3) analysis of space-charge-limited currents as shown in Figure 19 (Popov, 1996).

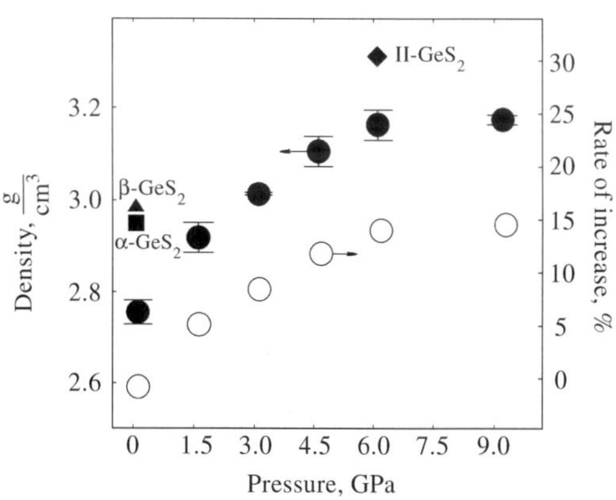

FIG. 17. Dependence of glassy GeS_2 density on pressure in permanent densification (Miyauchi et al., 2001).

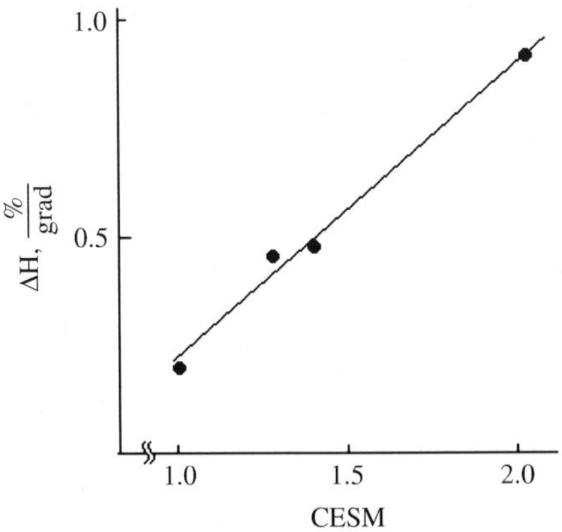

FIG. 18. Changes in microhardness of glassy chalcogenides as a result of changes in the thermal history of a material, ΔH, vs. the criterion of structural modification efficiency (CESM).

Despite application of the method of structural modification of properties to various non-crystalline semiconducting materials (chalcogenide glasses (Popov et al., 1983, 1988; Luksha, 1994; Miyauchi et al., 2001) thin films a-Si:H (Popov, 1996; Khokhlov, Mashin and Khokhlov, 1998; Mashin and Khokhlov, 1999), a-SiC:H (Ligachev, Svirkova, Filikov and Vasil'eva, 1996b) and a–C (Ligachev, Popov and Stuokach 1994) the issue of the applicability limits of one or another level of structural modification remains open. To answer this question, let us analyze the efficiency of structural changes at the different levels in covalent semiconducting elements, belonging to Groups IV–VI of the periodic table (carbon, silicon, germanium, phosphorus, arsenic, antimony, sulfur, selenium and tellurium) and also in a number of their systems.

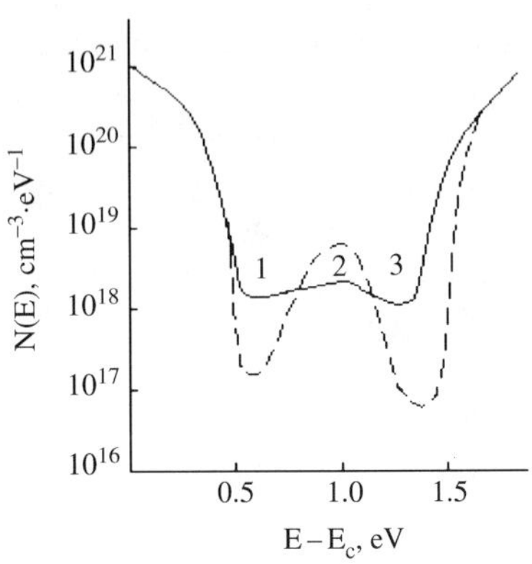

FIG. 19. Density of localized states N(E) in the mobility gap of a–Si:H before (solid line) and after (dashed line) exposure to ultraviolet radiation (dose 10^{19} cm^{-2}) (1 constant photocurrent method; 2 conductivity temperature dependence; 3 space-charge-limited currents).

5.2. Structural Changes at the Short-Range Order Level

The first level of structural modification involves pronounced changes in the short-range order; i.e., changes in the hybridization of electron orbitals of all (or most) atoms constituting the sample. Unique in this respect, carbon has long been considered the only element of those considered here that existed in allotropic crystalline modifications of diamond (sp^3 hybridization) and graphite (sp^2 hybridization). The discovery of carbines (sp hybridization) in 1960 (Sladkov, Korshak, Kudryavtsev and Kasatochkin, 1972) served as further evidence to the unique properties of carbon. At the same time, it should be noted that there have been reports that the short-range order might change substantially under certain conditions for other elements as well. For example, cubic α and β modifications of crystalline selenium with atom coordination numbers of, 4 and 6, respectively, were obtained (Andrievsky, Nabitovich and Krinykovich, 1959; Andrievsky and Nabitovich, 1960) in electron beam-induced crystallization of thin films. However, these reports failed to attract due attention of the scientific community at that time. The monopoly of carbon on the possibility of existence of forms with different hybridizations of electron orbitals and, consequently, with different atomic coordination numbers was radically broken up by a series of investigations (Khokhlov et al., 1998; Mashin and Khokhlov, 1999). New forms of silicon that appeared under certain conditions in films of amorphous silicon and in a-Si:H were discovered in these studies—silicine with sp hybridization of electron orbitals and atomic coordination of 2 and a form with sp^2 hybridization and atomic coordination of 3. However, it should be noted that, in these investigations devoted to silicon and in studies of selenium (Andrievsky et al., 1959; Andrievsky and Nabitovich, 1960), the new forms were obtained only under specific conditions.

Thus, in the considered group of covalent semiconducting materials (Periods 2–5, Groups IVA–VIA of the Periodic Table), structural changes at the short- range order level are observed for elements belonging to Group IVA, Periods 2 and 3 (C, Si), and Group VIA, Period 4 (Se) (Table VI), i.e., for three elements out of the considered nine. We should also note the increase in the first coordination number from 2 to 3 in tellurium melt (at 600 °C), with the covalent nature of chemical bonds being preserved (Fig. 20, Poltavzev, 1984). The mentioned elements show no fundamental distinctions in electron shell structure or other parameters from the rest of the considered elements. In view of the above, it seems reasonable to assume that structural changes at the short-range order level are not unique to carbon, being characteristic to all of the considered covalent semiconducting materials. Experimental evidence in favor of this assumption has been obtained for carbon, silicon and selenium, and obtaining it for the other elements is only a matter of time and attention devoted to the problem.

5.3. Structural Changes at the Medium-Range Order and Morphology Levels

At the second level of structural modification, the obtained differences in the atomic structure are due to changes in the medium-range order (distribution of dihedral angles

TABLE VI

SHORT-RANGE ORDER LEVEL

Periods	Columns|Coordination number (N_c)					
	IVA	N_c	VA	N_c	VIA	N_c
2	${}^{6}_{12}C$	2 3 4				
3	${}^{14}_{28}Si$	2 3 4	${}^{15}_{31}P$	3	${}^{16}_{32}S$	2
4	${}^{32}_{72.5}Ge$	4 4	${}^{33}_{75}As$	3	${}^{34}_{79}Se$	2 4 6
5			${}^{51}_{122}Sb$	3	${}^{52}_{128}Te$	2 3[a]

[a]melt

and their signs, degree of molecule polymerization, extent of disorder in alloys, etc.). In this case, the maximum possible structural changes can be evaluated by the CESM, expressed in a simplified form as (Popov et al., 1983)

$$\text{CESM} = (\text{VEC} - N)/[N(1 - I_c - M)] \tag{29}$$

where VEC is the average concentration of valence electrons, N the average coordination number, I_c the ionicity of chemical bonds and M the degree of bond metallization.

CESM varies between two for sulfur and selenium and zero for silicon. There exist some boundary values of CESM below which it becomes impossible to change the material properties by structural modification at this level. The question arises in this

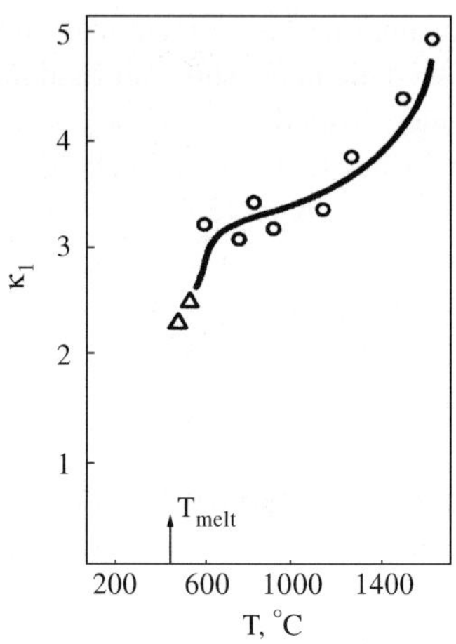

FIG. 20. Temperature dependence of the liquid tellurium first coordination number (Poltavzev, 1984).

connection is 'what determines the applicability limit of the structural modification at the medium-range order level and where does it lie?'

Qualitatively, it may be stated that this limit is determined by the rigidity of the structural network: for changes in the medium-range order to occur, the network must exhibit certain flexibility and flexible bonds are to be present. In view of the above, it seems reasonable to suggest that the limit in question corresponds to the rigidity threshold of the structural network (Phillips, 1979a, b, 1980) at which the average number of force constants per atom becomes equal to the number of degrees of freedom. Phillips (1979b) determined the critical coordination number corresponding to the rigidity threshold of the structural network for chalcogenide glasses: $N_c = 2.4$. For the systems Ge–Se, Ge–S, As–Se and As–S, the rigidity threshold is shown by the dashed line in Table VII. However, experimental studies of these systems (Yun, Li, Cappelletti, Enzweiler and Boolchand, 1989; Aitken, 2001) have shown that the rigidity threshold lies at higher average coordination numbers. It was demonstrated by Popov (1994) that the reason for the discrepancy between the calculated and experimental values of the rigidity threshold comes from the ionic component and metallization of chemical bonds being neglected. The rigidity threshold obtained for the above-mentioned systems with account of the ionicity of chemical bonds is shown in Table VII by solid line. In accordance with this suggestion, the line is the applicability limit of the second level of structural modification.

The morphology of films of non-crystalline materials means the presence of microheterogeneities in them (columns, globules, cones, etc.). A certain morphology (column structure) was observed in a–Si:H films by Knights and Lujan (1979). Danchenkov, Ligachev and Popov (1993); Vasil'eva (1997); Ligachev (1998); Filikov, Popov, Cheparin and Ligachev (1999); have demonstrated that the presence of heterogeneities is a characteristic feature of amorphous tetrahedral semiconductors (a–Si:H, a–C:H, a–Si_xC_{1-x}:H) and established a relationship between the averaged morphological parameters (cross dimensions of columns), spectra of electron states and material properties.

It should be noted that the necessary condition for obtaining any solid non-crystalline material is the thermodynamically non-equilibrium process of its synthesis. In conformity with the basic concepts of the theory of self-organization (synergetics), the non-equilibrium conditions of material formation result in the appearance of heterogeneities, or certain morphology. Thus, general considerations suggest the presence of macro-heterogeneities in all of the non-crystalline semiconductors in question. At the same time, while the presence of certain morphology in films of tetrahedral non-crystalline semiconductors has been firmly established (see references above), a 'structureless' smooth surface is commonly observed in vitreous materials. The most likely reason for this contradiction is the indefinite distinction between 'macroheterogeneities' and 'microheterogeneities'.

Experimental data mostly indicate the occurrence of certain morphology in films of tetrahedral amorphous semiconductors and the possibility of obtaining homogeneous 'structureless' films of vitreous semiconductors. Therefore, it can be assumed that (at least for the time being) the third level of structural modification is applicable to non-crystalline materials with a rigid structural network, and that the line corresponding to the rigidity threshold (Table VII) separates its applicability area from that of the second level of structural modification.

TABLE VII

Medium range order and morphology levels

Columns		$A^{IV}_x (A^V_x) B^{VI}_{1-x}$ systems Value of X							Column VI A
IV A	V A	0,60	0,50	0,40	0,33	0,25	0,20	0,11	
Ge **4** ***4***					**$GeSe_2$** **2,66** ***2,40***	**$GeSe_3$** **2,50** ***2,30***	**$GeSe_4$** **2,40** ***2,24***	**$GeSe_8$** **2,22** ***2,12***	**Se** **2** **2**
Ge **4** ***4***			**GeS** **3,00** ***2,73***	**Ge_2S_3** **2,80** ***2,49***	**GeS_2** **2,66** ***2,30***	**GeS_3** **2,50** ***2,23***	**GeS_4** **2,40** ***2,19***	**GeS_8** **2,22** ***2,10***	**S** **2** **2**
	As **3** ***3***	**As_3Se_2** **2,60** ***2,53***	**AsSe** **2,50** ***2,42***	**As_2Se_3** **2,40** ***2,31***	**$AsSe_2$** **2,33** ***2,25***	**$AsSe_3$** **2,25** ***2,19***	**$AsSe_4$** **2,20** ***2,16***	**$AsSe_8$** **2,11** ***2,09***	**Se** **2** **2**
	As **3** ***3***	**As_3S_2** **2,60** ***2,49***	**AsS** **2,50** ***2,38***	**As_2S_3** **2,40** ***2,26***	**AsS_2** **2,33** ***2,22***	**AsS_3** **2,25** ***2,17***	**AsS_4** **2,20** ***2,14***	**AsS_8** **2,11** ***2,08***	**S** **2** **2**

▬ ▬ ▬ - rigidity threshold in the case of purely covalent bonds;
▬▬▬▬ - rigidity threshold in the case of bond ionicity is taken into account;
2,25 - average coordination number, N_c;
2,25(italics) - $N_c(1 - I_c)$, where I_c – mean coefficient of bond ionicity.

5.4. Structural Changes at the Defect Subsystem Level

The fourth level of structural modification is associated with changes in the defect subsystem under the influence of either sample fabrication conditions or various external factors and is manifested in changes in the spectrum of localized states in the gap, which in turn leads to the changes in material properties.

Golikova, Kazanin, Kudoyrova, Mezdrogina, Sorokina and Babokhodjaev (1989); Golikova (1991) observed the effect of amorphous silicon pseudo-doping that varied with sample preparation conditions. A significant change in the spectrum of localized states in a–Si:H, resulting from a change in the relative content of Si–H, Si–H_2, and Si–H_3 complexes, was achieved by treating samples with ultraviolet radiation (Popov, 1996). The effect of weak electric and magnetic fields on quasimolecular defects and properties of vitreous selenium, arsenic triselenide, and materials of the selenium–tellurium system was observed by Dembovskii, Kozukhin, Chechetkina, Vikhrov, Denisov and Kobtzeva (1984); Dembovskii, Chechetkina and Kozukhin (1985). Dembovskii, Zyubin and Grigor'ev (1998) extended these results to sulfur and arsenic trisulfide.

Thus, the fourth level of structural modification is observed experimentally both in a tetrahedral material with rigid covalent structural network (a–Si:H) and in vitreous materials of group VI and V–VI chalcogenide glasses (Table VIII).

TABLE VIII

Defect Subsystem Level

Columns		Systems		Column
IV A	V A	$A_2^V B_3^{VI}$	$A^{VI} B^{VI}$	VI A
C				
Si	P			
Ge	As	As_2S_3	S–Se	S
		As_2Se_3	Se–Te	Se
		As_2Te_3		Te

The above suggests that the structural modification at the defect subsystem level is inherent, as is the structural modification at the short-range order level, in all the considered non-crystalline semiconducting materials.

It can thus be concluded that there exist at least four levels of structural modification of the properties of non-crystalline semiconductors that differ in the resulting structural changes. Control over properties of non-crystalline semiconductors by changing the structure at the short-range order level is, in principle, possible for all of the considered materials.

The fourth level of structural modification—control over properties by treating the defect subsystem is also applicable to all these materials.

Control over properties by changing the medium-range order is possible for vitreous semiconductors that have high CESM values, with the applicability limit of this level corresponding to the rigidity threshold of the structural network, calculated with the consideration of the ionicity of chemical bonds. On the other side of this boundary (materials with rigid structural network) lies the area of applicability of the third level of structural modification, involving changes in the morphology of semiconductor films.

5.5. Correlation Between Structural Modification and Stability of Material Properties and Device Parameters

Structural modification (changes of structure) can take place in the process of producing material as well as in the process of affecting material or devices on the basis of different factors (Table V). Moreover, atomic structure and consequently, properties of non-crystalline semiconductor can be changed in the process of using such devices because their operation as a rule is subjected to the influence on material of active part of device by electrical or electro-magnetic fields, high temperature and so on. Two conclusions follow from the above and these must be taken into consideration when designing devices on the basis of the observed materials.

The first conclusion consists of the possibility to improve parameters of devices by purposeful effect on a structure of material during their production or after completion. The present approach has been put into practice during the creation of electro-photographic photoreceptors on the basis of glassy selenium (thermal treatment and electromagnetic radiation) (Kotov, 1984; Popov, 1997), integrated circuit of reprogram constant memory matrix on the basis of chalcogenide glassy semiconductors (γ-emission)

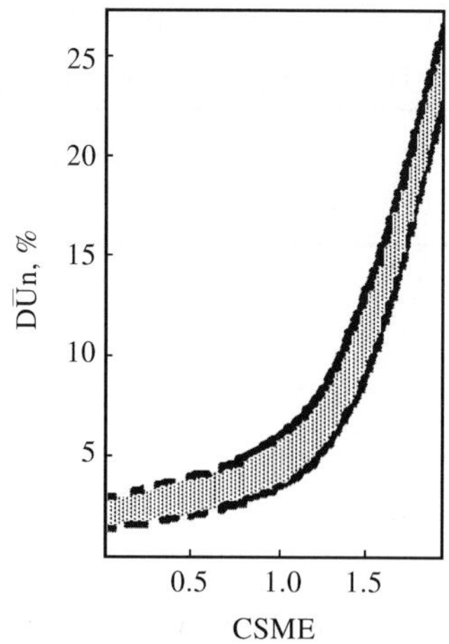

FIG. 21. Statistical variations of threshold voltage in various chalcogenide glasses as a function of the criterion for structural modification efficiency.

(Popov and Shemetova, 1982), photo-thermoplastic optical information memory units on the basis of chalcogenide glasses (thermal treatment) (Baratov et al., 1983), thin film transistors on the basis of hydrogenated amorphous silicon (ultraviolet emission) (Popov, 1996).

The second conclusion is that the possibility of changes in non-crystalline semiconductor structure during the device operation leads to statistical variations of device characteristics that cannot be avoided or eliminated. The value of these variations will grow in line with the level of structural changes possible in used non-crystalline semiconductors at the given levels of external factors. Popov et al. (1983), analyzed statistical variations of threshold voltage of switches on the basis of chalcogenide glassy semiconductors of different compositions according to a value of CESM (Fig. 21). As seen in Figure 21, the experimental data of statistical variations of threshold voltage obtained from 10 papers of different authors on the basis of various chemical substances maps well to power dependence on a value of criterion efficiency of structural modification of matters, calculated in the above-mentioned report.

In many cases, structural modification of properties of non-crystalline semiconductors is an effective method of device parameter control. In addition, it is also necessary to take into consideration that possibility of structural changes of non-crystalline semiconductors in the period of device operation defines the value of unavoidable variation of device parameters.

References

Adler, B.J. and Hoover, W.G. (1968) *Physics of Simple Liquids*, North-Holland, Amsterdam.

Aitken, B.G. (2001) Structure-property relationships in Ge–As selenide and sulphide glasses. *Proceedings International Congress on Glass, Edinburgh, Invited Papers*, pp. 135–140.

Aivasov, A.A., Budogyn, B.G., Vikhrov, S.P. and Popov, A.I. (1995) Disorded semiconductors (Neuporyadochennee poluprovodniki), Vesshaya shkola, Moscow (in Russian).

Anderson, P.W., Halperin, B.I. and Varma, C.H. (1972) Anomalous low-temperature thermal properties of glasses, *Philos. Mag.*, **25**(8), 1–9.

Andrievsky, A.I. and Nabitovich, I.D. (1960) Structure of β-cubic modification of selenium, *Sov. Phys. Crystallogr.*, **5**, 442–443.

Andrievsky, A.I., Nabitovich, I.D. and Krinykovich, P.K. (1959) Structure of selenium in thin layers, *Sov. Phys. Dokl.*, **4**, 16–18.

Bal'makov, M.D. (1975) *The Influence of Structural Changes on the Properties of Disorder Systems*, Voprose fisiki poluprovodnikov, Kaliningrad, Vol. 1, 19–24 (Russian).

Baratov, A.G. and Popov, A.I. (1990) Molecular structure and properties of chalcogens (sulphur, selenium, tellurium). Aistan, Erevan, (in Russian).

Baratov, A.G., Popov, A.I., Bekicheva, L.I. and Michalev, N.I. (1983) The influence of chalcogenide glass structure on the characteristics of photothermoplastic recording medium, *Electronnaya Teknika, Materiale*, **10**, 66–68 (in Russian).

Barreto, L.S., Alves, O.L., Mort, K.A. and Jackson, R.A. (2001) *Molecular Dynamic Simulation of the Structure and Properties of Lithium Acetate Glass*, Proceedings of the International Congress on Glass, Edinburgh, Extended abstracts, pp. 472–473.

Cherkasov, Yu.A. and Kreitor, L.G. (1974) Exiton absorption and photoconductivity in amorphous analogs of various crystalline forms of selenium, *Fizika Tverdogo Tela*, **16**, 2407–2410 (in Russian).

Cormack, A.N., Du, J. and Zeitler, T. (2001) *Molecular Modeling of Glass*, Proceedings of the International Congress on Glass, Edinburg, Invited papers, pp. 170–174.

Danchenkov, V.A., Ligachev, V.A. and Popov, A.I. (1993) Morphology, conductivity, and pseudodoping in amorphous and amorphous-crystalline C:H films, *Semiconductors*, **27**, 683–686.

Dembovskii, S.A., Chechetkina, E.A. and Kozukhin, S.A. (1985) Anomalous influence of weak magnetic fields on diamagnetic glassy semiconductors, *JETP Lett.*, **41**, 88–90.

Dembovskii, S.A., Kozukhin, S.A., Chechetkina, E.A., Vikhrov, S.P., Denisov, A.L. and Kobtzeva, Yu.N. (1984) *Investigation of Resonance Phenomena in Chalcogenide Glassy Semiconductors by Viscosimetry Method*, Proceedings of the International Conference Amorphous semiconductors-84, Vol. 1, Gabrovo, Bulgaria, pp. 183–185.

Dembovskii, S.A., Zyubin, A.S. and Grigor'ev, F.V. (1998) Modeling of hypervalent configurations, valence alternation pairs, deformed structure, and properties of a–S and a–As_2S_3, *Semiconductors*, **32**, 899–916.

Elliot, S.R. (1987) Medium-range order in amorphous materials: documented cases, *J. Non-Cryst. Solids*, **97 and 98**, 159–162.

Filikov, V.A., Popov, A.I., Cheparin, V.P. and Ligachev, V.A. (1999) Morphology formation in silicon-based thin amorphous films as self-organization manifestation, *Nano-Crystalline and Thin Film Magnetic Oxides*, Kluwer, The Netherlands, pp. 347–351.

Golikova, O.A. (1991) Doping and pseudodoping of amorphous silicin, *Sov. Phys. Semiconductors*, **25**, 915–938.

Golikova, O.A., Kazanin, M.M., Kudoyrova, V.Kh., Mezdrogina, M.M., Sorokina, K.L. and Babokhodjaev, U.S. (1989) Effect of amorphous silicon pseudodoping, *Sov. Phys. Semiconductors*, **23**, 1076–1079.

Greaves, G.N., Elliott, S.R. and Davis, E.A. (1979) Amorthous Arsenic, *Adv. Phys.*, **28**, 49–141.

Hogarth, C.A. and Popov, A.I. (1983) The control of the electrical conductivity of cooper phosphate glasses by the admixture of chlorine during preparation, *Int. J. Electron.*, **54**, 171–174.

Khokhlov, A.F., Mashin, A.I. and Khokhlov, D.A. (1998) New allotropic form of silicon, *JETF Lett.*, **67**, 675–679.

Knights, J.S. and Lujan, R.A. (1979) Microstructure of plasma-deposited a–Si:H films, *Appl. Phys. Lett.*, **35**, 214–216.

Kolomietz, B.T. (1960) Glassy state (Stekloobraznoe sostoynie), Moscow-Leningrad (in Russian).

Kotov, V.M. (1984). "Investigation of structure and properties of selenium layers for electrophotografic drums." Author's Abstract of Candidate's Dissertation, Moscow (in Russian).

Kozlov, V.M. and Krikis, U.U. (1978) Simulation of the chalcogenide glasses structure by the Monte Carlo method, *Isvestiya Akademee Nauk Latviyskoiy SSR*, **6**, 45–49 (in Russian).

Lacourse, W.C., Twaddell, V.A. and MacKenzie, J.D. (1970) Effects of impurities on the electrical conductivity of glassy selenium, *J. Non-Cryst. Solids*, **3**, 234–236.

Ligachev, V.A. (1998). "The morphology effect on the electronic spectrum, optical, and electrical properties of thin a–Si:H, a–C:H, and a–$Si_{1-x}C_x$:H films." Author's Abstract of Doctoral Dissertation, Moscow (in Russian).

Ligachev, V.A., Popov, A.I. and Stuokach, S.N. (1994) Condition of preparation structure and properties of hydrogenated films of amorphous carbon, *Semiconductors*, **28**, 1183–1188.

Ligachev, V.A., Popov, A.I., and Stuokach, S.N. (1996a). "Morphology, composition and properties of the amorphous hydrogenated carbon films." *In* Proceedings of the Conference "Optical Characterization Techniques in High Performance Device Manufacturing" (Austin, USA), SPIE, **2877**, pp. 46–50.

Ligachev, V.A., Svirkova, N.N., Filikov, V.A. and Vasil'eva, N.D. (1996b) Morphology and spectra of density of states in a–SiC:H films prepared by high frequency sputtering, *Semiconductors*, **30**, 834–844.

Long, M., Galison, P. and Alben, R. (1976) Structural model of amorphous selenium and tellurium, *Phys. Rev.*, **B13**, 1821–1829.

Lucovsky, G. (1987) Specification of medium-range order in amorphous materials, *J. Non-Cryst. Solids*, **97 and 98**, 155–158.

Lucovsky, G. and Galeener, F.L. (1980) Intermediate range order in amorphous solids, *J. Non-Cryst. Solids*, **35 and 36**, 1209–1214.

Luksha, O.V. (1994) *Structural Modification of Thin Amorphous Chalcogenide Films*, Proceedings of International workshop on advanced technologies of multicomponent solid films and structures, Uzhgorod, Ukraine, pp. 6–8.

Malaurent, J.C. and Dixmier, J. (1977) *A Random Chain Model for Amorphous Selenium*, The structure of non-crystalline materials, Cambridge University Press, Cambridge, UK, pp. 49–51.

Mashin, A.I. and Khokhlov, A.F. (1999) Multiple bonds in hydrogen-free amorphous silicon, *Semiconductors*, **33**, 911–914.

Michalev, N.I. (1983). "The influence of external factors on the structure and properties of the non-crystalline arsenic–chalcogen alloys." Author's Abstract of Candidate's Dissertation, Moscow.

Michalev, N.I., and Popov, A.I. (1987). "Interpretation of the diffraction examination results of glassy nonstoichiometric arsenic–chalcogen alloys." Structure and nature of metallic and nonmetallic glasses. Ijevsk (in Russian).

Miyauchi, K., Qiu, J., Shojiya, M. and Kawamura, N. (2001) Structural study of GeS_2 glasses permanently densified under high pressures up to 9 GPa, *J. Non-Cryst. Solids*, **279**, 186–195.

Morigaki, K. (1999). "Physics of Amorphous Semiconductors." Imperial College Press and World Scientific Publishing, Singapore.

Mott, N.F. and Davis, E.A. (1979) Electron processes in non-crystalline materials, Clarendon press, Oxford.

Nabitovich, I.D. (1970). "Examination of selenium layers structure." Author's Abstract of Doctoral Dissertation. L'vov (in Russian).

Ovshinsky, S.R. (1977) *Chemical Modification of Amorphous Chalcogenide*, Amorphous and Liquid Semiconductors, University of Edinburgh, Edinburgh, pp. 519–523.

Phillips, J.C. (1979a) Topology of covalent non-crystalline solids I: short-range order in chalcogenide alloys, *J. Non-Cryst. Solids*, **34**, 153–181.

Phillips, J.C. (1979b) Hollow claster models of atomic structures of chalcogenide glasses, *Inst. Phys. Conf. Ser.*, **43**, 705–708 (ch. 21).

Phillips, J.C. (1980) Chemical bonding, internal surfaces, and the topology of non-crystalline solids, *Phys. Status Solidi*, **B101**, 473–479.

Poltavzev, Yu.G., (1984). "Structure of Semiconductor Melts." Metallurgiya, Moscow (in Russian).

Poluchin, V.A. and Vatolin, N.A. (1985) Simulation of the amorphous metals structure, Nauka, Moscow, (in Russian).

Popov, A.I. (1978) The reasons of the influence of isoelectron impurity on the electrical conductivity of amorphous selenium, *Isvestiya Akademee Nauk SSSR, Neorganicheskie materiale*, **14**, 236–238.

Popov, A.I. (1980a) *Structural Modification—The Properties Control Method of Some Chalcogenide Glassy Semiconductors*, Proceedings of the International Conference Amorphous Semiconductors-80, Kishinev, pp. 150-153 (in Russian).

Popov, A.I. (1980b) The influence of the structural changes on viscosity of selenium, *Fisika i Khimia Stecla*, **6**, 307–311.

Popov, A.I. (1994) Determination of the rigidity threshold of the structural network in covalent glasses, *Glass Phys. Chem.*, **20**, 553–554.

Popov, I.A. (1996) Structural modification of amorphous hydrogenated silicon films with help of ultra-violet radiation, *Semiconductors*, **30**, 258–265.

Popov, A.I. (1997) Materials for electrophotographic recording medium, *Materialovedenie (Material Sciences Transactions)*, **2**, 27–34 (in Russian).

Popov, A.I., Domoryad, I.A. and Michalev, N.I. (1988) Structural modification of arsenic chalcogenide glasses under γ-radiation, *Phys. Status Solidi*, **A106**, 333–337.

Popov, A.I., Geller, I., Karalunets, A. and Ipatova, N. (1980) Effect of doping on molecular structure and VAP states of amorphous selenium, *J. Non-Cryst. Solids*, **35 and 36**, 871–876.

Popov, A.I., Ligachev, V.A., Vasil'eva, N.D. and Stuokach, S.N. (1995) *Preparation of Thin Hydrogenated Carbon Films by the HF Sputtering Method*, Proceedings of Ukrainian Vacuum Society, Kiev, Vol. 1, pp. 248–251 (in Russian).

Popov, A.I., Michalev, N.I. and Shemetova, V.K. (1983) Structural modification of some glassy chalcogenides, *Philos. Mag.*, **B47**, 73–81.

Popov, A.I. and Shemetova, V.K. (1982) Threshold voltage stability of chalcogenide glass switches, *Isvestiya Akademee Nauk SSSR, Neorganicheskie materiale*, **12**, 1997–2000 (in Russian).

Popov, A.I. and Vasil'eva, N.D. (1990) Ordering criteria for the atomic structure of non-crystalline semiconductors, *Fisika Tverdogo Tela*, **32**, 2616–2622 (in Russian).

Popov, A.I., Vasil'eva, N.D. and Khalturin, V.P. (1981) *Mathematics Simulation Method of the Non-Crystalline Semiconductor Structure*, Proceedings of the Moscow Power Engineering Institute, Moscow, 537, pp. 105-110 (in Russian).

Popov, A.I., Vorontsov, V.A. and Popov, I.A. (2001) Levels of structural modification of noncrystalline semiconductors and limits of their applicability, *Semiconductors*, **35**, 637–642.

Renninger, A.L., Rechtin, M.D. and Averbach, B.L. (1974) Monte Carlo models of atomic arrangements in arsenic-selenium glasses, *J. Non-Cryst. Solids*, **16**, 1–14.

Richter, H. and Breiting, G. (1971) Verschiedene Formen von amorphen Selen, *Z. Naturforsch*, **26a**, 1699–1708.

Rubish, V.M. and Kuzenko, Y.P. (1993) *Local Structure of Chalcogenide and Oxide Glassy Alloys on the Base of Arsenic and Antimony*, Proceedings of the Moscow Power Engineering Institute, Moscow, Vol. 667, pp. 31–45 (in Russian).

Skreshevsky, A.F. (1980). "Structural study of liquids and amorphous solids." Vesshaya shkola, Moscow (in Russian).

Sadkov, A.M., Korshak, V.V., Kudryavtsev, Yu.P. and Kasatochkin, V.I. (1972) USSR Discoverer's Certificate No 107, *Byull. Izobret.*, **6**(3).

Smorgonskaya, E.A. and Tsendin, K.D. (1996) Atomic and electronic structure of chalcogenide glassy semiconductors, *Electronic Phenomena in Chalcogenide Glassy Semiconductors*, Nauka, Sant-Peterburg, pp. 9–33 (in Russian).

Spear, W.E. and Le Comber, P.G. (1977) Doped amorphous semiconductors, *Amorphous and Liquid Semiconductors*, University of Edinburgh, Edinburgh, pp. 309–322.

Vasil'eva, N.D., (1989). "Molecular structure of non-crystalline selenium and its influence on electrographic layers properties." Author's Abstract of Candidate's Dissertation, Moscow (in Russian).

Vasil'eva, N.D. and Khalturin, V.P. (1986) *Atomic Structure Simulation of Selenium Electrographic Layers*, Proceedings of the Moscow Power Engineering Institute, Moscow, Vol. 103, (in Russian) 92–95.

Vasil'eva, N.D. and Popov, A.I. (1995) Effect of deposition conditions on the atomic structure of hydrogenated amorphous carbon films, *Inorg. Mater.*, **31**, 417–422.

Vasil'eva, N.D., Popov, A.I. and Khalturin, V.P. (1982) *Application of Subgradient Method to the Non-Crystalline Semiconductor Structure Simulation*, Proceedings of the International Conference Amorphous semiconductors-82, Bucharest, pp. 50–52 (in Russian).

Vasil'eva, N.D., Popov, A.I. and Shutova, I.V. (1997) Morphology and atomic structure of hydrogenated amorphous carbon films, *Vestnic Moskovskogo Energeticheskogo Instituta*, **3**, 77–81 (in Russian).

Vinogradova, G.Z. (1984) Glass formation and phase equilibrium in chalcogenide systems, Nauka, Moscow, (in Russian).

Voyles, P.M., Zotov, N., Nakhmanson, S.M., Drabold, D.A., Gibson, J.M., Treacy, M.M.J. and Keblinsky, P. (2001) Structure and physical properties of paracrystalline atomic models of amorphous silicon, *J. Appl. Phys.*, **90**, 4437–4451.

Yun, S.S., Li, H., Cappelletti, R.L., Enzweiler, R.N. and Boolchand, P. (1989) Onset of rigidity in $Se_{1-x}Ge_x$ glasses: ultrasonic elastic module, *Phys. Rev.*, **B39**, 8702–8706.

CHAPTER 3

EUTECTOIDAL CONCEPT OF GLASS STRUCTURE AND ITS APPLICATION IN CHALCOGENIDE SEMICONDUCTOR GLASSES

V. A. Funtikov

KALININGRAD STATE UNIVERSITY, UNIVERSITETSKAYA STREET, 2 KALININGRAD, 236040, RUSSIA

1. The Role of Stable Electronic Configurations in the Creation of a Glass-Forming Ability of Chalcogenide Alloys

ABSTRACT

A model is proposed, in which the possibility of glass formation in elemental substances and their alloys is related to specific features in the electronic structure of atoms, such as stable electron configurations p^0, p^3, d^0, d^5, d^{10}, f^0, f^7, and f^{14}. Glass formation in alloys is promoted by the structural–configurational equilibriums likely to exist in the melt at the synthesis temperature between clusters that form because the electron configurations of atoms in the chemically bound states are close in terms of energy and that vary widely in the degree of polymerization. Glass formation parameters are defined quantitatively and they characterize t he capacity of the atoms of the chemical elements that make up the melt to form a vitreous network. The dependence of these parameters on the nuclear charge of the elements exhibits a primary and a secondary periodicity in the case of sulfide, selenide, telluride, and oxide systems. The electron configuration model has proved applicable to halide, diamond-like, and metallic systems as well.

As noted in Goryunova and Kolomiets (1958), Winter-Klein was the first to draw attention to the relation between the capacity of substances to form glasses and the number of valence p-electrons per effective atom. According to Winter-Klein's criterion, the tendency to glass formation would be at its strongest in alloys with atoms having two to four valence p-electrons. It was found, however, that the criterion did not hold in general, but only in specific cases. Unfortunately, Winter-Klein's ideas have received no further development. A way out, we believe, can be found if the valence capabilities of various atoms are analyzed considering not only partly filled valence orbitals, but also vacant and completely filled orbitals close to their valence counterparts in terms of energy.

ISBN 0-12-752187-9
ISSN 0080-8784

The pivotal point of our hypothesis is: one of the principal conditions for glass formation by substances is the structural–configurational equilibrium between the low- and high-molecular-weight forms of atomic groups in melts (solutions) at the synthesis temperature; this equilibrium is related to the electron configuration equilibriums in the atoms that make up all of these groups. The most salient features of the proposed approach are provided with reference to selenide and telluride systems in Funtikov (1977, 1984, 1987, 1989). Here, we report new findings that confirm the applicability of our approach to a wide range of glass-forming substances.

In our analysis of the relation that exists between the electronic structure of atoms and their ability to form a vitreous matrix, it appears reasonable to start from the elemental substances. We might single out a fragment of the Periodic System that consists of four elements that most easily vitrify, and at the same time, readily form different modifications that can be denoted as the low-molecular (LM) and high-molecular (HM) ones (Funtikov, 1987). Sulfur, selenium, phosphorus, and arsenic were found to be such elements. Precisely, such simple substances in the molten state can produce both types of molecular groups possessing the same free energy and existing in equilibrium with one another. For sulfur and selenium, experimental evidence has been gathered for the existence of cyclic X_8 and chain X_n molecules in the molten and vitreous states (Addison, 1964; Feltz, 1983). The equilibrium between cyclic and chain molecules in the chalcogens is only a special case of $n\,\mathrm{LM} \Leftrightarrow \mathrm{HM}$ equilibriums. In other instances these may not only be molecules, but also the products of their ionic and radical decomposition. But in all such cases, we think the particles involved in the equilibriums must significantly differ in the degree of polymerization and the character of medium-range order because the formation of kinetic barriers between them is most probable. The latter is a mandatory condition for an increase in the relaxation time for the corresponding equilibriums, and as a consequence, for a decreased possibility that the substance would crystallize. From the viewpoint of solution theory, glass-forming melts have much in common with solutions of high-molecular substances in low-molecular solvents. A significant distinction between glass-forming melts and classical solutions of high-molecular compounds is the possibility of intraconversion between low-molecular and high-molecular particles. Moreover, the structural groups show a wider range of distribution in glass-forming melts. This stems from the more general treatments, such as the multiminimum concept of the vitreous state (Bal'makov, 1989).

The above structural–configurational equilibrium may exist in various substances, especially in the elemental ones, only if the corresponding atoms can readily rearrange their electron configurations under synthesis conditions. In the case of the chalcogens, for example, the structural–configurational equilibrium $n\,\mathrm{LM} \Leftrightarrow \mathrm{HM}$, which in our hypothesis leads to glass formation may be determined by electron configuration equilibriums of the types $\mathrm{s^1p^{\mathit{x}}} + (5 - x)\mathrm{e^-} \Leftrightarrow \mathrm{s^2p^4}$, $x = 1, 2, 3$. From oxygen to tellurium, this kind of equilibrium should shift to the right when considered under identical conditions. The shift of equilibrium will occur in a non-monotone manner due to the orbitals of free sublevels, $\mathrm{d^0}$ and $\mathrm{f^0}$. Among the Group VI elements of the periodic system, they first appear in sulfur ($\mathrm{3d^0}$) and tellurium ($\mathrm{4f^0}$). This is dealt in detail in Funtikov (1977, 1984, 1987, 1989), and we will dwell on the energy aspects of the processes. Evidence that we have chosen a correct way to analyze atomic states leading to glass formation under suitable conditions of substances can be found in Samsonov (1965) who examines the more general aspects of how stable electron configurations affect the properties of chemical elements and compounds. In isolated atoms, it is known, the outer s- and p-orbitals differ in energy rather strongly. For example, the difference is 8.1 eV in carbon, 11.4 eV in nitrogen, and 18.9 eV in oxygen. In sulfur and selenium the difference decreases to 10.0 and 10.1 eV, respectively (Akhmetov, 1975). Conversion of these values to units more familiar in chemistry yields 1823 kJ per (mol at) for oxygen. None the less, this is no obstacle for the oxygen atom to exist in the sp^3 hybrid state in the water molecule. Obviously, in multiparticle systems the sub-levels split up, and as a consequence, the energy gap between them decreases. The energies of the various electron configurational states come closer to one another as well. For example, from thermochemical data it can be calculated that the transition of diamond (sp^3) to graphite (sp^2) is accompanied by the release of a mere 1.8 kJ mol^{-1} of energy, and for the transition of graphite (sp^2) to carbyne (sp) the figure is 33.5 kJ mol^{-1} (Akhmetov, 1988). For the chalcogens the single-bond energies (in kJ per mol) are as follows: 211 for S_2, 229 for S_3, 239 for S_4, 249 for S_5, 256 for S_6, 257 for S_7, 260 for S_8, 261 for S_9, and 263 for S_{10}; 165 for Se_2, 161 for Se_3, 170 for Se_5, 157 for Se_6, 174 for Se_7, and 186 for Se_8 (Shukarev, 1974). Clearly, the single-bond energy increases with increasing number of atoms per molecule of a chalcogen. Indeed, the increase is just a few kJ per mol; this may be attributed to a gradual change in the type of hybrid state. As the chalcogens change from cyclic to chain molecules, the valence orbitals probably change from the sp^3 to the p-electronic states. This follows from the fact that chain molecules become stable in selenium and tellurium, which show almost no tendency to sp^x hybridization. Further evidence is provided by the decrease in the bond angle in chain molecules: from 107° for S_n to 105° for Se_n, to 102° for Te_n, (Addison, 1964; Glazov, Chizhevskaya and Glagoleva, 1967). We can judge from the bond angle that sulfur tends to change to the sp^3 hybrid state to form S_8 molecules under ordinary conditions. Closeness in energy between different molecular species in a glass-forming melt should lead to their distributions by size and other parameters. In turn, this should produce in the glass a wide range of reference and defect fragments of the vitreous network and fluctuations of the network structure (Bal'makov, 1988).

We propose two quantities as the parameters that characterize the glass-forming capacity of the elements making up multicomponent alloys. One is the limiting concentration $P_{\rm lim}$ of the element in the glass former. It is the amount expressed as an atomic percentage at which the vitreous state is still preserved. The other is the fraction S expressed likewise as a percentage that the glass formation region takes up in the entire concentration space. In Funtikov (1987) we presented a graph relating the glass-forming capacity of alloys in the A^{III-V}–E systems (E = S, Se, Te) to the ordinal number of p-elements in Groups III, IV, and V. It shows that, on passing to the alloys containing Period 5 elements, a sudden decrease occurs in their glass-forming capacity. This correlates with the fact that the atoms of this period acquire a free acceptor inner 4f sublevel ($4f^0$). It is responsible for the metallization of the chemical bond and for a proportionate decrease in the tendency of alloys to form glass. In this and other cases, we calculated the glass formation parameter using the data reported in Borisova (1972, 1983), Vinogradova (1984), Mazurin, Strel'tsina and Shvaiko-Shvaikovskaya (1973,

1975, 1980), Nedoshovenko, Turkina, Tver'yanovich and Borisova (1986) and Tver'yanovich, Borisova and Nedoshovenko (1986).

An increase in the number of components in an alloy is accompanied by an increase in its glass-forming capacity. This happens because the fragments that make up the vitreous network of atoms turn up in a number of alternative types. Consequently, any specific features in the electronic structure of atoms will affect the glass-forming capacity most strongly in ternary alloys. This can be readily illustrated by noting, for example, the manner in which the glass formation parameter P_{lim} varies with the ordinal number of the element in the As_2Se_3–E system in groups and rows (Funtikov, 1977). Note that the Group IV–VI elements in the 3rd and 4th periods show a similar behavior in this case. Likewise, a similarity in behavior is observed between the elements in the 5th and 6th periods. This observation correlates with the appearance of stable electron configurations $3d^0$ and, accordingly, $3d^{10}$ in the atoms of the 3rd and 4th periods, and stable electron configurations $4f^0$ and $4f^{14}$ in the elements of the 5th and 6th periods. The manner in which the glass-forming capacity of ternary alloys varies with the ordinal number of the element remains unchanged upon the replacement of a 4th-period chalcogen (selenium) by a 3rd-period chalcogen (sulfur), or upon the replacement of the glass-former, a Group 5 element (arsenic) by a Group 4 element (germanium). This is confirmed by a comparison of the As–Se–E^V and Ge–S–E^V, and As–Se–E^{VII} and As–S–E^{VII} systems (Fig. 1). For the telluride systems the glass-forming capacity of their alloys varies with the ordinal number of the elements in the same manner as in the case of alloys in the selenide and sulfide systems.

This can be illustrated by referring to the Ge–S–E^{III}, Ge–Se–E^{III}, Ge–Te–E^{III}, Si–Te–E^{III}, As–S–E^{III}, As–Se–E^{III}, and As–Te–E^{III} systems (Fig. 2). The dependence cited above confirms that the stable electron configurations d^0, d^{10}, f^0, and f^{14} play a special role in glass formation (Funtikov, 1977). Alloys in the TeO_2–$E_2^{III}O_3(Tl_2O)$ oxide systems behave similarly (Fig. 2).

So far we have been dealing with chalcogenide and oxide glasses based on p-elements. It is possible to identify the key electron configuration equilibriums that lead to

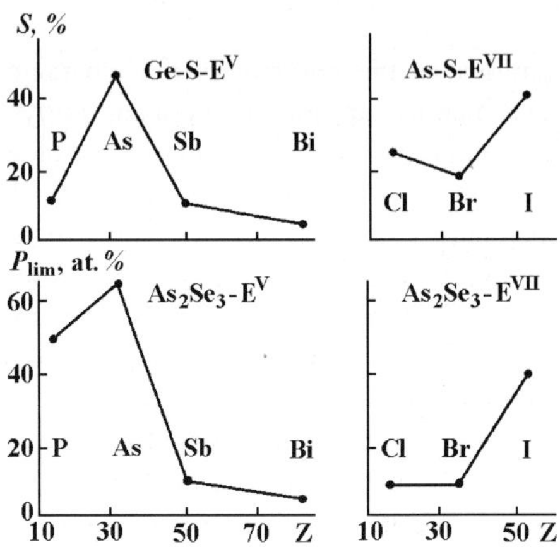

FIG. 1. Glass-forming capacity of alloys in the germanium- and arsenic-containing chalcogenide systems plotted against the nuclear charge Z of Group V and VII p-elements.

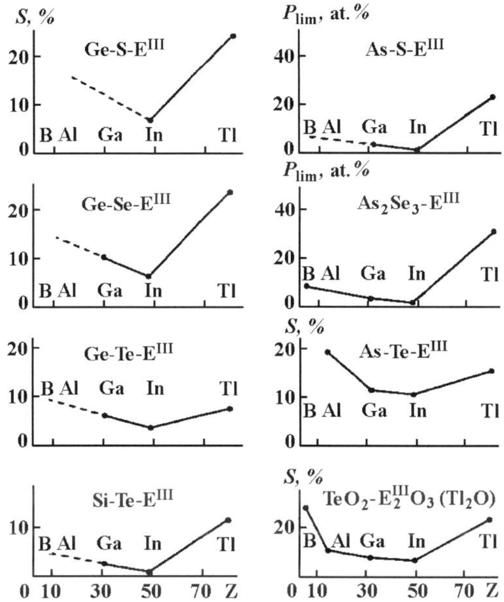

FIG. 2. Glass-forming capacity of alloys in the germanium- and arsenic-containing chalcogenide systems and $TeO_2-E_2^{III}O_3(Tl_2O)$ system plotted against the nuclear charge of Group III p-elements.

structural–configurational equilibriums of the type nLM $\Leftrightarrow$ HM in glass-former Group VI atoms

$$s^2p^{2(3,4)} \Leftrightarrow s^1p^3 + 0(1,2)e^- \tag{1}$$

$$s^1p^3 \Leftrightarrow s^1p^2 + e^- \tag{2}$$

$$s^1p^2 \Leftrightarrow s^1p^1 + e^- \tag{3}$$

The behavior of the glass-formation parameter with the ordinal number of an s-element for the $As_2Se_3-E^I$, $Sb_2S_3-E_2^IS$, $GeS_2-Ga_2S_3-E^ICl$, $GeS_2-E_2^IS$, $SiO_2-s_2^IO(E^{II}O)$, and $TeO_2-E_2^IO$ systems is shown in Figure 3. From lithium to sodium, a significant increase occurs in the glass-forming capacity of the alloys. This correlates with the changes in the structure of the valence level in the alkali element atom, namely, with the appearance of a free d sublevel ($3d^0$). This is not a mere coincidence. That much is confirmed by the fact that the alkali metals produce the most stable forms of oxides and sulfides (the oxides are formed on burning) (Akhmetov, 1975)

$$\begin{array}{lll} 2s^1 2p^0 & Li_2O & Li_2S_n\ (n = 1, 2) \\ 3s^1 3p^0 3d^0 & Na_2O_2 & Na_2S_n\ (n = 1\text{–}5) \\ 4s^1 3d^0 4p^0 4d^0 & KO_2 & K_2S_n\ (n = 1\text{–}6) \\ 5s^1 4d^0 5p^0 5d^0 & RbO_2 & Rb_2S_n\ (n = 1\text{–}6) \\ 6s^1 4f^0 5d^0 6p^0 6d^0 & CsO_2 & Cs_2S_n\ (n = 1\text{–}6) \end{array}$$

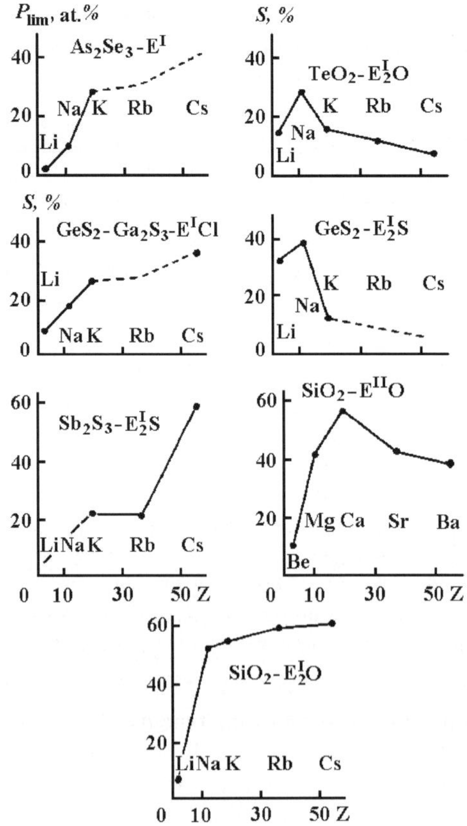

FIG. 3. Glass-forming capacity of alloys in chalcogenide-, oxide-, and halide-based systems plotted against the nuclear charge of Group I and II s-elements.

As modifiers in vitreous alloys, the Group I and II s-elements (Fig. 3), act solely to induce the tendency to chain formation in the Group VI p-elements, and as a consequence, cause a shift in the corresponding electron configuration equilibriums and in the structural–configurational ones. Unfortunately, the space available in an article is limited and we cannot dwell in detail on an interpretation of the observed relations. However, the aim of this work is to generalize these relations to as large a number of vitreous alloys of widely varying nature as possible.

Diamond-like vitreous semiconductors may also fit into our model, and they too acquire an electron configuration equilibrium of the type $s^2p^2 \Leftrightarrow s^1p^3$ that leads to the structural–configurational equilibrium $n\,\mathrm{LM} \Leftrightarrow \mathrm{HM}$ in melts. This is borne out by the formation of two types of atomic microgroups in the vitreous $CdGeAs_2$ compound, in which germanium exists in two valence states (2 and 4) (Turaev, Seregina and Kesamanly, 1984).

Halide glasses have been studied in less detail than oxide and chalcogenide glasses. They are mainly beryllium fluoride-based glasses. The compound that is most stable in the vitreous state is beryllium difluoride, BeF_2. It is able to form tetrahedral structural elements, BeF_2. It is likely that in the beryllium fluoride melt, the low-molecular and high-molecular clusters are formed in a similar manner as the silica melt, and a similarity in structure should exist between vitreous SiO_2 and BeF_2. This may be associated with the electron configuration equilibrium $s^1p^1 + 2e^- \Leftrightarrow s^1p^3$ that involves the electrons of the silicon and beryllium atoms, and is responsible for the formation of corresponding O bonds. There is ample reason to predict that SiO_2- and BeF_2-based glass-forming systems should be equivalent in terms of some characteristics. Notably, they should be similar in manifestations of primary and secondary periodicity.

From a study into the nature of atomic radial distribution functions for metallic glasses in the Fe–P, Fe–B, Fe–C, and Pd–Si systems, Boglaev, Il'in, Kraposhin, Matveeva and Ushakov (1985) deduced that these non-crystalline materials form in effect a class of 'ultradisperse eutectics,' in which the inclusions have a characteristic dimension of about 10^{-7} cm. This observation proves that the melts of these systems are dominated by certain structural–configurational equilibriums and, therefore, fit well into the framework of our approach.

It is of interest to look into the behavior of the transition elements in metallic, semiconductor, and dielectric glasses. According to how they find their way into chalcogenide glasses, the d-elements may be divided into two groups: (1) the d-elements whose concentration in the glass ranges from a few hundreds to 1% and (2) the d-elements whose concentration in the glass may be as high as 30 at.% (Funtikov, 1977). The first group includes the elements for which the electron configuration of atoms in the isolated state is d^n $(1 \leq n \leq 9)$. The second group includes the elements for which the stable configuration is d^{10}. In addition, it turns out that when they are present with the atoms of d-elements together, the d^{10} and f^{14} configurations have a positive effect on the concentration in chalcogenide glasses. To a first approximation, when they are part of chalcogenide glasses, it may be predicted that f-elements will behave similarly to d-elements. The point is that the partly filled f sublevel is close in energy to the free d sublevel, and it is therefore possible for the valence electrons to transfer from the f to the d sublevel. However, chalcogenide glasses fulfill this prediction only in part. In the Ln_2S_3–Ga_2S_3 systems (Ln = lanthanides), the glass formation region monotonically decreases from the lanthanum-containing system ($S = 25\%$ upon quenching from 1100 °C in water) to the gadolinium-containing system ($S = 0\%$), excluding the europium-containing system (Vinogradova, 1984). This proves that the open stable electron configuration f^7 has a negative effect.

In contrast to their behavior in chalcogenide glasses, the d- and f-elements exhibit entirely different traits in metallic glasses. Among other aspects, an inversion takes place in the way the electron configurations d^0, d^n, d^{10}, f^0, f^m, and f^{14} affect the character of glass formation. An analysis of all kinds of transition-element atoms (Guntherodt and Beck, 1981; Beck and Guntherodt, 1984) indicate that most metallic glasses contain atoms of d and/or f elements with open valence electron configurations d^n $(1 \leq n \leq 9)$, f^m $(1 \leq m \leq 13)$, and especially with stable d^5 and f^7 configurations.

Conclusion

A hypothesis is proposed, according to which the glass formation in alloys is promoted by the existence of a great number of equilibriums between low-molecular and high-molecular clusters in their melts, correlating with the corresponding electron configurational equilibriums between the component atoms. The specific features have been identified in the correlation between the electronic structure of atoms and glass formation in alloys containing s, p, d, and f elements. Analysis has proved the applicability of the proposed approach to chalcogenide, oxide, halide, diamond-like, metallic, and mixed alloys. The impact that stable d^0, d^5, d^{10}, f^0, f^7, and f^{14} configurations and unstable d^n and f^m configurations have on the glass-forming capacity of dielectric, semiconductor, and metallic alloys has been established. The influence that open d^n $(1 \leq n \leq 9)$, f^m $(1 \leq m \leq 13)$, and, especially, stable d^5 and f^7 electron configurations have on glass formation in alloys has been found to undergo an inversion from dielectrics and semiconductors to metals. The electron-configurational approach makes it possible to predict the compositions of new glass-forming systems and to interpret the physico-chemical properties of vitreous alloys from the fundamental characteristics of atoms (Funtikov, 1994).

2. Features of Chemical Bonds in Chalcogenide Vitreous Semiconductors

ABSTRACT

A theoretical approach to the structure of chalcogenide glasses is proposed, which is based on the electronic configuration model of glass formation, and the formation of additional chemical bonds in the chalcogenides. The energy of π_{d-p} chemical interaction is evaluated from photoluminescence, complex permittivity, and extrapolation. It is found to be more than 100 kJ mol^{-1} for sulfur and tellurium, 100–115 kJ mol^{-1} for selenium, 120 kJ mol^{-1} for As_2Se_3, and 80–90 kJ mol^{-1} for As_2S_3. A scheme is proposed for explaining the formation of 'intrinsic' defects from the viewpoint of inorganic chemistry. Under this scheme, the concept of lone p-electron pairs appears invalid. The concept of isoelectronic structural elements is introduced, whereby the intrinsic defects that come about in an equilibrium way in the melt lose their fundamental differences from reference structural elements. A non-traditional view on the structure of chalcogenide glasses is proposed, in which vitreous selenium is considered to be equivalent to a glass in the As–Se–Br system. The proposed approach may be applied to chalcogenide systems as well as other systems.

Advances in the theory of the crystalline state have brought along a number of concepts and notions specifically introduced to deal with crystals, but which are also used in the theoretical treatment of the non-crystalline, notably, vitreous states. To some extent, this has been prompted by the closeness in certain characteristics between vitreous and crystalline alloys. The debate still goes on between the proponents of the homogeneous and the microhomogeneous structure of vitreous alloys (Appen and Galakhov, 1977; Mazurin and Porai-Koshits, 1977), although experiments bring in more and more evidence in favor of the microinhomogeneous structure of glasses. On their part, the proponents of the homogeneous structure believe that there must exist, at least in principle, a defect-free homogeneous glass (Zakis, 1984). They ascribe all deviations from the ideal mainly to imperfections in glass synthesis. Bal'makov (1988) examined the difficulties that arise in defining the notion of a defect in glasses. He also pointed out that it is difficult to draw a line between defects and fluctuations of the structure. Problems arise even when attempting to define the concept of a reference structural element.

This chapter sets forth an approach to the problem of 'defects' in glasses, which has as its point of departure in the electronic configuration model of glass formation proposed in Funtikov (1989). I have also introduced the notion of isoelectronic structural elements, thus suggesting a non-traditional look at the 'reference' and 'defect' structural elements. To simplify matters, fluctuations of a structural element are treated as its energy modifications, provided its topological characteristics remain unaltered. Chalcogenide glasses were chosen as model objects because these inorganic vitreous alloys have a high degree of covalence and therefore allow one to analyze short- and medium-range order in the glass structure without any particular qualifications.

According to the charged defect model proposed by Street and Mott and based on Anderson's idea about the effective negative correlation energy of electrons, there comes about an equilibrium of the type (the SM model) (Feltz, 1983)

$$2D^0 \Leftrightarrow D^- + D^+ \tag{4}$$

Kastner, Adier, and Fritzsche have refined the model by introducing the concept of valence-alternation pairs (the KAF model) (Feltz, 1983)

$$2C_3^0 \Leftrightarrow C_1^- + C_3^+ \tag{5}$$

where C is a chalcogen.

In Popov (1980), he proposes a quasimolecular defect model based on the formation of tri-center, tetra-electron chemical bonds. As will be recalled, tri-center, tetra-electron bonds are hypervalent bonds, HV–I, if one uses p-electron pairs. The central atoms of hypervalent bonds must have low electronegativity and the ligands must have a high electronegativity. Thus, hypervalent bonds can be formed only if the chemically bound atoms differ greatly in electronegativity. This condition does not hold for chalcogenide glasses. Therefore, as believed, the quasimolecular defect model does not apply to vitreous chalcogenides, but might prove useful for halogenide vitreous alloys.

An attempt is made here to elaborate upon the ideas of charged defects and reference structural elements from the viewpoint of inorganic chemistry. Note that in the KAF model defects C_3^+ are assumed to be formed by a donor–acceptor mechanism. This is highly improbable because the original unstable defects C_1^+ are actually radicals. Moreover, the defects C_1^- are presumed to be unstable because a possibility is allowed for the more stable negatively charged defects C_3^- to be formed by the donor–acceptor mechanism. As a suitable prototype for a negatively charged 3-fold coordinated structural element C_3^-, one might consider an iodine ion J_3^-, which is produced exothermally from a J_2 molecule and a J^- ion ($\Delta H = -16.7$ kJ mol^{-1}) (Nekrasov, 1973). Before analyzing the structural features of vitreous chalcogenides, we first consider the character of chemical bonding in these alloys. As is known, the atoms of the elements in the main subgroups, beginning from the 3rd period, are able to form additional π_{d-p} bonds by the donor–acceptor mechanism through the use of vacant d-orbitals of appropriate symmetry and p-electron pairs. For example, the replacement of the carbon in carbon–halogen bonds by silicon causes the bond energy to rise by 20–80 kJ mol^{-1} (Akhmetov, 1975). Further evidence that such bonds can be formed is the dissociation energy for Me_2 molecules in the secondary and primary subgroups of Group I in the periodic system. The bond energy for Cu_2, Ag_2, and Au_2 molecules is four or five times that for K_2, Rb_2, and Cs_2 (Akhmetov, 1975).

Using published data (Huheey, 1983), we have plotted graphs that relate the energy of homopolar bonds for the elements in the main subgroups of Groups IV–VII (Figs. 4 and 5). As one moves from the second to the third Period, the energy of single homopolar bonds decreases in a regular fashion between the atoms of the elements in the 4th and 5th groups and increases likewise in a regular manner between the atoms of the elements in Groups VI and VII. This is easy to explain if one assumes that S, Se, Te, Cl, Br, and I atoms in the isolated state acquire vacant d-orbitals and form additional π_{d-p} bonds. By extrapolation (Fig. 5), the energy of the additional π_{d-p} bonds is estimated to be (a) 100 kJ mol^{-1} and (b) 110 kJ mol^{-1}.

From the quasimolecular approach, it is possible to surmise the energy diagrams of molecular orbitals for the Se–Se, As–Se, and As–S bonds (Fig. 6) and the model of band structure for vitreous Se, As_2Se_3, and As_2S_3. We have determined the energy levels of the bonding π_{d-p} and σ_{d-p} states from the frequency dependence of the imaginary part of the complex dielectric constant, ε_2, of the above alloys (Brodsky, 1979). From ε_2, two absorption maximals stand out. In the literature, they are assigned to two separate components of the valence band, and the upper component is associated with the lone valence p-electron pairs of chalcogen atoms. We believe, however, that the low-energy absorption maximum is associated with the formation of π_{d-p} bonds. In Figure 6, the position of the lone p-electron pairs, E_{Lp}, is shown by the dashed line. The free electron band is depicted, proceeding from the energy symmetry of bonding and loosening states.

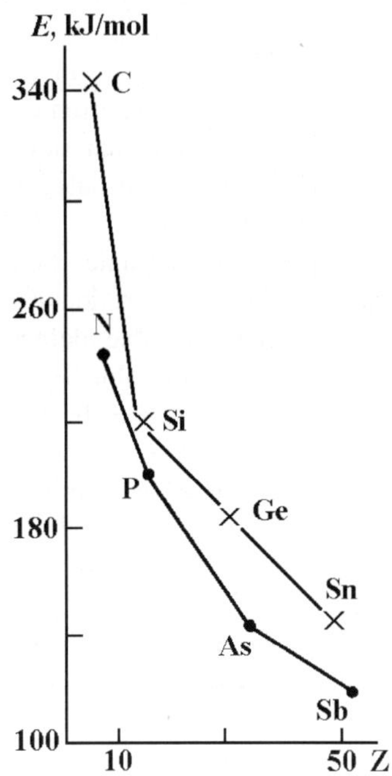

FIG. 4. Energies of single bonds $E^{IV}-E^{IV}$, $E^{V}-E^{V}$ plotted against nuclear charge Z.

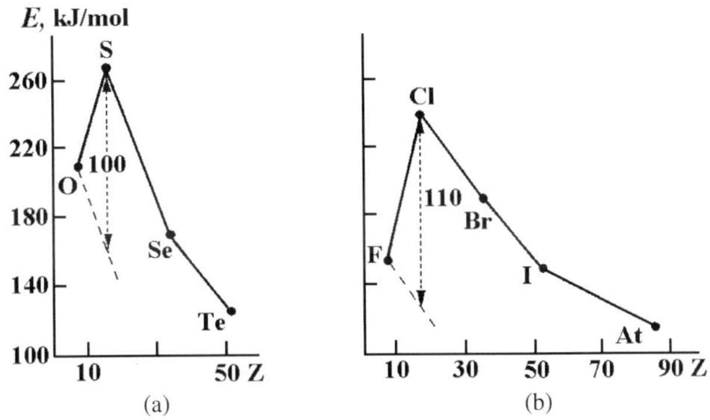

FIG. 5. Energies of single bonds (a) $E^{VI}-E^{VI}$ and (b) $E^{VII}-E^{VII}$ plotted against nuclear charge Z.

The lower edge of the free electron band is taken to correspond to the assumed zero energy level, because, when excited, valence electrons above would all populate the lower part of the free electron bond in the case of an intrinsic semiconductor. As has been found, the energy level corresponding to the p-electron pairs lies below the level of their lone state, E_{Lp}, by 100 kJ mol^{-1} for Se–Se, 120 kJ mol^{-1} for As–S, and 80 kJ mol^{-1} for As–Se. This depression can be associated with the formation of additional π_{d-p} bonds.

It is of interest to consider some characteristics of the photoluminescence of vitreous As_2Se_3 and Se (Feltz, 1983). As will be recalled, the maximum of the excitation spectrum of chalcogenide glasses displays an appreciable Stokes shift with respect to the maximum of the emission spectrum. This shift is 1.2 eV for Se and 0.9 eV for As_2Se_3. Characteristically, the maximum of the absorption spectrum occurs at the optical absorption edge. We regard the process of photoluminescence in chalcogenide glasses as an intrinsic one, primarily associated with the reference structural elements. It may be presumed that excitation is accompanied by the destruction of some reference structural elements ($C^0_{2(\pi)}$) both at the π_{d-p} bonds ($C^0_{2(\pi)} \Rightarrow C^0_2$) and at the σ_{p-p} bonds ($C^0_2 \Rightarrow 2C^0_1$),

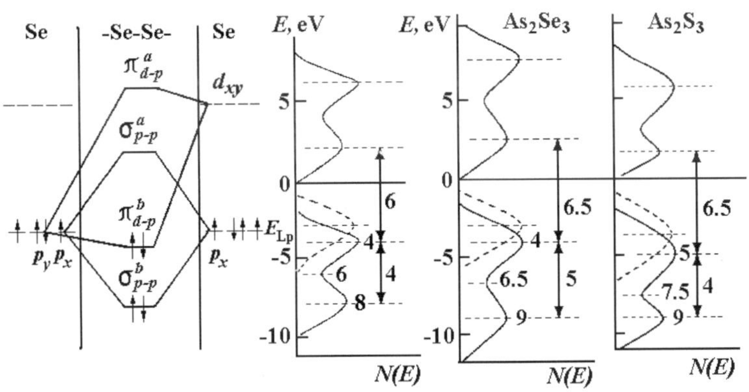

FIG. 6. Energy diagrams of molecular orbitals for Se–Se bonds and band structure models for vitreous Se, As_2Se_3, and As_2S_3.

and that radiative recombination proceeds as far as C_2^0 ($2C_1^0 \Rightarrow C_2^0$). If one assumes that the defects C_1^0 are situated near the conduction band, it will be obvious that the Stokes shift of absorption and emission maximum is equal to the energy of the additional π_{d-p} bonds, namely, 115 kJ mol^{-1} for Se–Se bonds and 90 kJ mol^{-1} for As–Se bonds.

As shown in Table I, the energies of the π_{d-p} bonds, evaluated in three independent ways, including extrapolation (Fig. 5), show satisfactory agreement.

Our analysis has proved the view held in the literature on lone p-electron pairs is not valid. From the chemical point of view it is more logical to assume that structural defects are formed in chalcogenide glasses by the following mechanism:

$$\begin{array}{c} 3C_{2(\pi)}^0 \Leftrightarrow 3C_2^0 \\ C_2^0 \Leftrightarrow 2C_1^0 \\ C_2^0 \Leftrightarrow C_1^+ + C_1^- \\ C_1^+ + 2C_1^0 \Leftrightarrow C_3^+ \\ \underline{C_1^- + C_2^0 \Leftrightarrow C_3^-} \\ 3C_{2(\pi)}^0 \Leftrightarrow C_3^+ + C_3^- \end{array} \tag{6}$$

The scheme proposed above for vitreous chalcogens may be adapted to vitreous chalcogenides as well. For vitreous selenium, the following overall equation may be written:

$$5[SeSe_{2/2}]^0 = 2[SeSe_{3/2}]^+ + 2[SeSe_{2/2}]^- \tag{7}$$

where $[SeSe_{3/2}]^+ \equiv$ 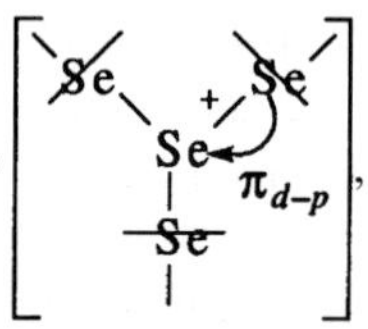, $[SeSe_{2/2}]^- \equiv$ 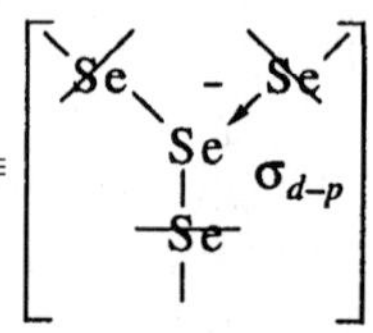

In analyzing the chemical bonding in chalcogenide structural elements, we use d–p donor–acceptor interactions of the π and σ types introduced for chalcogenide glasses in

TABLE I

ENERGY OF ADDITIONAL π_{D-P} BONDS (KJ MOL^{-1})

Bond	Method of calculation		
	Extrapolation	Photoluminescence	Complex dielectric constant
Cl–Cl	≥110	–	–
Br–Br	≥110	–	–
I–I	≥110	–	–
S–S	≥100	–	–
Se–Se	≥100	115	100
Te–Te	≥100	–	–
As–S	–	–	120
As–Se	–	90	80

(Funtikov, 1977). π_{d-p} bonds can be formed as both interatomic and intermolecular bonds; π_{d-p} bonds can be formed only as intermolecular ones. The chemical interaction of the d–p type in chalcogenide glasses can act as additional bonds responsible for medium-range as well as short-range order, and their consideration enables one to readily interpret many of the physicochemical properties of chalcogenides. Our studies show chemical $\pi(\sigma)_{d-p}$ bonds cannot be formed in non-crystalline semiconductors of the {Ge} type in principle, their Si–Si and Ge–Ge bonds are broken homolytically and, in consequence, an ESR signal appears due to unpaired electrons. In chalcogenide glasses, however, the bond breaking associated with the additional chemical $\pi(\sigma)_{d-p}$ interaction proceeds by the heterolytic mechanism, and glasses of the {Se} type do not generate an ESR signal in the dark 'equilibrium state.'

By the electronic configuration model of glass formation (Funtikov, 1989), a structural–configurational equilibrium should establish itself in a vitrifying melt between the low- and high-molecular structural elements (nLM ⇔ HM), say, $n/8\ Se_8 \Leftrightarrow Se_n$. Obviously, the short-range reference structural elements in elemental substances are the same for both the LM and the HM elements. Therefore, along with short-range structural elements (SRSEs), one should consider medium-range structural elements (MRSEs). For the crystalline state, one should further consider long-range structural elements (LRSEs). An apt example is vitreous As_2S_3, for which: SRSE ≡ $AsS_{3/2}$, …, MRSE = As_4S_6, $(As_6S_6S_{6/2})_m$,….

The vitreous state results from the practically relaxation-free cooling of the melts in which an equilibrium (nLM ⇔ HM) has established itself. Otherwise, crystallization occurs. Hence, even with elemental substances and readily vitrifying stable compounds, the molecular structural composition of glasses cannot be homogeneous for purely fundamental reasons, even if one leaves out defect formation. Undoubtedly, intrinsic structural defects are not formed in the vitreous state; instead, this happens in the vitrifying melt, where an appropriate equilibrium establishes itself between the reference and defective molecular structural elements. The situation is complicated by the fact that both the low- and the high-molecular groups experience partial equilibrium decomposition. For simplicity, we will limit ourselves to short-range order, because in most cases SRSEs are the same for both the low- and the high-molecular elements. Therefore, an ideal glass should display a broad set of equilibriums between the reference low- and high-molecular atomic groups, which on their part are in equilibrium with their defective structural elements.

In this sense, the use of the term 'defect' with regard to equilibrium structural defects impedes further development of ideas about glass. In view of this, we propose the concept of isoelectronic short-range structural elements (ISRSEs). By this, we mean structural elements with the same number of electrons in all atoms that make up an SRSE. From the standpoint of isoelectronic structural elements, there is no fundamental difference between the following ones:

$$[SeSe_{3/2}]^{+} \equiv [AsSe3/2]^{0}$$

$$[SeSe_{3/2}]^{-} \equiv [Br_{1/2}SeSe_{2/2}]^{0}$$

$$[AsSe_{2/2}]^{-} \equiv [SeSe_{2/2}]^{0}$$

$$[AsSe_{4/2}]^{+} \equiv [GeSe_{4/2}]^{0}$$

$$[GeSe_{3/2}]^+ \equiv [GaSe_{3/2}]^0$$

$$[GeSe_{3/2}]^- \equiv [AsSe_{3/2}]^0$$

Obviously, Eq. (7), which describes the formation of defects in vitreous selenium, should be recast:

$$5[SeSe_{2/2}]^0 \Leftrightarrow 2[SeSe_{3/2}]^+ + 2[SeSe_{2/2}]^- \equiv 2[AsSe_{3/2}]^0 + 2[Br_{1/2}SeSe_{2/2}]^0 \quad (8)$$

This, in turn, allows one to assert that elemental vitreous selenium contains structural elements analogous to those based on atoms of arsenic and bromine. Vitreous selenium is therefore an analog of As–Se–Br glass, whose composition depends on synthesis conditions. Moreover, structural elements produced in an equilibrium way stabilize the vitreous state of selenium. This can be confirmed by the fact that in vitreous selenium the low-frequency dielectric constant ε_1 is higher than its high-frequency counterpart, n^2. Apparently, this is because the chemical bonds are partly ionized.

The concept of isoelectronic structural elements provides a means for the researcher to consider the intrinsic defective elements of glasses not as defective qualities, but also to regard the structure and properties of vitreous elemental substances and their alloys from the viewpoint of multi-component systems. As a further illustration of the proposed idea, we can consider the way defects are formed in the most typical chalcogenide structural elements

$$2[AsSe_{3/2}]^0 \Leftrightarrow [AsSe_{4/2}]^+ + [AsSe_{2/2}]^- \equiv [GeSe_{4/2}]^0 + [SeSe_{2/2}]^0 \quad (9)$$

$$4[GeSe_{4/2}]^0 \Leftrightarrow 2[GeSe_{3/2}]^+ + 2[GeSe_{3/2}]^- + [SeSe_{2/2}]^0$$

$$\equiv 2[GaSe_{3/2}]^0 + 2[AsSe_{3/2}]^0 + [SeSe_{2/2}]^0 \quad (10)$$

The approach we have used with regard to chalcogenide glass may be extended to other glass-forming systems, such as oxides:

$$2[SiO_{4/2}]^0 \Leftrightarrow [SiO_{3/2}]^+ + [O = SiO_{3/2}]^- \equiv 2[AlO_{3/2}]^0 + [O = PO_{3/2}]^0 \quad (11)$$

This implies that the resultant structural elements can be further modified, depending on the potential capabilities of the atoms that make their composition.

CONCLUSION

From the standpoint of the electronic configuration glass formation model, it can be shown that vitreous melts cannot be structurally homogeneous for fundamental reasons, even if they are elemental in composition. It has been established that a special role in the formation of reference and defective structural elements in chalcogenide glass belongs to additional interatomic chemical π_{d-p} bonds and intermolecular π_{d-p} and σ_{d-p} bonds. The energies of the additional bonds have been found to lie in the range 80–120 kJ mol^{-1}. The concept of isoelectronic structural elements has been introduced, allowing one to formulate a non-traditional approach to an analysis of structural elements in vitreous melts. Consideration of isoelectronic structural elements removes fundamental differences between reference and defective structural elements (Funtikov, 1993).

3. Geometrical and Topological Aspects of Structure Formation in Chalcogenide Semiconductor Glasses

ABSTRACT

The report states the topological approach of analysis of processes of structure in glass formation. Our theoretical approach is based on the premise that extremely small volumes of a space permit the formation of the continuous ordered structure for all possible elements of the symmetry, with first on infinite orders. According to our topological model, only the large power-generating barrier of the transition from microscopic particles, having infinite set elements of the symmetry, in the crystalline state with the limited symmetry and permits substance to pass in the glassy state from the melt and in the amorphous film from a vapor.

The main defect of many structural models of a substance's glassy state consists of a predominant crystallographic or the formal topological approach. It is developed most obviously in a crystallite concept, offered by Frankenheim (1835). The initial crystallite concept is not used. It is possible to discover the increase of attention to purely geometrical and topological aspects of the formation of a structure in glasses. The boundary between geometry and topology is almost elusive. In the case of glassy substances, the formation of ideal geometrical figures is excluded, and characteristic types of connectedness of atoms have a preference. Phillips has applied the topological approach for the first time for the analysis of a formation of a grid of a glass instead of the geometrical approach. From this position, it is shown that a grid of glass is not casually formed. For optimization of a substance's glass-forming tendency, it is necessary to optimize dimensions of different clusters, which will correspond to mechanical stability of a glass (Phillips, 1981). The last condition is possible if

$$N_c \cdot N_t = N_c \cdot N_d \tag{12}$$

where N_c is the middle number of atoms in a cluster, N_t is the number of force communications on the atom, N_d is the dimensionality of space and $N_c \cdot N_d$ is the dimensionality of configurational space, describing the situation of atoms. Phillips has shown that in systems As–S(Se) connections $As_2S(Se)_3$ extremum of mechanical and chemical stability have on the combinations $As_2S(Se)_3$. Unfortunately, results of calculations differ from the experimental results. It follows that the maximum ability towards glass formation do not correspond to a given system on the combinations $As_2S(Se)_3$, or on eutectic alloys. The eutectic alloys crystallize with great difficultly.

We offered eutectoidal model of formation of structures in glasses (Funtikov, 1995). The eutectoidal model of the structure of glasses is offered that is based on the Smits's idea about pseudobinary systems (Smits, 1921). All glasses, including elementary glasses, are analyzed as a variety of eutectics, formed by the interaction among themselves of 'pseudophases.' Pseudophase is equivalent of the nucleus of the ordered particle that appears in the supercooling melt. According to our eutectoidal model even the ideal glass is non-uniform and can be submitted as the ensemble of sticking microspheres.

The eutectoidal model requires the elucidation of the gear formation of a middle order in glasses. With this purpose we offer topological model of a structure of glassy substances, based on following hypotheses:

(1) The elemental structural elements of glassy alloys, responsible for the formation of glasses and being centers of glassy nuclei, arise in the closed space, homomorphous to sphere S^3, and have in limiting geometrical space R^3 the uncrystallographical axes of a symmetry $L_5, L_7, L_8, L_9, \ldots, L_\infty$.
(2) The belonging of dominating structural elements of glasses to closed space leads up to the impossibility of formation of an unbroken glassy grid.
(3) In the case of weak interactions the ideal structural elements of glasses correspond to structural elements, which are homomorphous to the icocahedron. In case of strong interactions, they correspond to the dodecahedron.
(4) The equilibrium is established between clusters with the crystallographic and non-crystallographic axes of a symmetry. The structural elements with any axis of symmetry should exist in glassy alloys for this reason.
(5) The crystallization of the glasses is accompanied by a final transition in the stable condition through set of stages, in that in the beginning icocahedronal (for metals) (Shechtman, Blech, Gratias and Cahn 1984) or dodecahedronal (for semicoductors and dielectrics) quasicrystals and then metastable and stable crystals are formed.

We propose the following concepts of nucleouses of a crystallization and vitrification. The clusters, possessed of a closed spherical space S and to the homomorphous geometrical structural elements with the axes of a symmetry $L_{1-4,6,}$ we shall identify with nuclei of crystallization. The clusters, possessed close to a spherical space S and to the homomorphous geometrical structural elements with the axes of a symmetry $L_{5,7,8,9,\ldots,\infty,}$ we shall identify with nuclei of a vitrification. The set of clusters in a melt W is possible to separate two subsets: (1) W_{cr}—a subset of nuclei of a crystallization, (2) W_{gl}—subset nuclei of a vitrification. Obviously, a transition from closed space S in a open space R in the case nucleouses of a crystallization can occur continuously without a rupture of a structure, and in the case of nucleouses a vitrification can take place only with a rupture of a structure.

Equilibriums exist at a temperature of synthesis:

$$2A \Leftrightarrow A_2, A_2 + A \Leftrightarrow A_3, A_3 + A \Leftrightarrow A_4, \ldots, A_{n-1} + A \Leftrightarrow A_n \tag{13}$$

In a general kind:

$$A_n + A_m \Leftrightarrow A_{n+m}, \ldots, A_n + A_k \Leftrightarrow A_{n+k}, \ldots, A_k + A_m \Leftrightarrow A_{k+m} \tag{14}$$

Most importantly for the vitrification equilibriums, it is possible to allocate equilibriums between clusters with the crystallographic and non-crystallographic axes of a symmetry, which is between nucleouses of a crystallization and vitrification:

$$A_n^{\text{cryst.}} \Leftrightarrow A_n^{\text{vitrif.}} \tag{15}$$

The greater the propensity of a melt toward glass formation, and the closer the temperature to the optimum temperature for glass synthesis, the closer the share of nucleouses of a vitrification approaches to a maximum. In our opinion, type two kinds of clusters exist in liquids of any structure, but in the case of easily crystallizing liquids, the power barrier to transition by anticrystalline clusters in crystalline clusters is extremely small, and crystallization occurs easily. The balance (15) can displace to the right in the case of a reduction of dimensions of clusters in a melt. The vitrification is possible by the cooling of melts only on the condition that a relaxation of an equilibrium (15) to the left will be practically impossible at the expense of a high-power barrier and by the large diversity of nucleouses of a vitrification. According to our topological model, just the large power-generating barrier of the transition from microscopic particles, having infinite set elements of the symmetry, in the crystalline state with the limited symmetry and permits substance to pass in the glassy state from the melt and in the amorphous film from vapor. The transitions of glasses and amorphous films in the crystalline state must pass through a stage of the formation of quasicrystals. From the topological model that follows, such quasicrystals should have axis of symmetry of 5th order and the icosahedral structure in the case of metal alloys and the dodecahedral structure in case of semiconductors and dielectrics. In the structure of all glasses, the structural elements should dominate, having in utmost geometrical variant by all elements of the symmetry of icosahedron and dodecahedron ($6L_5 10L_3 15L_2 15$PC). The given conclusions are confirmed by the experimental reception of the quasicrystals at a tempering of metal glasses. It is possible to predict, that such phases should occur at a tempering of the semiconductor and dielectric glasses. From the topological model follows that even the ideal glass non-uniformly and can be submitted as the ensemble of sticking spheres. Direct experimental confirmation consists in a supervision by means of the electron microscope the many-range structure typical of one-component glasses after their chemical etching. The schematic graphs of a dependence of a free Gibbs energy from dimensions are shown in Figure 7.

Proceeding from our concept, it is possible to make conclusions about the structural elements of a middle order in glasses, and it is illustrated on covalent chalcogenide glasses. The structural elements of a medium-range order for As, S and As_2S_3 in non-crystalline alloys are shown in Figure 8. The predominance in glasses of such structural elements should lead up to a many-range structure with a spherical symmetry, that results in formation of a blistery fracture in glasses at their destruction.

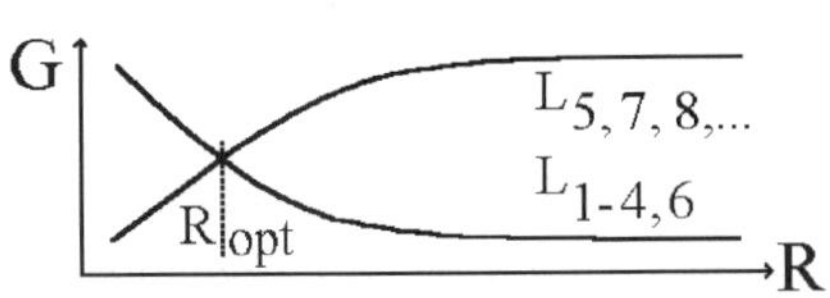

FIG. 7. The schematic graphs of a dependence of a free Gibbs energy from dimensions of nuclei of a crystallization ($L_{1-4,\ 6}$) and of nuclei of a vitrification.($L_{5,7,8,9,\ldots,\infty}$).

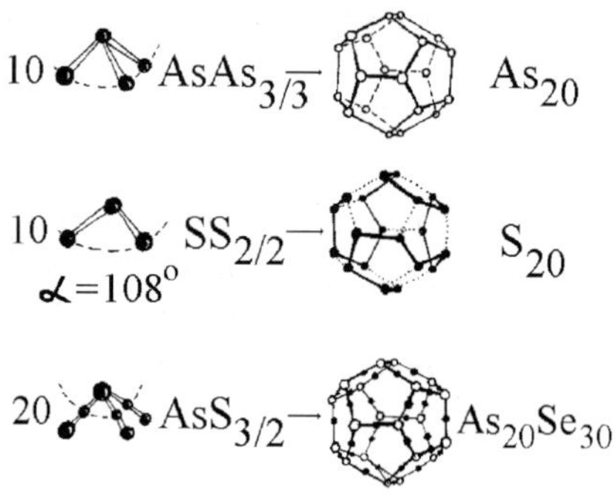

FIG. 8. The structural elements of near- and medium-range orders of As, S, and As_2S_3, in non-crystalline alloys.

CONCLUSION

The topological approach to the description of a structure of glasses is offered, and only extremely small volumes of space are needed for the formation of a continuous ordered structure, which permit all possible elements of the symmetry with first on infinite orders. The concept of nucleouses of a vitrification is entered, possessed to close a spherical space S and to the homomorphous geometrical structural elements with the uncrystallographic axes of a symmetry $L_5, L_7, L_8, L_9, \ldots, L_\infty$. The higher the propensity of a melt toward glass formation and the closer the temperature to optimum temperature of synthesis of a glass, the closer the share of nucleouses of a vitrification reaches a maximum. From a position of our topological approach, a microheterogeneity is not only a feature of a real glass, but should be included and in concept of an ideal glass (Funtikov, 1996a).

4. Stable and Metastable Phase Equilibriums in Chalcogenide Systems

ABSTRACT

A physicochemical approach to the analysis of structural transformations in semiconducting vitrifying melts during the semiconductor–metal transition is proposed. The pseudobinary model for this transition is based on the concept of quasi-components and the principles of physicochemical analysis. It is postulated that the character of interactions between components in the semiconductor–metal systems is identical to that in the melts in the temperature range of a semiconductor–metal smeared phase transition. According to the proposed model, the number of components in the melts increases and the phase-separated microinhomogeneity is formed in the region of smeared phase transition.

The semiconductor–metal transition in vitrifying melts is of particular interest, primarily, from the viewpoint of modeling the formation of medium-range order in vitreous alloys with covalent and metallic bonds, because, in this case, both covalently bound and metallic glasses can be produced from the same melt. Glasses of the former type are formed at a cooling rate of 180 K s^{-1}. Metallic glasses can be produced from the completely metalized melts at extremely high rates (of an order of 10^6 K s^{-1}) of cooling from temperatures above the upper boundary (7) of the temperature range of the semiconductor-metal transition.

Until recently, the semiconductor-metal transition has been considered in the literature from a strictly physical standpoint. In this work, in order to gain a deeper insight into this phenomenon, we applied a physicochemical approach. It is currently known that, under isobaric conditions, there are two types of semiconductor-metal transitions that manifest themselves in a number of properties such as electric conductivity, thermal emf, relative density, magnetic susceptibility, viscosity, and heat capacity (Glazov et al., 1967; Mustyantsa, Velikanova and Mel'nik, 1971; Shmuratov, Andreev, Prokhorenko, Sokolovskii and Bal'makov, 1977; Tver'yanovich, Borisova and Funtikov, 1986; Tver'yanovich and Gutenev, 1997; Tver'yanovich, Tver'yanovich and Ushakov, 1997). The semiconductor–metal transitions of the first type (for example, melting of germanium and silicon) occur at specific temperatures (Glazov et al., 1967). The transitions of the second type—the semiconductor–metal transitions smeared over a sufficiently wide range of temperatures—are observed, for example, in chalcogenide glass-forming melts (Mustyantsa et al., 1971; Shmuratov et al., 1977; Tver'yanovich et al., 1986; Tver'yanovich and Gutenev, 1997; Tver'yanovich et al., 1997).

In Tver'yanovich and Gutenev (1997) and Tver'yanovich et al. (1997), Tver'yanovich et al. discussed the question as to whether the semiconductor–metal transition can be referred to as a smeared first-order phase (polymorphic) transition. This inference was corroborated by the example of the As_2Te_3 glass-forming melt, for which these authors managed to measure the thermal effect of the semiconductor–metal transition. It was shown that the semiconductor–metal transition involves elementary transformations within the microregions, which consist of several dozens of atoms and comprise only several nanometers across. Note that the estimated average number of atoms in a microregion of the elementary structural transformation was 33–60 (Tver'yanovich et al., 1997). With differential scanning calorimetry, it was found that the heat is released over the entire range of the semiconductor–metal transition occurring in the As_2Te_3 melt, and the maximum thermal effect is observed at a temperature close to the midpoint (780 K) of this range (Tver'yanovich and Gutenev, 1997). The total thermal effect is equal to 25 kJ kg^{-1} (0.6 kcal mol^{-1}). The fact that the semiconductor–metal transition in the arsenic telluride melt is the first-order phase transition is evidenced by the thermal effect ($\Delta H_{S-M} \neq 0$). The second-order phase transitions imply no release or absorption of the heat (Karapet'yants, 1975).

The lower boundary of the colloidal state corresponds to a range of particle dimensions <1 nm. In the case under consideration, the microinhomogeneity region corresponds to a particle size of >1 nm, which, at high temperatures, can be associated with the early stage of microemulsion formation in a melt owing to the phase separation processes. A prerequisite to the formation of the emulsion is the complete or partial immiscibility of liquid phases, which stems from their strong difference in chemical bonding types. In our case, these are metallic and covalently bound phases. As a rule, the emulsions are coarsely disperse systems, because small drops rapidly disappear due to the isothermal distillation. Only in two cases the microemulsions arise spontaneously and represent rather stable formations (Frolov, 1982; Fridrikhsberg, 1984). In a binary system (free of an emulsifying agent), this can be realized at temperatures somewhat below the critical temperature. Under these conditions, the interfacial tension is so small ($<0.1 \times 10^{-3}$ J m^{-2}) that it is completely compensated for by the entropy factor. Judging from the particle sizes and a total absence of the third component, it can be assumed that the critical microemulsions with particle sizes varying from several nanometers to several dozens of nanometers are formed in the temperature range of the semiconductor-metal transition in a melt, which is in agreement with the average particle sizes estimated in (Tver'yanovich et al., 1997). The critical microemulsions represent a variety of the lyophilic colloidal systems, because they arise spontaneously (Shchukin, Pertsov and Amelina, 1982). The sizes of individual particles are so small that the interfaces between them are smeared, and, thus, the critical emulsions can appropriately be named as the quasi-heterogeneous systems. However, in any case, the micro-inhomogeneity on the medium-range level is retained with an alternation of the medium-range orders of the covalently bound and metallic phases.

In order to avoid a number of questions in considering the pseudobinary model of the semiconductor–metal transition in a melt, it should be reminded that physicochemical analysis is formed on the three fundamental principles: continuity, correspondence, and compatibility (Goroshchenko, 1978). The continuity principle implies that the composition and properties of alloys in a system change continuously until a new phase arises. The correspondence principle consists in the fact that a particular geometrical image or a combination of such images in the phase diagram corresponds to each chemical individual, phase of variable composition, and phase transformations of various types. The compatibility principle has hitherto received little use, even though its application holds much promise. According to this principle, any set of components, irrespective of their number and properties, can constitute a physico-chemical system.

As is known, the components of a system are called the substances if variations in their masses are independent and determine all possible variations in the composition of a system (The Collected Works of G. Willard Gibbs, 1928; Anosov, Ozerova and Fialkov, 1976; Khimicheskaya entsiklopediya, 1992). For systems with chemical transformations, the number of components is equal to the difference between the number of particle types in a system and the number of independent reactions. In general, the number of components depends on the conditions under which a system occurs. If the reversible chemical reactions do not proceed in a system, the number of components is equal to the number of substances. For example, simple substances with different compositions and molecular structures, such as molecular oxygen and ozone, are independent components (Goroshchenko, 1978). Each component should consist of particles that are kinetically independent of other particles and, under certain conditions, capable of forming a particular phase. In essence, the metastable components (quasi-components) are no different from the stable components, but they participate in the metastable phase equilibriums.

The classical physicochemical analysis involves consideration of the equilibrium reversible systems under the assumption that the initial components are universally unaltered. Under close examination of the actual compounds, it can be concluded that only a few number of systems possess the above feature. All the actual systems can be separated into two groups: (1) the systems in which the total number of components remains constant with any variations in temperature and pressure, but the components can undergo a chemical modification and (2) the systems in which the number of components and their chemical nature change upon transition from one state of aggregation to another state and even within the same state of aggregation.

By the chemical modification of components, we mean the change in chemical bonding and structure of particles that are treated as components. This problem does not arise in the thermodynamics—a division of science that covers only the macroscopic parameters of systems and does not examine the structure of substances. Unfortunately, this approach holds good only for ideal systems in which, under any conditions, the interactions involve only components that exhibit an absolute minimum of the free energy. In actual practice, the presence of several structural–chemical modifications of a particular component can give rise to the metastable phase equilibriums, and the corresponding phase diagram will differ substantially from that for the stable phase equilibriums.

Table II presents the classification of physicochemical systems and chemical modifications of the initial components for different states of aggregation for alloys in a system.

In melts of systems 3–6 (Table II), the initial components can exist in diverse chemical forms due to the processes like dissociation, association, etc. The chemical equilibrium is attained between new and initial forms of components. As long as this equilibrium occurs in a melt and is easily reestablished with variations in temperature and pressure, the number of components in systems 3 and 4 remains unchanged. However, upon rapid 'freezing' of the equilibrium, all the molecular forms can produce new components, and we have $K > 1$. This is possible because even the sole, chemically independent molecule (unlike a chemical individual whose identification requires the phase formation) can become a component. Substances in the molten state, which can undergo an abrupt semiconductor-metal first-order phase transition should be assigned to systems of type I (Table II). Materials that exhibit smeared phase transitions belong to systems of type II, because the microparticles with covalent and metallic chemical bonds, which coexist in

TABLE II

CLASSIFICATION OF PHYSICOCHEMICAL SYSTEMS

	N_i, state of aggregation		
	Crystalline	Liquid	Gaseous
I			
1. $K = \text{const} = K^0$	$N_i = 1$	$N_i = 1$	$N_i = 1$
2. $K = \text{const} = K^0$	$N_i = 1$	$N_i = 1$	$1 < N_i < \infty$
3. $K = \text{const} = K^0$	$N_i = 1$	$1 < N_i < \infty$	$1 < N_i < \infty$
4. $K = \text{const} = K^0$	$1 < N_i < \infty$	$1 < N_i < \infty$	$1 < N_i < \infty$
II			
5. $K \neq \text{const} < K^0$	$1 < N_i < \infty$	$1 < N_i < \infty$	$1 < N_i < \infty$
6. $K \neq \text{const} > K^0$	$1 < N_i < \infty$	$1 < N_i < \infty$	$1 < N_i < \infty$

Note: K^0 is the number of the initial components in a system and N_i is the number of the structural–chemical modifications of the ith component in crystalline, liquid, and gaseous states.

the transition range, represent nuclei of two phase modifications corresponding to two different components.

In the analysis of semiconductor-metal transition, as for the eutectoidal model of the vitreous state, we will use the generalized definition of the component (Funtikov, 1995). Simple substances (or chemical compounds) correspond to one component in melts until the melts remain homogeneous and all the products of their dissociation are in a chemical equilibrium. If the microinhomogeneous structure with different types of medium-range order arises in a melt and the resulting microregions coexist, it is incorrect to argue for the one-component melt ($K = 1$). In the case when the transition occurs between the phase characterized by covalent medium-range order and the phase with metallic medium-range order in the microregions of a melt at temperatures corresponding to the semiconductor-metal transition, the melt should be considered a two-component melt ($K = 2$). Therefore, the formally one-component melt possesses two (covalent and metallic) components. If the semiconducting (covalent) or metallic modification of a substance is stable only at certain temperatures, this does not imply that it is impossible to choose such conditions at which the temperature range of their steady-state existence can be extended. Of course, when these conditions are removed, the corresponding modification either exists in the metastable state or immediately transforms into a thermodynamically more stable phase.

In the case under study, the interaction between components with covalent and metallic bonds can be described by a pseudobinary phase diagram with phase separation. The phase diagrams for the systems, in which the initial components are represented by metallic simple substances and semiconducting chemical compounds, can serve as a phenomenological basis of the subsequent discussion. Typical examples of these systems are as follows: $Cu-Cu_2S$, $Cu-Cu_2Se$, $Cu-Cu_2Te$, $Ag-Ag_2S$, $Ag-Ag_2Se$, $Ag-Ag_2Te$, $Tl-Tl_2S$, $Tl-Tl_2Se$, etc. (Abrikosov, Bankina, Poretskaya, Skudnova and Chizhevskaya, 1975). All the above particular systems are characterized by a cupola-shaped phase diagram that corresponds to the liquid–liquid phase separation. We postulate that the character of interactions between components in the semiconductor-metal systems is identical to that in the melts in the temperature range of a semiconductor-metal smeared phase transition.

The temperature dependence of the magnetic susceptibility χ for vitrifying melts in the temperature range of the semiconductor–metal transition is schematically represented in Figure 9a. Such a dependence in every detail is observed for high-conductivity semiconductors as As_2Te_3 and $TlAsTe_2$ (Tver'yanovich and Gutenev, 1997). The semiconductor–metal transition temperature, which is maximum at the value of $d\chi/dT$ and 50% of bonded electrons are delocalized, is designated as T_{S-M}. The starting and final temperatures of the semiconductor–metal transition are denoted by T_b and T_f, respectively.

Figure 9b depicts the hypothetical phase diagram for a semiconductor-metal pseudobinary system. The line ABC corresponds to the actually observed states of a melt in the range of the semiconductor–metal transition. An *S*-shaped form of this line is explained by the fact that, in reality, it represents the inverse dependence of the fraction of delocalized bonded electrons as a function of the melt temperature in the range of the semiconductor-metal transition (Tver'yanovich and Gutenev, 1997). In the phase diagram (Fig. 9b), line *M* indicates a monotectic equilibrium when two liquid and one

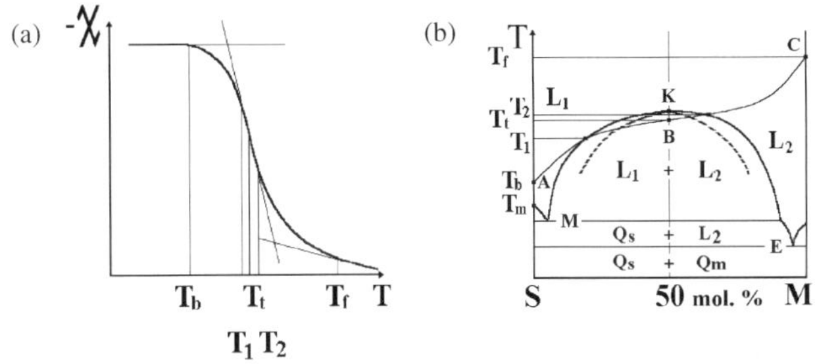

FIG. 9. (a) Temperature dependence of the magnetic susceptibility χ for vitrifying melts in the semiconductor-metal transition range and (b) hypothetical phase diagram for a semiconductor-metal pseudobinary system. Line ABC corresponds to the actually observed states of a melt in the semiconductor-metal transition range; T_b and T_f are the temperatures of the onset and completion of the semiconductor–metal transition, respectively; T_{S-M} is a temperature of the semiconductor-metal transition at which $d\chi/dT$ reaches a maximum; T_1 and T_2 are the boundary temperatures of the range in which the melt structure is assumed to be microinhomogeneous; M indicates the line of a monotectic equilibrium; E represents the line of an eutectic equilibrium; L_1 and L_2 are the covalently bound liquid and metallic liquid phases, respectively; and Q_S and Q_M are the covalently bound crystalline and metallic crystalline phases, respectively. Spinodal is shown by the dashed line.

solid phases are in equilibrium ($L_1 \Leftrightarrow L_2 + Q_s$) (Anosov, 1976). The line of an eutectic equilibrium when two solid and one liquid phases are in equilibrium ($L_2 \Leftrightarrow Q_s + Q_m$) (Anosov, 1976) is denoted by E. The covalently bound liquid, metallic liquid, covalently bound crystalline, and metallic crystalline phases are designated by L_1, L_2, Q_s, Q_m, respectively. The phase diagram is displayed at $T_m < T_b$, where T_m is the melting temperature of a simple substance (a chemical compound) or the liquidus temperature of an intermediate alloy. The ratio between the melting temperatures of semiconducting ($T_m(S)$) and metallic ($T_m(M)$) phases is immaterial for our purposes. In the case under discussion, $(T_m(M)) < (T_m(S))$ and $T_m(S) < T_b$. Unlike the states above, the critical point K, where the homogeneous melt structure is stable, the reference point lying between the binodal curve (a cupola of phase separation) and the monotectic line M corresponds to the phase separation of a liquid into two liquid phases L_1 and L_2. The points A, B, and C match the equilibrium states of an alloy at specified constant pressure and other possible external conditions. The point B lies somewhat below the critical point K and corresponds to the condition for the spontaneous formation of critical microemulsions. This is also favored by the location of the B point inside the spinodal (a dashed line), where the metastable states of homogeneous melts cannot occur. Hence, it follows that the pseudobinary model of transition makes possible the prediction of the microinhomogeneous structure of melts in a narrow temperature range from T_1 to T_2 (Fig. 9b), at which the ABC line intersects the binodal. Moreover, it is clear that the particle sizes should reach a maximum at the semiconductor–metal transition temperature T_{S-M}. As a first approximation, we can assume that the T_1 and T_2 temperatures should correspond to the initial and end points of virtually linear portions of the temperature dependences of some properties of the melts (e.g., density and magnetic susceptibility) near the transition temperature T_{S-M} (Fig. 9a).

A question arises as to which practical inferences can be drawn from the proposed model. First, we can make the conclusion that an increase in pressure should bring about a decrease in the binodal size in the phase diagram for a pseudobinary system and a gradual transition to the state when the critical point K lies below the point 5, and the semiconductor–metal transitions occur through a continuous change in the melt structure rather than through a phase transformation. Second, it is possible to solve the problem by preparing the metallic phase in a metastable crystalline state. The fact is that, if one attempts to produce this phase (e.g., for the $TlAsTe_2$ compound) by ultrafast cooling of the melt from temperatures above T_f, the crystallization of the resulting metallic glass most likely cannot cease at the stage of an intermediate metallic phase because of very low energy barrier and considerable heat release upon crystallization. In order to produce and investigate the metallic modification, it is necessary to decrease the sizes of crystallized particles, to introduce stabilizers inhibiting their rapid growth, and to reduce the heat released upon crystallization. The crystallization of melts in the temperature range corresponding to the semiconductor–metal transition meets the above requirements. With a cooling rate intermediate between 10^2 and 10^6 K s^{-1}, it is possible to produce a microheterogeneous material whose semiconducting vitreous matrix incorporates metallic crystalline inclusions of the metastable modification with sizes quite suitable for the X-ray powder diffraction analysis.

How can a semiconductor–metal smeared transition in an alloy consisting of several simple substances or compounds be interpreted? Investigations of the ternary systems, in which two of the three components are immiscible, but each of them are completely miscible in pairs with the third component, indicate that all the above conclusions and reasoning are also true for ternary systems (Shchukin et al., 1982). In this case, a system possesses only a larger number of degrees of freedom, and the critical state can be approached in going from a binary system due to the change in both temperature and composition of a system. Reasoning from the principle of compatibility of different physicochemical systems, the aforementioned inferences can also be extended to more multicomponent systems. In this respect, it is easy to explain why the dependences of the properties for melts composed of the As_2Te_3, $TlAsTe_2$, and $TlAsSe_2$ compounds in the temperature range of the semiconductor-metal transition are not radically different from those for intermediate melts in the $TlAsTe_2$–$TlAsSe_2$ system (Tver'yanovich et al., 1986). Indeed, the initial components of the latter system are completely miscible in the liquid state.

Conclusion

The pseudobinary model of a semiconductor–metal smeared phase transition in vitrifying melts is developed within the context of the physicochemical approach. The model is based on an idea of the variable number of components in a system. According to the proposed model, the melts in the temperature range close to the semiconductor–metal transition point T_{S-M} exhibit a microinhomogeneous structure, and the particle sizes are maximum at the temperature T_{S-M} (Funtikov, 1998a).

5. Eutectoidal Model of Glassy State of Substance

Abstract

The model of the structure of glasses is offered that is based on the Smits's idea about pseudobinary systems. All glasses, including one-element glasses, are analyzed as a variety of eutectics formed by the interaction among themselves of 'pseudophases.' Pseudophase is the equivalent of the nucleus of the ordered particle that appear in the supercooling melt.

The debate still goes on between the proponents of the homogeneous and the microinhomogeneous structure of vitreous alloys, although experiments bring in more and more evidence in favor of the microinhomogeneous structure of glasses. According to our eutectoidal model, even the ideal glass is non-uniform and can be submitted as the ensemble of sticking microspheres.

The equilibrium between low- and high-molecular clusters is a necessary condition of the stable glass formation, according to the electronic configuration model of the glass formation (Funtikov, 1989). These clusters may be neutral and have charge particles (Funtikov, 1993). It follows from model that the availability of a large number of internal reactions and appropriate chemical equilibriums promotes the formation of glasses. It is characteristic as for one-element substances as it is for their compounds. In case of elementary substances and steady compounds, the interaction between clusters of the different degrees of the polyregularity predominates. We assume that co-existing low- and high-molecular clusters can be considered as quasi-independent components. These components are capable of existing in the stable and metastable alloys. It is obvious that any forming glass melts are multicomponent alloys. The eutectoidal model of the glassy state of the substance is based on the Smits's idea about pseudobinary systems (Smits, 1921). All glasses, including one-element glasses, are analyzed as a variety of eutectics formed by the interaction among themselves of 'pseudophases.' Pseudophase is the equivalent of the nucleus of the ordered particle that appear in the supercooling melt. The conception of pseudophase is introduced by Poryi-Coshitz (1983). The simple pseudobinary systems are the systems based on sulfur and selenium. In the case of sulfur, pseudobinary system consists of the molecules S_8 and S_n ($n = \text{const}$). We have shown that the system $S_8{-}S_n$ and similar systems are the simple eutectic systems with a limited solubility in the solid state. It is shown that conditions (P, T) must exist for the formation of the eutectic melt. The components of this melt must lose a degree of the freedom during solidification in the case of the slow relaxation of the corresponding equilibriums. Thus the glasses synthesized from melts are microdispersed metastable eutectic alloys even in case of the elementary substances. The hypothetic diagram of states of the pseudobinary system based on elementary substance is shown in Figure 10. It is suggested that sets of equilibriums between low- and high-molecular clusters ($n\,\text{LM} \Leftrightarrow \text{HM}$) in the melt at the temperature of the synthesis of the glass T_S determine the structure of glass. The high viscosity of such melts is determined by the large power-generating barrier of the transition between the low- and high-molecular clusters. During the cooling of the melt from temperature T_S, relaxation of the chemical equilibriums becomes extremely weak, and true solution (melt) L_T is turned into lyophilic colloid L_c

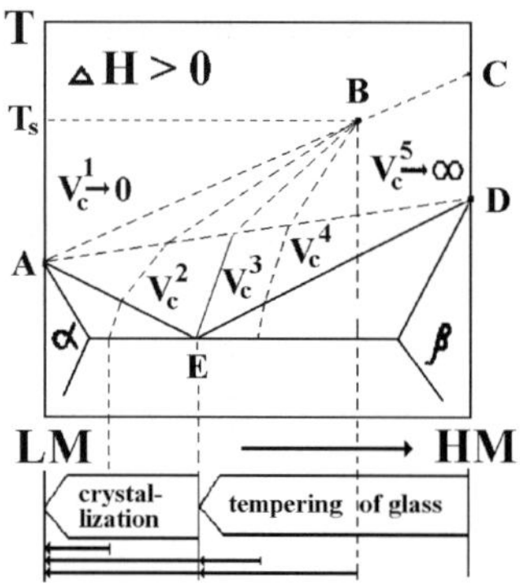

FIG. 10. The hypothetic diagram of states of the pseudobinary system based on the elementary substance. (a) The equilibrium between low- and high-molecular clusters (n LM ⇔ HM) is determined for the temperature of the melt T_S, a change of the enthalpy of the direct reaction $\Delta H > 0$, a change of the free energy $\Delta G = 0$. (b) Low-molecular clusters are more stable than high-molecular clusters for $T = 298$ K ($G_{LM} < G_{HM}$); the velocity of a cooling of melts $V_c^1 < V_c^2 < V_c^3 < V_c^4 < V_c^5$, the optimum velocity of cooling of melts $V_c^{opt} = V_c^3$ corresponds to the maximum dispersity.

(Fig. 11). In the case of the optimum velocity of cooling of melts $V_c^{opt} = V_c^3$, the maximum dispersity of particles of the glass is formed owing to equilibrium between L_T, L_c and solid solutions S'_α and S'_β. The tempering of glasses must stimulate relaxation of the chemical equilibriums and give the same result.

These conceptions agree with the topological approach to the analysis of processes of the formation of the structure in glasses (Funtikov, 1996a). Our topological approach is based on the idea that extremely small volumes of space for the formation of the continuous ordered structure permit all possible elements of the symmetry with first

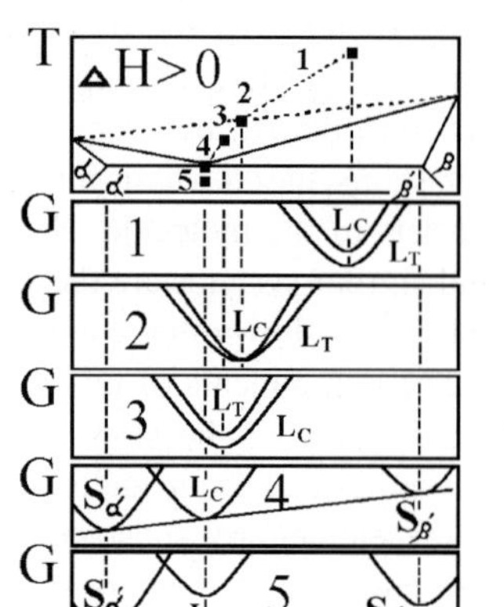

FIG. 11. The schematic diagram of dependences of the Gibbs's free energy G of a true solution (melt) L_T, lyophilic colloid L_c, solid solutions S_α' and S_β' from composition of alloys of low- and high-molecular components of the pseudobinary system based on the basis of the elementary substance. Figures 1–5 correspond to the points 1–5 on the hypothetic diagram of states of the pseudobinary system based on the basis of the elementary substance.

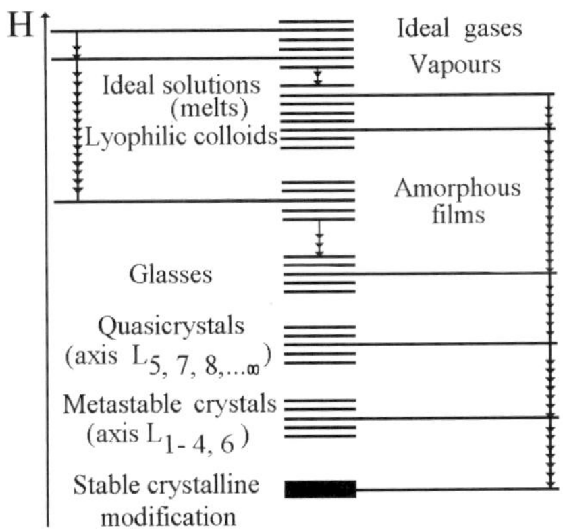

FIG. 12. The integrated enthalpy (H) diagram of different states of substances, inclined to formation of glasses and amorphous films.

infinite orders. The conventional approach to the analysis of the structure of glasses is based, as a rule, on the laws of crystallography. According to these laws (the elements of symmetry of macrostructure), only 1, 2, 3, 4, and 6 orders are permitted, and from that theoretical opportunity, the formation of the ideal homogeneous glass follows. According to our topological model, only the large power-generating barrier of the transition from microscopic particles, having infinite set elements of the symmetry, to the crystalline state with the limited symmetry permits substance to pass in the glassy state from the melt and in the amorphous film from a vapor. The diagram of the transitions of glasses and amorphous films in the crystalline state through a stage of the formation of quasicrystals is offered (Fig. 12). From topological model it follows that such quasicrystals should have an axis of symmetry of fifth order and the icosahedral structure in case of metal alloys and the dodecahedral structure in case of semiconductors and dielectrics. In the structure of all glasses, the structural elements having utmost geometrical variant all elements of the symmetry of icosahedron and dodecahedron ($6L_5 10L_3 15L_2 15$PC) should dominate. The given conclusions are confirmed by the experimental reception of the quasicrystals at a tempering of metal glasses.

It is possible to predict, that such phases should occur during tempering of the semiconductor and dielectric glasses. From the topological model it follows that even the ideal glass non-uniformly can be submitted as the ensemble of sticking spheres. The direct experimental confirmation consists of supervision by means of the electronic microscope; the multiple-range structure typical of one-component glasses after their chemical etching.

CONCLUSION

The eutectoidal model of glassy state of substance is offered, which allows to investigate the mechanism of a glass transition and to predict compositions of new glassy systems (Funtikov, 1995, 2000).

6. Experimental Proof of the Eutectoidal Nature of Glasses

ABSTRACT

The presence of nanostructures in glasses and their eutectic interaction has been shown experimentally. The correctness of the eutectoidal model of a glassy state of substance is as follows:

At present all advanced models of a glassy state of substance do not take into account the microscopic aspect of process of glass formation. For this reason it is impossible to create a uniform idea about glass, which is not concerned to a nature of a chemical bond and other features of vitrescent substances. We suggest the eutectoidal model of a glassy state of substance, which takes into account the microscopic aspect of a glass transition. It also offers the method of physicochemical model operation of process of a glass transition, which allows to construct glasses chemically. The method of the differential thermal analysis is used for the investigations of the metastable chalcogen-based systems for experimental verification of the suggested principles.

The modern idea of glass formation is based on the kinetic theory of a glass transition related with relaxation processes in vitrescent melts. Using the kinetic theory, the glass is the supercooled fluid having so high viscosity that it acquires the properties of a solid body. The concept mentioned above is macroscopic in all advantages and does not allow making one-valued deductions about processes occurring at an atomic–molecular level. At present, the discussion about the mechanisms of formation of structure of glass remains unfinished, and there is not as yet a generally accepted definition regarding the glassy state of a substance. The modern idea of glass formation breaks the development of representations about glass and principles of a series of technological processes in the production of glasses. First of all, it concerns new classes of glasses such as chalcogenide and metal glasses. For development of principles of formation of a glassy state of substance, taking into account the macroscopic and microscopic aspects and not concerning with a concrete method of reception of glasses, we advanced the eutectoidal model of a glass transition of substances (Funtikov, 1995). According to this model, the glasses are a type of the ultradispersed multicomponent eutectoidal structures, which are not structurally homogeneous material in the limits of the medial order. The medial order in glasses is the topological order at a level of second, third, and other spheres of interaction of atoms. The upper sphere of such interaction of atoms is determined by the sizes of nanostructures, which join the composition of a glassy network. We suppose that these nanostructures should not be less than two types in glasses. In the case of simple substances and stable chemical combinations, a topological order within a frame of one nanostructure should be homeomorphous with geometrical order in crystals of some modification of a substance (Funtikov, 1996a).

The presence of nanostructure in chalcogenide glasses is shown by a method of chemical and electrochemical equivalent measurement (Funtikov, 1998b). It is shown that the composition of nanostructures in glasses can correspond to the composition of simple substances, stable, and metastable compounds. The experimental results show, that nanostructures can be formed by selenium, stable compounds $GeSe_2$, As_2Se_3, by metastable compound $GeTe_2$ and other compounds (Funtikov, 1996b–e, 1998b) for binary glass systems As–Se, As–Te, Ge–Se, Ge–Te and 3-fold systems Tl–Ge–Te, Tl–As–Te and other systems as an example. We are offered chemical and electrochemical equivalent measurements (Funtikov, 1998b). The new method is based on the law of equivalents and on the analysis of a dependence of a chemical equivalent of the glasses and the ceramics from their composition. The glassy and ceramic insulators, the glassy semiconductors, and the amorphous semiconductor films are microheterogeneous alloys. Therefore, they can be investigated by the selective dissolution of the separate fragments of alloys in the solutions of acids, alkalis, oxidizers, reductants, organic compounds, and by the electrochemical method.

The dependence of the experimental CE^{exp} and theoretical sizes CE_{I}^{theor}, CE_{II}^{theor}, CE_{III}^{theor} of a molar weight of a chemical equivalent of glasses of a system Ge–Se from a composition of glasses by dissolving them in 1 M KOH solution is shown in Figure 13. The theoretical sizes CE_{I}^{theor}, CE_{II}^{theor}, CE_{III}^{theor} are calculated for three variants of the gear of a dissolution of glasses: (1) a sediment of $GeSe_2$ is formed and Se is dissolved to Se^{2-} and SeO_3^{2-}; (2) a sediment of $GeSe_2$ is formed and Se is dissolved to Se_4^{2-} and SeO_3^{2-}; (3) a sediment of Se or Ge is formed and $GeSe_2$ is dissolved to $GeOSe_2^{2-}$. The shift of a gear of dissolution of the glasses is realized at the alloys with the eutectic combinations 10 at.% Ge. The obtained results specify the presence of nanostructures on the basis of a compound $GeSe_2$ and selenium in glasses of a Ge–Se system (Funtikov, 1996c).

The glasses of a system Tl–Ge–Te on a section GeTl–Te with the shortage of tellurium are dissolved electrochemically in 1 M solution of NaOH. The graphs are submitted for the dependence of the mole weight of a chemical equivalent of glasses, received unstationary and stationary methods by electrochemical dissolution in 1 M solution of NaOH and as well as calculated to next gears: (1) by a loss in the sediment

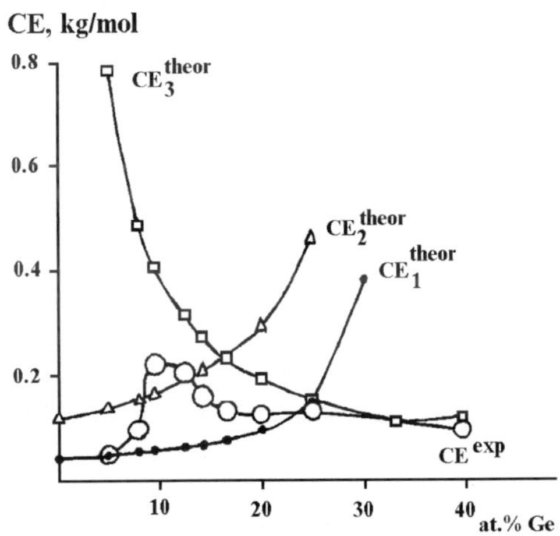

FIG. 13. The dependence of the experimental CE^{exp} and theoretical sizes CE_{I}^{theor}, CE_{II}^{theor}, CE_{III}^{theor} of a molar weight of a chemical equivalent of glasses of a system Ge–Se from a composition of glasses by dissolving them in 1 M KOH solution. The theoretical sizes CE_{I}^{theor}, CE_{II}^{theor}, CE_{III}^{theor} are calculated for three variants of the gear of a dissolution of glasses: (1) a sediment of $GeSe_2$ is formed and Se is dissolved to Se^{2-} and SeO_3^{2-}; (2) a sediment of $GeSe_2$ is formed and Se is dissolved to Se_4^{2-} and SeO_3^{2-}; (3) a sediment of Se or Ge is formed and $GeSe_2$ is dissolved to $GeOSe_2^{2-}$.

of a compound $GeTe_2$; (2) by a loss in the sediment of a compound GeTe; (3) by a complete dissolution of glasses. The fragments of the metastable compound $GeTe_2$ are discovered by the unstationary method of electrochemical equivalent measurement in glasses of a system Ge–Te–Tl (Funtikov, 1996b).

Using the principles of the eutectoidal model of the structure formation in glasses, the author simulated the process of a glass transition to experimentally prove the validity of the eutectoidal concept. The chalcogens, i.e., sulfur, selenium, and tellurium, are the most successful model objects for this purpose. The metastable systems $Se_8(Se_{monocline})$-$Se_n(Se_{hexagonal})$, $Se_8(Se_{red\ amorphous})-Se_n(Se_{black\ amorphous})$, $Se_8(Se_{monocline})-Te_n$ $(Te_{hexagonal})$, $Se_8(Se_{red\ amorphous})-Te_n(Te_{black\ amorphous})$, $Se_8(Se_{monocline})-S_8(Se_{rhombic})$, $Se_8(Se_{monocline})-I_2$ and the stable system $S_8(S_{rhombic})-Se_n(Se_{hexagonal})$ are investigated by the method of differential thermal analysis. Besides, the glassy selenium is studied by the method of differential thermal analysis. The obtained results show that the choice of physicochemical model operation for the study of the process of glass transition of substances is a rather perspective method.

According to the eutectoidal model of a glassy state of substance, the formation of glasses is connected with a variable number of components of the vitrescent melts and with the special role of metastable phase equilibriums (Funtikov, 1995). The simple metastable physicochemical system is required for the experimental and theoretical modeling of processes, enabling to reveal a role of chemical and phase equilibriums in the formation of the medium order and physical and chemical properties of glasses. The selenium-based systems (Se_8- Se_n) are the best samples for this condition. Selenium is one of the four chemical elements (P, As, S, Se), which can form the one-element glasses. All four elements exist in the form of low- and high-molecular modifications. The chemical equilibrium between the monomeric and polymeric molecules is established in their melts. It is supposed that such molecules are Se_8 and Se_n in the case of selenium. The chained molecules (Se_n) form a basis of the stable phase of the gray hexagonal selenium. The cyclic molecules (Se_8) form three modifications of the metastable red selenium, of which the monocline modification is considered the most stable modification. Previously, we had designed a new method of making of red monoclinic selenium before the examination of physicochemical interaction of stable and metastable components on the basis of selenium.

The differential thermal analysis was applied for the research of a physicochemical interaction between crystals of the gray and red selenium. The present experiment is planned on the assumption that the structure and physical–chemical properties of glassy selenium are determined by the special role of metastable phase equilibriums between gray and red modifications of selenium. To realize this type of a physicochemical interaction, it is necessary that the chemical equilibrium between the molecules Se_8 and Se_n is infringed at the time of experiment. This condition will be created by sharp cooling of a vitrifying melt.

For the first time, the diagram of fusibility of a system on the basis of components of the chemical composition of selenium is investigated. The experimental results obtained by us confirm the eutectoidal model of a glassy state of a substance. In particular, it is stated that the performances of the curves of the differential thermal analysis of alloys of system $Se_8(Se_{monocline})-Se_n(Se_{hexagonal})$ change non-linearly at transferring from low-molecular weight to a high-molecular component.

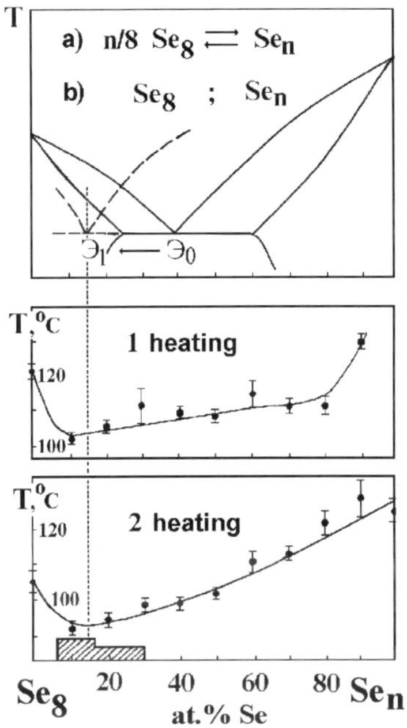

FIG. 14. The prospective diagram of states of the pseudo-binary system Se_8(β–Se)–Se_n(γ–Se) and the concentration dependence of a temperature at the beginning of melting of crystalline alloys of a specified system for first and second heatings. The shaded area corresponds to a complete glass formation of alloys after the second heating.

The minimum temperature of the initial melting of alloys is observed in the range of 10–20 initially, of given mass % of hexagonal gray selenium (Se_n) (Fig. 14). The alloys of the compositions mentioned above seem completely vitrified objects after the experiment, which was concerned with the special role of metastable phase equilibriums between stable and metastable components on the basis of selenium in the glass formation of a melt of selenium. The noted phenomenon can be connected with the eutectic interaction of modifications of selenium.

In order to magnify the dispersity of the starting stable and metastable (quasi-) components, we used the red and black amorphous modifications of selenium. Experimental data show that the structure of red and black amorphous modifications of selenium differs significantly. It is demonstrated that the red amorphous selenium consists of cyclic molecules Se_8, and black amorphous selenium from chained molecules Se_n.

In the system $Se_8(Se_{red\ amorphous})–Se_n(Se_{black\ amorphous})$, in the case of surplus red modification, it appears that the curves of the differential thermal analysis obtained during contact melting of two amorphous modifications of selenium coincide with the curve of the differential thermal analysis obtained during heating of the glassy selenium. This fact is fundamental and it allows planning the theoretical and experimental process of a glass transition.

Conclusion

Thus, for the first time, the metastable phase equilibriums between stable and metastable physicochemical components have been studied using differential thermal analysis. Moreover, for the first time, the physicochemical model operation of a glass transition process has been made, using chalcogen as an example (Funtikov and Funtikova, 2000, 2001).

7. Physicochemical Analysis of Vitreous Semiconductor Chalcogenide Systems

Abstract

A theoretical approach to the analysis of the structure of vitreous alloys is proposed, which interprets glasses as intermediate in their characteristics between homogeneous and heterogeneous systems. This approach is based on the eutectoidal hypothesis for the glass structure, which provides a way of describing the mechanism for the formation of medium-range order groupings in a vitrifying melt and, correspondingly, in a glass. In terms of the eutectoidal hypothesis, the 'composition–property' diagrams of vitreous systems can be analyzed in the same manner as the corresponding diagrams in physicochemical analysis of heterogeneous systems.

It is well known that alloys close in composition to the metastable eutectics exhibit a high glass-forming ability (Comet, 1976). Knowledge of metastable phase diagrams, where the compositions of eutectic alloys can differ from those of equilibrium eutectics, is of prime importance in treating the nature of glass formation.

It can be assumed that the above-mentioned tendency implies something more than a mere correlation between a particular composition of alloy and its glass-forming ability. In this respect, we proposed the eutectoidal hypothesis for structure formation in glasses. Under this hypothesis, melts of all substances prone to glass transition should feature a quasi-eutectic structure (Funtikov, 1990, 1991, 1995). In these works, it was shown that glasses of any composition (even elemental) could be considered as a modification of the multicomponent, highly disperse eutectics. The eutectoidal hypothesis is based on the Smith idea of pseudobinary systems and the notion of pseudophases introduced by Porai-Koshits (1985) for glasses.

To answer a number of questions, there is a need to define the term 'components.' As is known, the components of a system are called substances if variations in their masses are independent and determine all possible variations in the composition of a system (The Collected Works of G. Willard Gibbs, 1928; Anosov, 1976; Khimicheskaya entsiklopediya, 1992). For systems with chemical transformations, the number of components is equal to the difference between the number of particle types in a system and the number of independent reactions. In general, the number of components depends on the conditions under which a system occurs. If the reversible chemical reactions do not proceed in a system, the number of components is equal to the number of substances. For example, simple substances with different compositions and molecular structures, such as molecular oxygen and ozone, are independent components (Goroshchenko, 1978).

The number of components in a system can vary from 1 to ∞. Each component should consist of particles that are kinetically independent of other particles and are, in principle, capable of forming a certain phase. It follows that ions in solutions or melts cannot serve as components. However, molecules with different degrees of polymerization and involving atoms of the same type can fulfill the role of components in the melts, provided that their interconversion is impossible. The foregoing is very readily illustrated by the behavior of sulfur. Under equilibrium conditions, sulfur should be treated as the sole component. This corresponds only to one equilibrium modification. At temperatures higher than 159 °C, the molten sulfur is considered as a solution of polymer in monomer; in this case, the ratio between these forms is temperature-dependent (Addison, 1964). If the melt is rapidly cooled to temperatures in the range 159–119 °C, the number of components in the melt increases sharply, because all the molecular forms of sulfur kinetically become virtually independent. Upon further rapid cooling of the molten sulfur, the components should interact even in a multi-component system. To put it differently, under certain conditions, even the systems involving the sole element can become multicomponent. In the vitrifying melt, the number of quasi-components is equal to the number of different molecules that hold their individuality during the cooling of a melt. In order for the homogeneity of the vitrifying melt to be at maximum upon freezing, it is necessary that the particles corresponding to different quasi-components grow simultaneously. From the standpoint of physicochemical analysis, this can be achieved only by a eutectic mechanism for the growth of ordering nuclei. As is known (Anosov, 1976), the eutectic of a system from any number of components can be defined as a solution that is in congruent equilibrium with solid phases whose number is equal to the number of components in a system. In the case of vitrifying melts, because of a considerable number of quasi-components, actual phases cannot, in principle, be formed, due to a deficit of material. As a result, the ordering ceases at the pseudophase forming stage. In the context of the eutectoidal model, the optimum glass transition will be observed in the case when, upon cooling, N components form an N-component, metastable, highly disperse eutectic. The particles of this eutectic are so small that they are more likely to be liquid than solid. This is quite consistent with the kinetic theory of glass transition. The number of quasi-components N in a melt of a good glass-former should be comparable to the number of structural elements. This is why the actual phases are not formed and the glass transition occurs. It is apparent that all the nuclei of chemical ordering should grow virtually simultaneously for the glass homogeneity to be at maximum. This can be realized only in the case of eutectoidal mechanism for the solidifying of a melt. From the aforementioned, it follows that the ideal glass is a multicomponent eutectic in which the number of components is comparable in order of magnitude to a feasible total number of structural elements forming the short-range order.

The eutectic melts are an example of lyophilic colloidal solutions in which the disperse particles are completely solvated by a dispersion medium (Zalkin, 1987). Hence, it can be inferred that among other aspects glasses can also be considered, as modification of the frozen lyophilic colloidal solutions. The vitrifying melts can correspondingly be treated as lyophilic colloidal solutions, where spherical micelles can be formed at the first stage. As the concentration of solutions increases, these micelles are able to transform into anisometric (liquid-crystalline) micelles (Shchukin et al., 1982). Table III demonstrates the feasible structural–configurational equilibriums in melts of the binary A–B systems,

TABLE III

STRUCTURAL–CONFIGURATIONAL EQUILIBRIUMS RESPONSIBLE FOR EUTECTOIDAL INTERACTIONS BETWEEN COMPONENTS AND GLASS TRANSITION IN MELTS OF THE BINARY A–B SYSTEMS

	A		B	
	$2A_1 \Leftrightarrow A_2$ $A_2 + A_1 \Leftrightarrow A_3$ $A_3 + A_1 \Leftrightarrow A_4$ $A_{n-1} + A_1 \Leftrightarrow A_n$		$2B_1 \Leftrightarrow B_2$ $B_2 + B_1 \Leftrightarrow B_3$ $B_3 + B_1 \Leftrightarrow B_4$ $B_{n-1} + B_1 \Leftrightarrow B_n$	
I. $(A$–$A)$ $(A_n)_m \cdot kA_1$ $\Delta H_{\mathrm{I}}^{\mathrm{form}} < 0$		II. $(A$–$B)$ $(B_n)_m \cdot kA_1$ $\Delta H_{\mathrm{II}}^{\mathrm{form}} < 0$ $\lvert\Delta H_{\mathrm{II}}^{\mathrm{form}}\rvert > 1/2\lvert\Delta H_{\mathrm{I}}^{\mathrm{form}} + \Delta H_{\mathrm{III}}^{\mathrm{form}}\rvert$		III. $(B$–$B)$ $(B_n)_m \cdot kB_1$ $\Delta H_{\mathrm{III}}^{\mathrm{form}} < 0$

Note: $\Delta H_{\mathrm{I}}^{\mathrm{form}}$, $\Delta H_{\mathrm{II}}^{\mathrm{form}}$, and $\Delta H_{\mathrm{III}}^{\mathrm{form}}$ are the formation heats for $(A_n)_m \cdot kA_1$, $(B_n)_m \cdot kA_1$, $(B_n)_m \cdot kB_1$, and solvates, respectively.

where the homosolvates and heterosolvates can be formed. One can assume that the heterosolvates are more stable, because, in principle the interconversion between the disperse phase and dispersion medium is impossible. Both neutral molecules and all the molecular products of their dissociation can be served as quasi-components (Table IV). All the charged products of this dissociation are not quasi-components and, in a way, only complicate the resulting glass structure.

It is obvious that 'heterogeneous' interactions should predominate among a wide variety of interactions. In a series of systems, a melt, and equilibrium crystal can be different in component composition. Tables V and VI present the enthalpy diagrams of the transitions between equilibrium crystals and their melts that already occur along the non-equilibrium pathway at the price of retention and even formation of the metastable crystalline phases. At high temperatures, these phases correspond to rather stable

TABLE IV

STRUCTURAL–CONFIGURATIONAL EQUILIBRIUMS IN MELTS OF THE VITRIFYING ELEMENTAL SUBSTANCES AND COMPOUNDS

Elemental substances and dystectic compounds	Peritectic compounds
$n\mathrm{LM}_i^0 \Leftrightarrow \mathrm{HM}_i^0$	$\mathrm{HM}_j^0 \Leftrightarrow \mathrm{LM}_k^0 + \mathrm{HM}_p^0$
$2\mathrm{LM}_i^0 \Leftrightarrow \mathrm{LM}_i^+ + \mathrm{LM}_i^-$	$2\mathrm{LM}_k^0 \Leftrightarrow \mathrm{LM}_k^+ + \mathrm{LM}_k^-$
$2\mathrm{HM}_i^0 \Leftrightarrow \mathrm{HM}_i^+ + \mathrm{HM}_i^-$	$2\mathrm{HM}_p^0 \Leftrightarrow \mathrm{HM}_p^+ + \mathrm{HM}_p^-$
$n(\mathrm{HM}_i^{0(+,-)}) + m(\mathrm{LM}_i^{0(+,-)})$	$n(\mathrm{HM}_p^{0(+,-)}) + m(\mathrm{LM}_k^{0(+,-)})$
$\Leftrightarrow (\mathrm{HM}_i^{0(+,-)})_n \cdot m(\mathrm{LM}_i^{0(+,-)})$	$\Leftrightarrow (\mathrm{HM}_p^{0(+,-)})_n \cdot m(\mathrm{LM}_k^{0(+,-)})$

Note: LM and HM designate the low- and high-molecular neutral (or charged) particles, respectively.

TABLE V

ENTHALPY DIAGRAMS FOR DIFFERENT STATES OF ELEMENTAL SUBSTANCES AND DYSTECTIC COMPOUNDS (*H* IS THE ENTHALPY OF THE CORRESPONDING STATE; LM AND HM DENOTE THE LOW- AND HIGH-MOLECULAR PARTICLES, RESPECTIVELY)

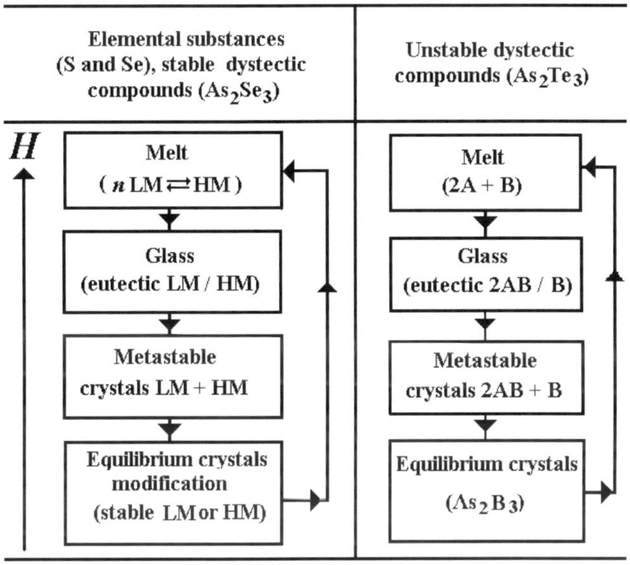

TABLE VI

ENTHALPY DIAGRAM FOR DIFFERENT STATES OF UNSTABLE COMPOUNDS (*H* IS THE ENTHALPY OF THE CORRESPONDING STATE; LM AND HM DENOTE THE LOW- AND HIGH-MOLECULAR PARTICLES, RESPECTIVELY)

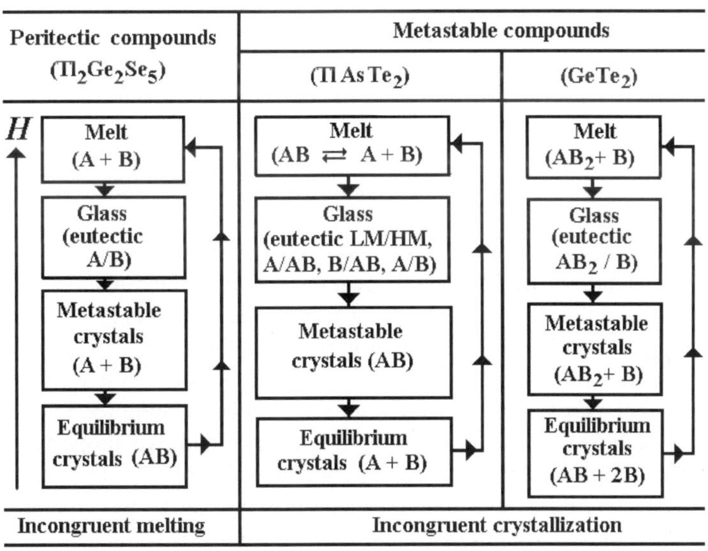

components, which are metastable under normal conditions. This can be illustrated by the metastable compound $GeTe_2$, which is absent in the equilibrium phase diagram, but can arise in the vitreous alloys. In the study of the electrochemical dissolution of the Tl–Ge–Te glasses, we showed that fragments of the $GeTe_2$ compound are incorporated into these glasses (Funtikov, 1996e).

From the above reasoning, it is clear that in terms of an eutectoidal hypothesis, the composition–property diagrams of vitreous systems can be treated both in the framework of the quasi-homogeneous systems (as it was most consistently performed by Gutenev (1993) and in the context of quasi-heterogeneous systems. However, we believe that the last approach is more informative.

Analysis of the composition–property diagrams (isotherms of properties) for the thermodynamically equilibrium multicomponent homogeneous and heterogeneous systems is theoretically unambiguous, if true equilibrium alloys are investigated. Many difficult solubility problems arise for the thermodynamically non-equilibrium systems. Such systems predominate among real systems and can correspond to crystalline, liquid, and vitreous states. Inorganic vitreous alloys cannot, in principle, exist in a thermodynamically equilibrium state. Because of this, different authors interpret the composition–property diagrams very ambiguously. Everything depends on the model used for the glass transition and glass structure. However, before proceeding to our point of view on this problem, we consider the correct techniques to construct the above diagrams. In this case, it is assumed that the same synthesis procedure is used for all glasses of a particular system. The pseudomolar, volume-additive, and mass-additive properties can be recognized. The pseudomolar properties are most appropriate for the correct physicochemical analysis. For simplicity, these properties subsequently will be called molar properties, because it is virtually impossible to calculate truly molar properties for thermodynamically non-equilibrium systems. For glasses, these can be properties calculated per one mole of atoms of a glass, as accurate stoichiometric compositions are not typical of vitreous alloys. Molar properties are an advantage in the case of systems where the components do not chemically interact; there, these properties linearly depend on the composition in terms of mole fractions. A value of mass-additive property is most simply converted into mole-additive modification. To accomplish this, a value of mass-additive property is multiplied by the mass of one mole of glass atoms. For the volume-additive property, its value should be multiplied by a molar volume, which, in millimeter, is calculated from a molar mass and experimentally determined density of substance. Therefore, in the last case, the parameter that is the product of values of two properties capable of reflecting the chemical processes that proceed in a system will be investigated as a function of mole fraction. As a result, three types of the composition–property diagrams are possible: (1) all features of the diagrams (extremums and inflections) become more pronounced, (2) all features become less pronounced, and (3) all features of the diagrams are completely mutually neutralized. The last case is most dangerous, because it leaves room for incorrect conclusion on a system. Therefore, we suggest that in some cases, more objective information concerning the interactions in a system can be obtained in the analysis of the volume-additive property (density ρ, refractive index n, optical density D, integrated intensity of the absorption bands B, dielectric constant ε, etc.)—mole fraction diagrams. We derive the conditions for the correct application of such diagrams for the physicochemical analysis.

We consider the binary A–B system. The following designations are used: N^{v} is the volume fraction, N^{m} is the mole fraction, V^{m} is the molar volume, P^{v} is the volume-additive property, and P^{m} is the molar property. Let $V_{\mathrm{A}}^{\mathrm{m}} = V_{\mathrm{B}}^{\mathrm{m}} \cdot k$. It is evident that

$$P^{\mathrm{v}} = P_{\mathrm{A}}^{\mathrm{v}} N_{\mathrm{A}}^{\mathrm{v}} + P_{\mathrm{B}}^{\mathrm{v}} N_{\mathrm{B}}^{\mathrm{v}} = P_{\mathrm{A}}^{\mathrm{v}} + (P_{\mathrm{B}}^{\mathrm{v}} - P_{\mathrm{A}}^{\mathrm{v}}) N_{\mathrm{B}}^{\mathrm{v}} \tag{16}$$

If the chemical interaction between A and B is weak,

$$N_{\mathrm{B}}^{\mathrm{v}} = (V_{\mathrm{B}}/(V_{\mathrm{A}} + V_{\mathrm{B}})) = N_{\mathrm{B}}^{\mathrm{m}}(1/(N_{\mathrm{B}}^{\mathrm{m}}(1-k)+k)) = N_{B}^{m}\gamma \tag{17}$$

where $\gamma = 1/(N_{\mathrm{B}}^{\mathrm{m}}\,(1-k)+k)$.

Thus,

$$P^{\mathrm{v}} = P_{\mathrm{A}}^{\mathrm{v}} + (P_{\mathrm{B}}^{\mathrm{v}} - P_{\mathrm{A}}^{\mathrm{v}}) N_{\mathrm{B}}^{\mathrm{m}} \gamma \tag{18}$$

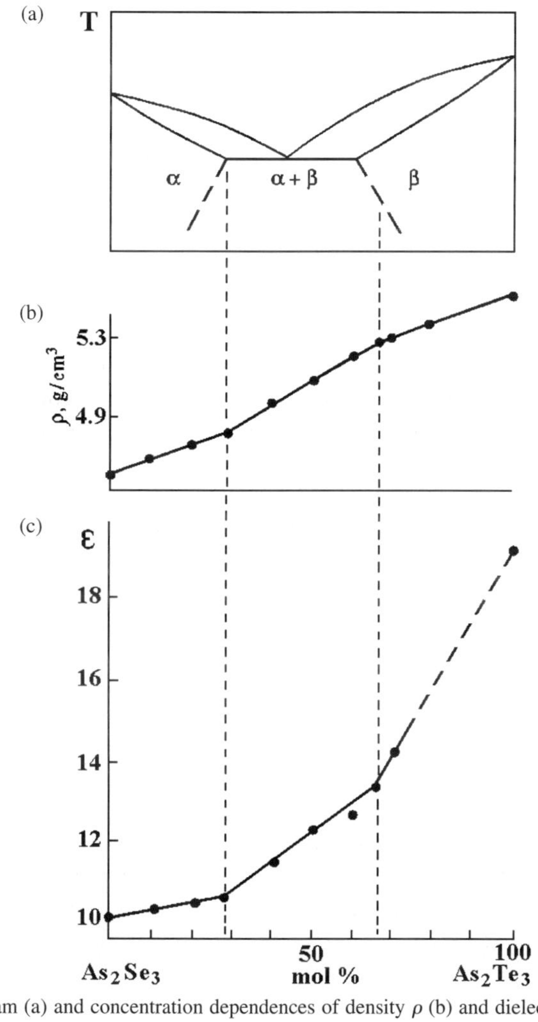

FIG. 15. Phase diagram (a) and concentration dependences of density ρ (b) and dielectric constant ε (c) for vitreous As_2Se_3–As_2Te_3 alloys.

The function obtained will be linear if $\gamma = 1$ at $k = 1$. For composition with $N_{\mathrm{B}}^{\mathrm{m}} = 0.5$ and $k = 1.1$, the deviation from the additivity is about 0.3%; in the case of $k = 1.4$, this deviation is about 3% (this is in agreement with the experimental error for the determination of the volume-additive properties). The foregoing approach can be illustrated by the As_2Se_3–As_2Te_3 system. According to the equilibrium phase diagram of this system, the regions of the solid α- and β-solutions, as well as the region of their coexistence, can be recognized in the crystalline state (Fig. 15) (Vinogradova, 1984). Using the data on density and dielectric constant that were obtained by Kasparova et al. (1984), one can see that the straight line portions in the concentration dependences of these parameters just correspond to the above three regions in the equilibrium phase diagram. This indicates that the interactions between equilibrium components differ little from those between predominant metastable components in the system under consideration. The points of inflections in the corresponding curves can be explained by the change in the type of glass microinhomogeneity.

Conclusion

In the context of the eutectoidal hypothesis for the structure formation in the vitreous systems, it is demonstrated that, in principle, vitreous alloys cannot be truly homogeneous. According to this model, the glasses can be treated as a modification of metastable highly disperse multicomponent eutectics, or frozen lyophilic colloidal solutions. An ideal glass is a multicomponent eutectic in which the number of components is comparable in the order of magnitude to a feasible total number of structural elements of the short-range order. It is shown that physicochemical analysis of vitreous systems should be considered as physicochemical analysis of quasi-heterogeneous systems (Funtikov, 1996f,g).

References

Abrikosov, N.Kh., Bankina, V.F., Poretskaya, L.V., Skudnova, E.V. and Chizhevskaya, S.N. (1975) Poluprovodnikovye khal'kogenidy i splavy na ikh osnove (Semiconducting Chalcogenides and Alloys on Their Base), Nauka, Moscow.

Addison, W.E. (1964) *The Allotropy of the Elements*, Oldbourne, London.

Akhmetov, N.S. (1975) Neorganicheskaya Khimiya (Inorganic Chemistry), Vysshaya Shkola, Moscow.

Akhmetov, N.S. (1988) Obshchaya i Neorganicheskaya Khimiya (General and Inorganic Chemistry), Vysshaya Shkola, Moscow.

Anosov, V.Ya., Ozerova, M.I. and Fialkov, Yu.Ya. (1976) Osnovy fiziko-khimicheskogo analiza (The Foundations of Physicochemical Analysis), Nauka, Moscow.

Appen, A.A. and Galakhov, F.Ya. (1977) On inhomogeneities in glass and the fundamental principle of its structure, *Fiz. Khim. Stekla*, **3**, 402–408.

Bal'makov, M.D. (1988) Defects and fluctuations of structure in disordered systems, *Fiz. Khim. Stekla*, **14**, 801–809.

Bal'makov, M.D. (1989) On the variety of structures, *Fiz. Khim. Stekla*, **15**, 293–295.

Beck, H. and Guntherodt, G.J. (Eds.) (1984) *Glassy Metals, II: Atomic Structure and Dynamics, Electronic Structure, Magnetic Properties*, Springer, Berlin.

Boglaev, I.P., Il'in, A.L., Kraposhin, V.S., Matveeva, L.A. and Ushakov, N.G. (1985) The nature of singularities in the atomic radial distribution functions of metallic glasses, *Fiz. Khim. Stekla*, **11**, 641–646.

Borisova, Z.U. (1972) Khimiya Stekloobraznykh Poluprovodnikov (Chemistry of Vitreous Semiconductors), Leningr. Gos. Univ., Leningrad.

Borisova, Z.U. (1983) Khal'kogenidnye Poluprovodnikovye Stekla (Chalcogenide Semiconductor Glasses), Leningr. Gos. Univ., Leningrad.

Brodsky, M.H. (Ed.) (1979) *Amorphous Semiconductors*, Springer, Berlin.

Comet, J.B. (1976) *The Eutectic Law for Binary TE-Based Systems: A Correlation between Glass Formation and the Eutectic Composition*, Struktura i svoistva nekristallicheskikh poluprovodnikov (Structure and Properties of Noncrystalline Semiconductors), Nauka, Leningrad, pp. 72–77.

Feltz, A. (1983) Amorphe und Glasartige Anorganische Fest-korper, Akademie-Verlag, Berlin.

Frankenheim, M.L. (1835) Die Lehre von der Cohäsion, umafassend die Elastizität der Gase, die Elastizität und Cohärenz der flüssigen und festen Korper, und die Krystallkunde, A. Schulz und Komp., Breslau.

Fridrikhsberg, D.A. (1984) Kurs kolloidnoi khimii (A Course of Colloidal Chemistry), Khimiya, Leningrad.

Frolov, Yu.G. (1982) Kurs kolloidnoi khimii (A Course of Colloidal Chemistry), Khimiya, Moscow.

Funtikov, V.A. (1977) *Impact of Stable Electron Configurations on the Glass-Forming Capacity of Chalcogenide Semiconductor Alloys*, Amorfhye i Stekloobraznye Poluprovodniki (Amorphous and Vitreous Semiconductors), Kaliningradskaya Pravda, Kaliningrad, Russia, pp. 42–50.

Funtikov, V.A. (1984) *Effect of Stable Electron Configurations on the Formation of Chemical Bonds and Structure in Chalcogenide Vitreous Semiconductors*, Mezhdunarodnaya Konferentsiya Amorfhye Poluprovodniki-84 (Int. Conf. on Amorphous Semiconductors), Gabrovo, Bulgaria, pp. 165–167.

Funtikov, V.A. (1987) *Stable Electron Configurations and Properties of Vitreous Semiconductors* Vses. Seminar Novye Idei v Fizike Stekla (All-Union Seminar on New Ideas in Glass Physics), Moscow, pp. 141–148.

Funtikov, V.A. (1989) *An Electron Configuration Model of Glass Formation*, Mezhdunarodnaya Konferentsiya Nekristallicheskie Poluprovodniki-89 (Int. Conf. on Non-Cryst. Semiconductors-89), Uzhgorod, pp. 58–60.

Funtikov, V.A. (1990) *Eutectoidal Model for the Vitreous State of a Substance*, Tezisy konferentsii Stroenie, svoistva i primenenie fosfatnykh, ftoridnykh i khal'kogenidnykh stekol (Proc. Conf. on Structure, Properties, and Application of Phosphate, Fluoride, and Chalcogenide Glasses), Latv. Akad-Nauk, Riga, pp. 235–236.

Funtikov, V.A. (1991) *Eutectoidal Model for the Structure of Noncrystalline Solids*, Tezisy II Vsesoyuznoi konferentsii po fizike stekloobraznykh tverdykh tel (Proc. II All-Union Conf. on Physics of Vitreous Solids), Latv. Akad-Nauk, Riga, p. 39.

Funtikov, V.A. (1993) On the structure of chalcogenide glasses, *Glass Phys. Chem. (Transl.)*, **19**, 111–115.

Funtikov, V.A. (1994) Electron configurations of atoms as a factor affecting glass formation in elemental substances and their alloys, *Glass Phys. Chem. (Transl.)*, **20**, 492–496.

Funtikov, V.A. (1995) *Eutectoidal Model of Glassy State of the Substance*, Proceedings of XVII Int. Congress on Glass, Vol. 2, Beijing, China, pp. 256–261.

Funtikov, V.A. (1996a) *Topological Model of Formation of Structure of Glasses*, Proc. of Int. Symposium on Glass Problems, Vol. 2, Istanbul, Turkey, pp. 367–371.

Funtikov, V.A. (1996b) *Electrochemical Dissolution of Glasses*, Proc. of Int. Symposium on Glass Problems, Vol. 2, Istanbul, Turkey, pp. 105–108.

Funtikov, V.A. (1996c) *Use of Chemical Equivalent for Estimation of Chemical Stability of Glasses*, Proc. of Int. Symposium on Glass Problems, Vol. 2, Istanbul, Turkey, pp. 380–382.

Funtikov, V.A. (1996d) Chemical equivalent and its application as a parameter for estimating the structural features of chalcogenide glasses, *Glass Phys. Chem. (Transl.)*, **22**, 215–218.

Funtikov, V.A. (1996e) Influence of structure of chalcogenide glasses on their electrochemical dissolution, *Glass Phys. Chem. (Transl.)*, **22**, 207–209.

Funtikov, V.A. (1996f) *To Question About Physicochemical Analysis of Glassy Systems*, Proc. of Int. Symposium on Glass Problems, Vol. 2, Istanbul, Turkey, pp. 361–365.

Funtikov, V.A. (1996g) On the structure of glass and physicochemical analysis of vitreous systems, *Glass Phys. Chem.*, **22**, 210–214.

Funtikov, V.A. (1998a) A pseudobinary model of the semiconductor–metal transition in vitrifying melts, *Glass Phys. Chem. (Transl.)*, **24**, 427–431.

Funtikov, V.A. (1998b). Equivalent measurement of glasses. In *Proc. of the XVIII Int. Congress on Glass.* Section D8 of CD-ROM Proceedings, San Francisco, CA, USA.

Funtikov, V.A. (2000) *Eutectoidal Model of Glassy State of the Substance*, Bulletin of the V. Tarasov's Center of the Chemotronics of Glass, Vol. 1, Publishing House of Mendeleev's University of Chemical Technology of Russia, Moscow, pp. 24–26.

Funtikov, V.A. and Funtikova, A.V. (2000) *The Role of Metastable Phase Equilibriums in Formation of Glassy State of the Selenium*, Bulletin of the V. Tarasov's Center of Chemotronics of Glass, Vol. 1, Publishing House of Mendeleev's University of Chemical Technology of Russia, Moscow, pp. 75–78.

Funtikov, V.A. and Funtikova, A.V. (2001) *The Role of Phase Equilibriums in Formation of Glasses*, Proceedings of XIX International Congress on Glass, Vol. 2, Edinburgh, Scotland, pp. 902–903.

Glazov, V.M., Chizhevskaya, S.N. and Glagoleva, N.N. (1967) Zhidkie Poluprovodniki (Liquid Semiconductors), Nauka, Moscow.

Goroshchenko, Ya.G. (1978) Fiziko-khimicheskii analiz gomogennykh i geterogennykh sistem (Physicochemical Analysis of Homogeneous and Heterogeneous Systems), Naukova Dumka, Kiev.

Goryunova, N.A. and Kolomiets, B.T. (1958) Vitreous semiconductors, IV. On relationships involved in glass formation, *Zh. Tekh. Fiz.*, **28**, 1922–1932.

Guntherodt, G.J. and Beck, H. (Eds.) (1981) *Glassy Metals, I: Ionic Structure, Electronic Transport, and Crystallization*, Springer, Berlin.

Gutenev, M.S. (1993) On the use of physicochemical analysis with reference to vitreous studies, *Fiz. Khim. Stekla*, **19**, 375–383 ([*Glass Phys. Chem.* (*Transl.*), **19**, 186–189]).

Huheey, J.E. (1983) *Inorganic Chemistry*, Harper and Row, New York.

Karapet'yants, M.Kh. (1975) Khimicheskaya termodinamika (Chemical Thermodynamics), Khimiya, Moscow.

Kasparova, E.S., Gutenev, M.S. and Baidakov, L.A. (1984) Dielectric investigation of glasses in the $AsSe_{1.5}$–$AsTe_{1.5}$ system, *Fiz. Khim. Stekla*, **10**, 541–548.

Khimicheskaya entsiklopediya (Chemical Encyclopedia) (1992), Vol. 3. Bol'shaya Ross. Entsiklopediya, Moscow.

Mazurin, O.V. and Porai-Koshits, E.A. (1977) On the principles underlying the formulation of a general theory of the vitreous state, *Fiz. Khim. Stekla*, **3**, 408–412.

Mazurin, O.V., Strel'tsina, M.V. and Shvaiko-Shvaikovskaya, T.P. (1973, 1975, 1980) Svoistva Stekol i Stekloobrazuyushchikh Rasplavov, Spravochnik (A Handbook on Properties of Glasses and Glass-forming Melts), Nauka, Leningrad, 1973, Vol. 1; 1975, Vol. 2; 1980, Vol. 4, Part 1.

Mustyantsa, O.N., Velikanova, A.A. and Mel'nik, N.I. (1971) Electric conductivity of melts in the As–Te System, *Zh. Fiz. Khim.*, **45**, 1738–1739.

Nedoshovenko, E.G., Turkina, E.Yu., Tver'yanovich, Yu.S. and Borisova, Z.U. (1986) Glass formation and interaction of components in the GeS_2–Ga_2S_3–NaCl System, *Vestn. Leningr. Univ., Ser. 4, Fizika, Khimiya*, **2**, 86–91.

Nekrasov, B.V. (1973) Osnovy Obshchel Khimii (Fundamentals of General Chemistry), Vol. 1, Khimiya, Moscow.

Phillips, J.C. (1981) Topology of covalent non-crystalline solids II: medium—range order in chalcogenide alloys and A-Si(Ge), *J. Non-Cryst. Solids*, **43**, 37–77.

Popov, N.A. (1980) *Quasimolecular Defects in Chalcogenide Semiconductor Alloys*, Materialy Mezhdunarodnoi Konferentsii Amorfnye Poluprovodniki-80 (Proc. Int. Conf. on Amorphous Semiconductors–80), Kishinev, pp. 154–157.

Porai-Koshits, E.A. (1985) Structure of glasses: the struggle of ideas and prospects, *J. Non-Cryst. Solids*, **73**, 79–89.

Poryi-Coshitz, E.A. (1983) *Glassy State*, Science, Leningrad, pp. 5–10.

Samsonov, G.V. (1965) Influence of stable electron configurations on the properties of chemical elements and compounds, *Ukr. Khim. Zh.*, **31**, 1233–1247.

Shchukin, E.D., Pertsov, A.V. and Amelina, E.A. (1982) Kolloidnaya khimiya (Chemistry of Colloids), Mosk. Gos. Univ., Moscow.

Shechtman, D., Blech, J., Gratias, D. and Cahn, J.W. (1984) Metallic phase with long-range orientational order and no translational symmetry, *Phys. Rev. Lett.*, **53**, 1951–1953.

Shmuratov, E.A., Andreev, A.A., Prokhorenko, V.Ya., Sokolovskii, B.L. and Bal'makov, M.D. (1977) Thermoelectromotive force and electric conductivity of arsenic chalcogenide melts upon high-temperature transition to metallic conductivity, *Fiz. Tverd. Tela (Leningrad)*, **19**, 927–928.

Shukarev, S.A. (1974) Neorganicheskaya Khimiya (Inorganic Chemistry), Vol. 2, Vysshaya Shkola, Moscow.

Smits, A. (1921) *Die Theorie der Allotropie*, Barth, Leipzig.

The Collected Works of G. Willard Gibbs: In Two Volumes (1928). Longmans Green, New York; Translated under the title *Termodinamicheskie raboty*. (1950). Gos. Izd. Tekhniko-Teor. Lit., Moscow.

Turaev, E.Yu., Seregina, L.N. and Kesamanly, P.P. (1984) Short-range structure of vitreous $CdGeAs_2$ studied by Mossbauer and photoelectron spectroscopy, *Fiz. Khim. Stekla*, **10**, 721–724.

Tver'yanovich, Yu.S., Borisova, Z.U. and Funtikov, V.A. (1986) Metallization of chalcogenide melts and its relation to the glass formation, *Izv. Akad. Nauk SSSR, Neorg. Mater.*, **22**, 1546–1551.

Tver'yanovich, Yu.S., Borisova, Z.U. and Nedoshovenko, E.G. (1986) *Solid Vitreous Electrolytes* GeS_2*–*Ga_2*-*S_3*–MeCl (Me = Li, Na)*, Stekloobraznoe Sostoyanie (The Vitreous State), Nauka, Leningrad, pp. 389–390.

Tver'yanovich, Yu.S. and Gutenev, M.S. (1997) Magnetokhimiya stekloobraznykh poluprovodnikov (Magnetochemistry of Vitreous Semiconductors), St.-Peterb. Gos. Univ., St. Petersburg.

Tver'yanovich, Yu.S., Tver'yanovich, A.S. and Ushakov, V.M. (1997) Microregions of cooperative structural transformations at the semiconductor–metal transition in As_2Te_3 melt, *Fiz. Khim. Stekla*, **23**, 55–60 ([*Glass Phys. Chem.* (*Transl.*), **23**, 36–39]).

Vinogradova, G.Z. (1984) Stekloobrazovanie i Fazovye Ravnovesiya v Khal'kogenidnykh Sistemakh (Glass Formation and Phase Equilibria in Chalcogenide Systems), Nauka, Moscow.

Zakis, Yu.R. (1984) Defekty v Stekloobruvwm Sostoyunii Vesh-chestva (Defects in the Vitreous State), Zinatne, Riga.

Zalkin, V.M. (1987) Priroda evtekticheskikh splavov i effekt kontaktnogo plavleniya (The Nature of Eutectic Alloys and Effect of Contact Melting), Metallurgiya, Moscow.

CHAPTER 4

CONCEPT OF POLYMERIC POLYMORPHOUS-CRYSTALLOID STRUCTURE OF GLASS AND CHALCOGENIDE SYSTEMS: STRUCTURE AND RELAXATION OF LIQUID AND GLASS

V. S. Minaev

JSC "ELMA", RESEARCH INSTITUTE OF MATERIAL SCIENCE AND TECHNOLOGY, ZELENOGRAD, 124460 MOSCOW, RUSSIA

1. General Observations on Glass Formation

Existing theories, concepts, criteria, and models of glass formation can be divided into three main groups: kinetic, thermodynamic, and structural–chemical. Boundaries between these groups are indistinct and the elements of one group contain elements of the other (Uhlman, 1977). A satisfactory theory of glass formation cannot be created on the basis of a single aspect of glass formation alone (Rawson, 1967). As early as 1933, Tammann was one of the first who attempted to characterize the glass transition process and connected thermodynamic and kinetic descriptions with the first structural notions on glass (Tamman, 1933). In the triune concept of glass formation, the most important elements of each of these aspects are harmonically integrated, and it is now the most important task of contemporary glass science (Minaev, 1991).

The concept of 'glass formation' is wider than the concept of 'glass transition,' which describes the regularities of properties and the change of glass-forming melt at its transition to glass (Mazurin, 1978). Glass formation, in our opinion, includes processes of liquid formation, processes of relaxation including those at glass transition, and processes of relaxation in response to external conditions and impacts. One can never claim that glass is completely formed. Even after special careful annealing, glass continues to change, modifying its properties and structure in time over hours or centuries. Increased interest has recently arisen in the thinly researched study of glass aging (Nemilov, 2001).

Of the three main aspects of glass formation, structural–chemical aspect is likely to be the least developed at present, despite the fact that over the past several decades, diffractometric, spectroscopic, and other research techniques have aided the comprehensive study of glass and glass-forming liquids. The cause of developmental lag in the structural–chemical aspect of glass formation lies seemingly not in structural investigation techniques, but because of outdated concepts; in the absence of new ideas, the archaic paradigm of glass structure hinders the attempt to impartially interpret

ISBN 0-12-752187-9
ISSN 0080-8784

the experimental structural data obtained. Without new ideas regarding the objective principles of organization of glass and glass-forming liquid, it is impossible to fully understand glass formation processes, including processes that occur during relaxation of the glass-forming liquid at cooling and heating, during relaxation of glass taking place at heating at the temperature range of its softening in particular; it is impossible to understand the essence of processes taking place at crystallization of one-component glass in the form of different, depending on external conditions, crystalline polymorphous modifications (PMs).

2. Main Concept of Glass Structure

Consciously or unconsciously, investigators rely on and are guided by certain concepts and models of glass structure to analyze and interpret diffractometric and spectrometric data to study the structure and structural changes in glass and glass-forming liquid.

The most famous concepts of glass structure are the crystalline concept of Frankenheim (1835, 1851) and Lebedev (1921, 1924), in which for the first time an hypothesis was offered regarding glass formation and polymorphism; the concept of polymeric structure (Mendeleev, 1864; Sosman, 1927; Tarasov, 1959, 1979, and others); the concept of a continuous random network of Zachariasen (1932); the polymeric-crystallite concept of Porai-Koshits (1959), which with some success combines three previous concepts; and the concept of clusters of structural-independent polyforms of Goodman (1975), which develops the ideas of Frankelgeim–Lebedev.

To a certain extent, each of the above-mentioned concepts reflects an objective glass structure, but they are distinctive either in excessive generality and do not explain some experimental data (i.e., the concept of polymeric structure and, to a certain extent the Zachariasen's concept—even in its modern form: the chemically ordered continuous random network (COCRN)—Lucovsky and Hayes, 1979); or the concept contains theses that contradict experimental data (i.e., the crystalline concept; modern diffractometric investigations show that in well-synthesized glass, there are no crystallites that are defined by Porai-Koshits (1959) as smallest crystals consisting of a small number of elementary cells).

Nevertheless, beginning from the 1950s to 1960s, it is commonly accepted that experimental data conform more fully to the hypothesis of the random network than to the most modern modification of the crystal hypothesis (Porai-Koshits, 1977). The same opinion is held by Elliott (1984). In his analysis of Goodman (1975), Elliott (1984) writes that structure of amorphous solid matter, composed of different types of micro-crystals based on the 'micro-crystallite' model, is not in compliance with the commonly accepted opinion that amorphous covalent solid matter is better described in the random network terms. Nevertheless, Elliott thinks that there is a close relation between the glass-forming ability and the existence of several crystalline polyforms in the same material. He believes that a structural unit in amorphous material forming the continuous random network (the SiO_4 tetrahedron, in the SiO_2 case) can be in several conformations, similar with respect to energy, corresponding to the existence of crystalline polyforms (quartz, cristobalite, tridimite). It is obvious that Elliott's point of view is of significant value, although it does not pretend to be a completed conceptual resolution of the problem.

So, to quote Elliot, 'experimental data are in better compliance with...' the random network, and not the crystalline model, and '... amorphous matter is better described...' by the network as well; moreover, the glass-forming structural unit $SiO_{4/2}$ 'can be in several conformations.' Does this mean that the crystalline model is not completely acceptable to Elliot? Or that some of its features cannot be applied to glass formation? Is it worthwhile to consider once more the micro-crystallite concepts with their polyforms more attentively and try to find out a rational grain, with some respect similar to those Elliott addressed?

3. Relation Between Glass Formation and Polymorphism in One-Component Glass

The relation between polymorphism and its ability to form non-crystalline matter, vitreous in particular, has been discussed by many authors during the past 150 years. Most part of these works, connected with glass formation and polymorphism of substance, has been discussed by Minaev (1991).

The author of the first scientific hypothesis of glass structure, Frankenheim (1851) considered glass as consisting of very small crystals of different sizes. He connected unambiguously the tendency of substance to polymorphism with its tendency to glass formation, assuming the appearance of a mixture of modifications taking S, Se, and As_2O_3 as examples. Unfortunately, the Frankenheim's hypothesis, surpassed more than 80 years' appearance of similar conceptions in modern science (Lebedev, 1921, 1924), has been completely forgotten (Nemilov, 1995).

Lebedev (1921, 1924) was the first to conclude that modifications of properties of soda-lime glass in the critical region are closely connected with $\alpha-\beta$ transformation of quartz and that 'complexes of SiO_2 molecules can be formed in glass, located approximately in the same way as in tridimite or cristobalite crystals and having, consequently, the transition temperature close to that of these minerals' (117 °C in tridimite, 240 °C in cristobalite, and 575 °C in quartz).

Tudorovskaya (1938) experimentally determined that the refraction index of soda-lime glass with 23% Na_2O changes at temperatures increasing in three temperature ranges: 85–120, 145–165, and 185–210 °C corresponding to $\alpha-\beta$ and $\beta-\gamma$ transformations of tridimite and $\alpha-\beta$ transformation of cristobalite. These changes decreased with decreasing silica content in glass.

Deeg (1957) revealed jumps of the torsion vibration coefficient in fused quartz at 150, 250, and 570 °C that were connected with transformations of tridimite, cristobalite, and quartz.

Mackenzie (1960) determined that an increase of exposure time at the melted state of SiO_2 led to significant changes of micro-hardness vs. temperature curves. The sharp bend at the temperature of the $\alpha-\beta$ transition of quartz ($\approx$570 °C) still remained after a 1 h exposure at 1900 °C.

Brückner (1970, 1971) also observed anomalies of temperature dependences of properties of vitreous silica (volume, refractory index, dielectric constants, etc.) appeared near the points of transformation of crystalline modifications. He thought that there were regions in pre-ordered state in glass where SiO_4 tetrahedrons form cristobalite- and quartz-like structures.

Porai-Koshits (1942) studied leaden-silicate glasses with 'quartz' ($\sin \Theta/\lambda \approx 0.15$ Å^{-1}) and (after exposure at 1500–1600 °C) 'cristobalite' ($\sin \Theta/\lambda \approx 0.12$ Å^{-1}) diffraction maximums on the X-ray scattering curve at large angles. Alteration of glass composition upon heating (decrease of lead content) did not allow to the author to consider the experiment as faultless and report on revelation of phase transformations in glass. As Porai-Koshits (1992) writes, his latest data have confirmed this hypothesis, showing vitreous silica in two different modifications, as reported in 1942.

Research following Gerber, Himmel, Lorenz and Stachel (1988), referenced by Porai-Koshits (1992), has shown that in the process of densification of glassy SiO_2 ($P = 7$ GPa, $T = 700$ °C) and transformation of structure with cristobalite topology into structure with quartz topology taking place (density increase of ≈15%) through the X-ray diffraction method at large angles. The main intensity maximum of X-ray scattering of untreated cristobalite-like glass obtained from melt ($d = 2.2$ g cm^{-3}) is located at $s = 15.0$ nm^{-1}, shifts to greater values along with densification of samples reaching the value of 17.4 nm^{-1} for the sample with $d = 2.56$ g cm^{-3} that is a little less than that of crystalline quartz.

The described phase transformation in glass assumes the presence in the melting of two analogous phases which are identified as two different topological structures.

Golubkov (1992) has determined by the small-angle X-ray scattering method that a vitreous silica sample may exist in two different structural states: quartz- and cristobalite-like states. Of principal importance here is that 'the sample by the corresponding heat treatment can be converted from one state to another many times, mixed states being possible at insufficient heating' are present, as seen in works by Gerber et al. (1988).

Rawson (1967) has shown several examples that properties of glass depend on the crystalline form of its initial material. He analyzed in detail silica phases, on polymorphism of BeF_2, $ZnCl_2$, B_2O_3, GeO_2, P_2O_5, As_2O_3, Sb_2O_3, TeO_2, S, and Se. Rawson came close to concluding a cause-and-effect relation between polymorphism and glass formation. He approached this conclusion, but did not offer it. However, in 1975, Goodman made such a conclusion. He has proposed the concept of glass formation in which base clusters of 'structurally independent polyforms' are laid connected with each other by sterically strained interfaces.

The universal correlation between non-crystalline state and polymorphism for elements of the periodic table has been established by Wang and Merz (1977). They revealed that the greater the number of PMs possessed by an element, the easier the non-crystalline states (including vitreous) are formed on its base. The easiest formation in non-crystalline state is for sulfur, selenium, phosphorus, boron, and arsenic that can exist in 10, 5, 5, 5, and 4 PMs, respectively.

It is interesting to note that structural changes of the type of modification transformation in such glass-forming liquid as H_2O (Maeno, 1988) were supposed as early as (Bernal and Fowler, 1933).

As for chalcogenide glasses, Blinov (1985) and Sugai (1986, 1987) were likely the first who indicated the relation between polymorphism and glass formation.

During the past 150 years, many investigators have revealed one or more relations between polymorphism of individual chemical substances (ICSs) of various chemical classes and the demonstration of those substances' glass-forming abilities.

Sharma, Mammone and Nicol (1981) compared Raman spectra of vitreous SiO_2, quartz, and koecite. Bates (1972) has compared the Raman spectrum of cristobalite. Comparison of these spectra shows that the most intensive spectrum lines of PMs are located in the region of the most intensive broad spectrum band of v-SiO_2 (250–520 cm^{-1}), allowing to interpret the latter as the approximate superposition of vibration bands of quartz, cristobalite, and koecite. In its turn, it can be interpreted as participation of structural fragments of quartz, cristobalite, and koecite in the organization of structure of vitreous SiO_2.

Thus a paradox can be observed: while it was indisputably proved long ago that carefully produced glass does not contain even the smallest crystals—crystallites (Warren, 1937; Mozzi and Warren, 1969), Porai-Koshits, 1977), Golubkov, 1992); yet glass properties and structure are simultaneously connected to the properties and structure of two (or even three for SiO_2 (Deeg, 1957)) crystalline PMs of the given substance (Minaev, 2000a).

Unfortunately, the existing paradox has not been completely resolved, and this has delayed the development of the structural–chemical aspect of the glass formation process for many decades.

4. Short-Range Order Definition and Its Consequences

The existence of a 60-year-old paradox that the properties of crystalline polyforms are manifested in glass without crystals (Minaev, 2002a) is explained, in our opinion, by the strong disagreement shown by glass investigators (including the author) towards the concept that appeared approximately a half century ago. The short-range order (SRO) in vitreous and crystalline ICS—for example, in SiO_2, P_2O_5, Se, etc.—is just the same. This thesis was confirmed in widely known works by Mott and Davis (1979), Feltz (1983), and many others, as presented at the XIX International Congress on Glass in 2001.

In this seemingly obvious thesis, the deadlock situation for solution of the glass structure problem was created long ago.

What is SRO actually?

The modern definition of SRO goes back to the works of Ioffe and Regel (1960) and Glazov, Chizhevskaya and Glagoleva (1962). The SRO is a local arrangement of atoms around a certain atom taken as a reference point. SRO is characterized by the coordination number (CN) and chemical nature of atoms located in the first coordination sphere of the atom taken as a reference point and the geometry of their arrangement: inter-atom distance values and interbond angle values. Common acceptance of this definition is confirmed by Lucovsky and Hayes (1979) and Mott and Davis (1979) (without reference to chemical nature of atoms).

What can be said about crystalline SiO_2 from the point of view of such definition? It can and must be said that there are two SROs in SiO_2: one is around a silicon atom and the other is around an oxygen atom (Minaev, 1996). These SROs are different in types of atoms in the first coordination sphere (oxygen and silicon, correspondingly), by the CN (4 and 2, correspondingly) and, naturally, by different interbond angles (O–Si–O and Si–O–Si, correspondingly). SRO around a Si atom (SRO-I) is described by the structural unit $SiO_{4/2}$, SRO around an O atom (SRO-II) is described by the structural unit $OSi_{2/4}$. Characteristics of SRO-I in all PMs of SiO_2 are almost the same (Leko, 1993). One of the SRO-II characteristics—the interbond angle Si–O–Si—has its individual value for each PM: 143.7° in quartz, 146.8° in cristobalite, 137.3, 148.5, and 180° in koecite. Association of SROs-I (practically the same in all PMs) with SROs-II (different in different PMs) creates different intermediate-range orders (IROs) and different LROs that are characteristic for quartz, cristobalite, tridimite, etc. (Minaev, 1994, 1996). IRO reflects the structure of a fragment of crystal structure in which the structural units of SRO-I of one PM are joined between themselves by structural units of SRO-II of the same PM, just as in the crystal of the same PM, but where there is no LRO yet. Atoms in such fragments are arranged in compliance with regularities of their arrangement in one of the PMs excluding one regularity—translation symmetry, i.e., LRO. This structural fragment is very much like a crystal, but nevertheless it is not a crystal—it is a crystalloid (Minaev, 1996). The quartz crystalloid is a bearer of SRO-I and SRO-II of quartz, a bearer of the IRO of quartz. The cristobalite crystalloid is a bearer of short-range and intermediate-range orders of cristobalite. The same can be said about tridimite, kitite, koecite, etc. The detailed definition of crystalloids is provided in this chapter.

From the commonly accepted definition of SRO, it follows that a crystalline ICS can contain more than one SRO in each PM; each PM contains its own particular IRO (Minaev, 1989, 1992, 1996). These conclusions completely disavow the thesis that SRO in glass is the same as SRO in crystal. And when these conclusions are applied to applicable data (Lebedev, 1921, 1924; Deeg, 1957; Brückner, 1971, etc.) showing dependence of structure and properties of one-component glass-forming substance on structure and properties of different PMs of this substance, the obvious inference arises: a vitreous one-component substance (a glass-forming ICS) is constructed from structural fragments (crystalloids) of its different PMs without LRO and it is characterized by SROs and IROs inherent to these PMs. On the analogy of the rejected thesis stating that SRO in glass is the same as SRO in crystal, one can claim that SROs and IROs in glass are similar to crystalline PMs taking part in glass formation. These very orders, their plurality, and their peculiarities are manifested in glass structure by diffractometric and spectroscopic methods as well as at fixation of extremum points on thermal dependences of glass properties, which are characteristic for different PMs in the above-mentioned works of Lebedev, Deeg, Brückner, and others.

These findings regarding short- and intermediate-range orders in crystalloids of different PMs as structural constituents of vitreous substance relate not only to SiO_2, but also to other one-component glassformers, including GeO_2, As_2O_3, P_2O_5, $BeCl_2$ as well as to Se, GeS_2, $GeSe_2$, AsSe, and other chalcogenide glassformers as described in works of Minaev (1991, 1996, 1998a, 2000b).

What can be said about the structure of one-component glass-forming liquid?

Associated glass-forming liquid is a 'forerunner' of glass from the structural point of view. Taking into account the fictive or structural temperature T_f, it is a structural analog of glass, according to Tool's (1946), kinetic glass formation theory, which was further developed in the works of several authors analyzed in Mazurin's (1986) monograph. Glass science has accumulated extensive experimental data in favor of this conclusion. The data have been analyzed by Rawson (1967), Minaev (1991, 1996) and Minaev, Timoshenkov and Tchernykh (2002).

v-SiO_2, obtained by quenching of melt, contains structural fragments that show a sharp bend on temperature dependences of properties typical for one, two or three crystalline PMs of SiO_2–quartz, tridimite, and cristobalite.

In a melt of stable hexagonal polymorphous selenium at high temperature, metastable molecules Se_8 are formed which 'are typical for structure of monoclinic selenium and can disintegrate on fragments at cooling according to Feltz's data (1983).' Molecular spectroscopy data indicate that melted selenium cooled to room temperature is a co-polymer consisting of structural fragments of *cis*- and *trans*-configurations typical for monoclinic and hexagonal PMs (Lucovsky, 1979).

For such glass-forming substances as $GeSe_2$, $SiSe_2$, $BeCl_2$, Raman spectral bands of liquid and glass are correlated with main vibrational modes of crystalline PMs.

X-ray diagrams of such glass-forming liquid as H_2O (Pauling, 1970) conform to the X-ray diagrams calculated for the mixture of micro-crystals of PMs ice I, ice II, and ice III with the ratio of 50:33:17. The density of water (1 g cm^{-3}) is of intermediate value between the densities of ice I, ice II, and ice III—0.94, 1.18, and 1.15, correspondingly (Pauling, 1970). The density of glass-like H_2O is 1.1 g cm^{-3} (Maeno, 1988).

Liquid is apparently different from glass in the degree of polymerization of substance. In essence, the structure of liquid at temperature T_f, which is characteristic for glass, is similar to the structure of glass.

The correlation between the intensity ratio of Raman-spectra bands of high-temperature (328 cm^{-1}) and low-temperature (274 cm^{-1}) PMs in glass and glass-forming liquid $BeCl_2$ vs. temperature has been investigated in the work of Pavlotou and Papatheodorou (2000). It is interesting that this ratio increases in the temperature range of melting temperature T_m (415 °C)–520 °C from ≈ 3 to ≈ 4. It also means that from the moment of high-temperature PM (HTPM) melting until the moment of the Raman-spectrum registration, a significant amount of structural fragments (crystalloids) of low-temperature PM (LTPM) appear, but their amount decreases as the temperature of liquid increases further.

During the increase of exposure temperature of SiO_2 melt (from the melting temperature 1723 to 1900 °C), a decrease of the $\alpha-\beta$ quartz transformation effect is observed on the micro-hardness vs. temperature dependence in glasses obtained from those melts (Mackenzie, 1960). With the increase of melt temperature, the concentration of structural fragments (crystalloids of LTPM, quartz) decreases. Its occurrence is due to the increase of the HTPM concentration—cristobalite crystalloids (Minaev, 1996).

The studies of Raman-spectrum taken immediately after HTPM show $GeSe_2$ films melt at temperatures higher than 712 °C. It has been shown that the intensity of the 201 cm^{-1} band typical for HTPM decreases, while the intensity of the 201 cm^{-1} band typical for LTPM increases. This indicates the formation of structural fragments of LTPM along with structural fragments of HTPM in liquid (Wang, Nakamura, Matsuda, Inoue and Murase, 1996).

Based on these data, as well as other research indicating the polymorphous-crystalloid structures of glass obtained by fast cooling of liquid (Minaev, 1996, 1998b, 2001a, 2001b; Minaev et al., 2002), the following conclusions can be made.

The process of formation of glass-forming liquid at the melting of HTPM of an ICS is the process of generating structural fragments of this PM without an LRO (crystalloids), their partial transformation into crystalloids of other PMs, and subsequent establishment of the concentration ratio of crystalloids of different PMs corresponding to each temperature.

In glass-forming liquid, as in glass, short-range and intermediate-range orders are the same as in the crystalline PMs of initial ICSs that take part in glass formation.

Let us sum up the above. To resolve the paradox mentioned earlier that the properties of crystalline polyforms are manifested in crystal-free glass, the brilliant experiments of Porai-Koshits (1942), Brückner (1970), Gerber (1988) and Golubkov (1992) were insufficient, due to the need to revise the paradigm of the one-component glass and glass-forming liquid structure; i.e., it was necessary to replace the outdated concepts of single SRO in glass, and to reconsider the idea of single IRO in glass. Efforts to establish a new paradigm, the concept of polymeric polymorphous-crystalloid structure of glass-forming ICS, and its liquid were made in the late 1980s by Minaev (1989, 1991). The development of this concept has been taking place during subsequent years (Minaev, 1996, 1998a, 2000a,b, 2001a,b, 2002; Minaev et al., 2002). We hope that this concept will make a significant contribution to the structural development of the future general theory of glass formation process, which will contain kinetic and thermodynamic aspects as well.

5. Main Theses of the Concept of Polymeric Polymorphous-Crystalloid Structure of One-Component Glass and Glass-Forming Liquid (CPPCSGL)

Based upon a critical analysis of the existing concepts of glass structure (see Section 2), including studies related to glass-forming liquid and glass structure by Bernal and Fowler (1933), Hagg (1935), Porai-Koshits (1942), Winter (1943), Rawson (1967), Brückner (1970, 1971), Vukcevich (1972), Landa and Nikolaeva (1979), Landa, Landa and Tananaev (1984), Blinov (1985), Blinov, Balmakov and Pochentsova (1988), Gerber et al. (1988), Golubkov (1992) as well as on other data analyzed by Minaev (1996, 1998a,b, 2000b) the following main theses of CPPCSGL have been established:

1. The process of one-component vitreous substance formation (an element or an chemical compound) is the process of generation, mutual transformation and co-polymerization of structural fragments of various PMs of crystal substance without an LRO (crystalloids) in disordered polymeric polymorphous-crystalloid structure (network, tangle of chains, ribbons, etc.) of glass.

The term 'crystalloid' was initially introduced by Graham (Encyclopedic Dictionary of Physics, 1961) and it described particles of substance in the state of molecular fragmentation which were able to crystallize from solution. It is evident that in our case this definition must be modified to some extent, as has been done by Minaev (1996).

2. The crystalloid is a fragment of crystal structure consisting of a group of atoms connected by chemical bonds according to stereometric ordering rules inherent to one of the crystal PMs of the substance and it has no translational symmetry of the crystal. There is no LRO of any kind in the crystalloid, i.e., even the smallest LRO defined as two neighboring elementary cells of crystal structure which are able to intertranslate.

3. The notion 'crystalloid' is directly connected with notions 'SRO' and 'IRO', 'short-range ordering' and 'intermediate-range ordering', applied both to non-crystalline and crystalline substances.

The SRO is a topologically determined composition of atoms, including the atom taken as the origin point and surrounding atoms of the first coordination sphere. SRO is characterized by the CN, by types of constituent atoms, by distances between atoms and by interbond angles.

The IRO is a stereometrically determined composition (topology) of SROs in the boundaries of crystalloid characterized by parameters of all SROs (CNs, distances between atoms, interbond angles) and dihedral angles.

The IRO, at least along one of the crystallographic axes, has dimensions less than two periods of the crystal lattice. Otherwise, the crystalloid becomes a crystallite—a minimum fragment of crystalline substance.

The short-range ordering is a combination of various SROs (in crystalline and non-crystalline substances).

The intermediate-range ordering is a combination of various IROs in a non-crystalline substance. In a crystalline substance the term 'intermediate-range ordering' coincides with the term 'intermediate-range order'.

4. In every non-crystalline substance there are two or more SROs, two or more IROs (plurality of SROs and IROs), and there is no LRO. The number of IRO types is equal to the number of PMs taking part in the formation of the non-crystalline substance. In crystalline substance there may be one, two, or more SROs and only one type of IRO and LRO.

5. Stereometrically ordered crystalloids of different PMs join together (polymorphous polymerization) in accordance with rules of one of the PMs (except for rules of translational symmetry); but because of statistical alteration inherent to them, they form disordered polymeric polymorphous-crystalloid structure of vitreous substance in which order (on the micro-level) and disorder (on the macro-level) organically co-exist.

Polymorphous polymerization is the necessary and sufficient condition of glass formation for ICSs.

6. Glass structure is not absolutely continuous, and there are separate broken chemical bonds and other structural defects.

7. Glass-forming liquid of ICS is also constructed of crystalloids of different PMs whose degree of co-polymerization decreases at an increased temperature. When a specific temperature is reached, effects include the disintegration of some crystalloids, the disappearance of IROs typical for certain PMs, and the formation of separate structural fragments characterized only by an SRO (for example, polyhedrons $SiO_{4/2}$, $GeSe_{4/2}$, $AsS_{3/2}$, etc.) begins.

In accordance with the above notions of the CPPCSGL, glass and glass-forming liquid structure and their properties are determined by the concentration ratio of crystalloids of different PMs inherent to the given glass (or liquid), and they are dependent on initial substance state, conditions of formation, and treatment of the vitreous material as well as external conditions affecting the investigated substance.

In glass-forming liquid and glass, substantial changes of structure and properties take place depending on concrete realization of the process of inter-transformations of crystalloids of different PMs.

$$A_k + B_l + C_m + \cdots + Z_x \overset{(T,P,\mathrm{Ph},E,H)}{\leftrightarrows} A_p + B_q + C_r + \cdots + Z_y \tag{1}$$

where A, B, C, ..., Z are crystalloids of different PMs of a substance with concentrations $k, l, m, \ldots, p, q, r, \ldots, x, y$ which are changed depending on conditions (temperature T, pressure P, irradiation Ph, electrical field E, magnetic field H, etc.) in the range from 0 to 100% (Minaev, 1989, 1996).

Expression (1) reflects the physical–chemical essence of relaxation processes in crystalline, vitreous, and liquid states of substance.

Thus, the CPPCSGL concept includes the following ideas from earlier concepts:

- polymeric structure (Mendeleev, 1864; Sosman, 1927; Tarasov, 1959);
- absence of an LRO in continuous glass-forming network (Zachariasen, 1932);
- chemical ordering of the network (Lucovsky and Hayes, 1979; Wright, Etherington, Desa, Sinclair, Connell and Mikkelsen Jr., 1982);
- influence of crystalline PMs on glass structure and properties (Frankenheim, 1851; Lebedev, 1921, 1924; Porai-Koshits, 1942, 1992; Goodman, 1975).

The CPPCSGL concept rejects the following ideas:

- the continuous random network consisting of randomly located separate atomic polyhedrons;
- presence of a single SRO;
- presence of a single IRO (having no strict definition);
- presence of micro-crystals (crystallites).

The CPPCSGL concept introduces the following new ideas:

- the crystalloid as a bearer of a strictly defined IRO of a concrete crystalline PM;
- presence in glass and glass-forming liquid of not less than two SROs and not less than two IROs pertaining to different crystalline PMs;
- polymorphous co-polymerization—polymerization of crystalloids of different PMs resulting in the formation of a continuous network that is disordered (random) on the macro-level and ordered on the micro-level (levels of short-range and intermediate-range orders);
- inter-transformation of crystalloids of different PMs and alteration of their concentration ratio in glass-forming liquid and glass under the influence of external impacts causing alteration of their structure and properties even up to their crystallization as this or that PM;
- formation of glass as a tangle of chain-like and ribbon-like or layer-like agglomerates consisting of fragments of different PMs of substance.

CPPSCGL represents the synthesis of existing concepts and new ideas into a single non-contradictory concept of glass structure. A new paradigm of glass structure has been created.

6. Influence of Polymorphous-Crystalloid Structure on Properties and Relaxation Processes in One-Component Chalcogenide Glass and Glass-Forming Liquid

The aim of this section is to consider structure and some properties of one-component chalcogenide glasses ($GeSe_2$, AsSe, Se, and others) as well as processes taking place in them under external impacts under the CPPCSGL concept.

6.1. Relaxation Processes in One-Component Condensed Substance—General Considerations

The nature of the vitreous state cannot be understood without the analysis of crystal and liquid states, their structure, the phase transformations that occur between these three states, as well as transformations inside them that take place under the effects of external impacts such as temperature, pressure, electromagnetic radiation, etc. Such transformations, including phase transformations as well as structural transformations occurring in liquid and vitreous states, represent relaxation[1)] of a condensed substance; i.e., the process of establishment of thermodynamic equilibrium (complete or partial) in a physical system consisting of a large number of particles (Prokhorov, 1982).

Phenomenological aspects of relaxation processes are considered in this section, particularly, in ICSs based on chalcogenides of IV and V main subgroups of the periodic table, as well as in chalcogene Se. Special attention is given to the well-documented germanium diselenide, $GeSe_2$. The processes of structural relaxation of these substances are much better known in their crystalline and vitreous states compared to their liquid states, of which information is quite limited.

The largest amount of experimental material on chalcogenides is collected in research related to the effects of temperature. In this section, substance relaxation will be considered mainly as establishment of equilibrium in substance as a result of temperature influence, i.e., as a result of 'placement' of substance from one temperature condition to the other. To understand the nature of glass formation, the processes connected with substance relaxation by placing the substance in higher temperature conditions (heating) are no less important than processes of cooling, even though a lion's share of experimental materials is devoted to research in cooling; note, for example, numerous publications on such problems as 'relaxation of glass-forming liquid upon cooling.'

Considering the relaxation processes, it must be kept in mind that a substance does not frequently achieve the equilibrium state because of an insufficient duration of conditions in which the system is kept. In the case where conditions continuously change, the potential equilibrium state to which the system aims at also changes.

Any alteration to the state of a system leads to the alteration of its properties. In accordance with the main formula of physical–chemical analysis, the formula of Kurnakov-Tananaev (Tananaev, 1972), substance properties are functionally dependent on composition, structure, and dispersion (the ratio of substance surface to its volume). At constant chemical composition and in the absence of significant dispersion, alteration of substance properties is directly related to the alteration of its structure as it takes place in glass and glass-forming liquid.

Knowledge of the structure of a substance is the most important condition for understanding its relaxation processes. As for ICSs, and chalcogenides in particular, there is a lack of data on structure of even the most simple crystal state, such as crystallographic parameters of low-temperature PMs for $GeSe_2$ and AsSe. There is no confidence that all PM of ICSs and chalcogenides, in particular, have been discovered.

[1] In this work we do not consider vaporous state of substance, analysis of transformation of which in condensed non-crystalline state, in particular in ultra-dispersed and vitreous, is very important in obtaining non-crystalline films of the chalcogenides.

The situation is more complicated with structure of vitreous substances, and glass-forming liquids in particular.

The founder of the kinetic theory of glass formation, Tool introduced the notion of 'the fictitious temperature' T_f, to describe the structure of liquid (melt). T_f is frequently called 'the structural temperature,' and according to Tool (1946), T_f is the temperature of melt at which the structure coincides with the structure fixed in this glass. But although structures of glass and liquid are similar at a certain temperature in accordance with the kinetic theory of glass formation, it says nothing about their peculiar concrete stereometry of atom arrangement in glass or liquid.

6.2. Germanium Diselenide $GeSe_2$

6.2.1. Crystalline PMs and Phase Transformations

Germanium diselenide $GeSe_2$ exists at normal pressure in two PMs: high-temperature (HTPM, β-modification), and low-temperature (LTPM, α-modification) (Dittmar and Schafer, 1975, 1976a,b; Popovic, Raptis, Astassakis and Jakis, 1998). The HTPM has two-dimensional layered structure in which $GeSe_{4/2}$ tetrahedrons, connected in vertexes (corner-shared tetrahedrons—CST), form chains connected by bridges from pairs of tetrahedrons, connected in edges (edge-shared tetrahedrons (ESTs)). In HTPM there are equal numbers of CSTs and ESTs. The HTPM has an yellow (Feltz, 1983) or orange color (Azoulay, Thibergen and Brenac, 1975) and has a melt temperature of 740 °C (Feltz, 1983).

The LTPM is constructed of tetrahedron chains arranged along directions [001] and [101] and connected by only CST in the three-dimensional network (Inoue, 1991). The LTPM can be obtained by photo-irradiation of thin vitreous samples by the He–Ne laser (Sugai, 1985), and by annealing at 325 °C of amorphous films deposited by vacuum evaporation (Inoue, 1991). A comprehensive X-ray determination of the LTPM structure is absent.

When heating glasses of Ge–Se system (with 15–30% Ge content) up to temperatures of 280–300 °C, Azoulay et al. (1975) discovered in all cases inclusions of the same phase consisting of small crystals of red-brown color with the approximate composition of $Ge_{30}Se_{70}$. At further heating, the orange phase $GeSe_2$ appeared, which co-existed with the red-brown phase. At heating to 400 °C and higher, the latter disappeared while the orange phase $GeSe_2$ remained.

In accordance with the eutectic type of the phase diagram Ge–Se, in the range Se–$GeSe_2$, at partial crystallization in the range of 15–30 at.% Ge only $GeSe_2$ phase can be formed. Therefore, Azoulay appears to be the first who discovered LTPM $GeSe_2$ and observed the phase transition LTPM → HTPM which can be interpreted as the process of relaxation of crystalline $GeSe_2$ at temperature increase. The confirmation of the fact that Azoulay dealt with the phase transition in $GeSe_2$ is the Inoue's experiment (Inoue, 1991), who at thermal annealing of amorphous film obtained the LTPM at 325 °C and the HTPM at 425 °C. The phase transition in the opposite direction HTPM → LTPM is registered (Popovic et al., 1998) when pressure is increased up to 7 GPa. The transition is realized via intermediate, completely disordered at 6.2 GPa, phase.

For $GeSe_2$, two PMs obtained at high pressures are also known (Shimada and Dachile, 1977).

As it is seen from the presented data, the information on the crystal state of $GeSe_2$ is far from being comprehensive. But the fact that stable HTPM and LTPM exists at normal pressure in certain temperature intervals itself says that $GeSe_2$ has the enantiotrope phase transition HTPM ↔ LTPM, which is characterized by a certain temperature of polymorphous transformation T_{tr}. Temperature of the enantiotrope transition T_{tr} is unknown. It is clear that only this transition is in the temperature range of 300–400 °C according to Azoulay et al. (1975), or 325–425 °C according to Inoue (1991). We did not come across any record of the HTPM → LTPM transition occurring at normal pressure. It appears that this transition is difficult to obtain due to the demand for sufficiently high temperatures. It is possible that it does take place, although slowly, at temperatures slightly less than the real T_{tr}. The transition HTPM → LTPM can be activated by high pressure (Popovic, 1998), which can be interpreted as a process of relaxation of crystalline $GeSe_2$ as a result of pressure increase, or as a result of presence of structural fragments (crystalloids) LTPM taking place in amorphous $GeSe_2$ film which consists, in accordance with the CPPCSGL as well as with data of Sugai (1986) and Inoue (1991), from fragments of structure of both PMs. Judgment may have to be reserved regarding a pure enantiotropic transition because, by definition, the enantiotropic transition must take place at normal pressure and in crystalline substance, but not in non-crystalline substance. Nevertheless, the presented facts give evidence that LTPM is significantly more stable at temperatures below T_{tr} than HTPM that allows to state that the discussed phase polymorphous transformation is an enantiotropic one. The mutual transformation HTPM ↔ LTPM in glass or liquid is a structural transformation, not a phase transformation, because it takes place in one phase without the formation of a phase interface. At the same time crystallization of glass or overcooled liquid as the result of the crystalloid concentration increase of one of the PMs to the critical concentration is the phase transformation 'glass → crystal' (or 'liquid → crystal'). Some results of the above discussion are shown in Figure 1.

6.2.2. *Structure of Vitreous $GeSe_2$*

Several structural models of vitreous $GeSe_2$ are known. The model of a continuous random network of Poltavtsev and Pozdnyakova (1973) is not distinctive in practical terms from Zachariasen's model (1932)—all $GeSe_{4/2}$ tetrahedrons are connected in vertexes through Se atoms.

Bridenbaugh, Espinosa, Griffits, Phillips and Remeika (1979) proposed a layered structure for v-$GeSe_2$ constructed from large ribbon-like structural units (clusters), 'outrigger-rafts,' as called by the authors. These structural units are obtained by extraction from the layer of HTPM. Along edges of ribbons Se atoms are joined in pairs by Se–Se bonds.

In the beginning of the 1980s, research by Lucovsky, Wong and Pollard (1983) and Nemanich, Galeener, Mikkelsen, Connell, Etherington, Wright and Sinclair (1983) was published in which the authors proposed models of the structure containing both

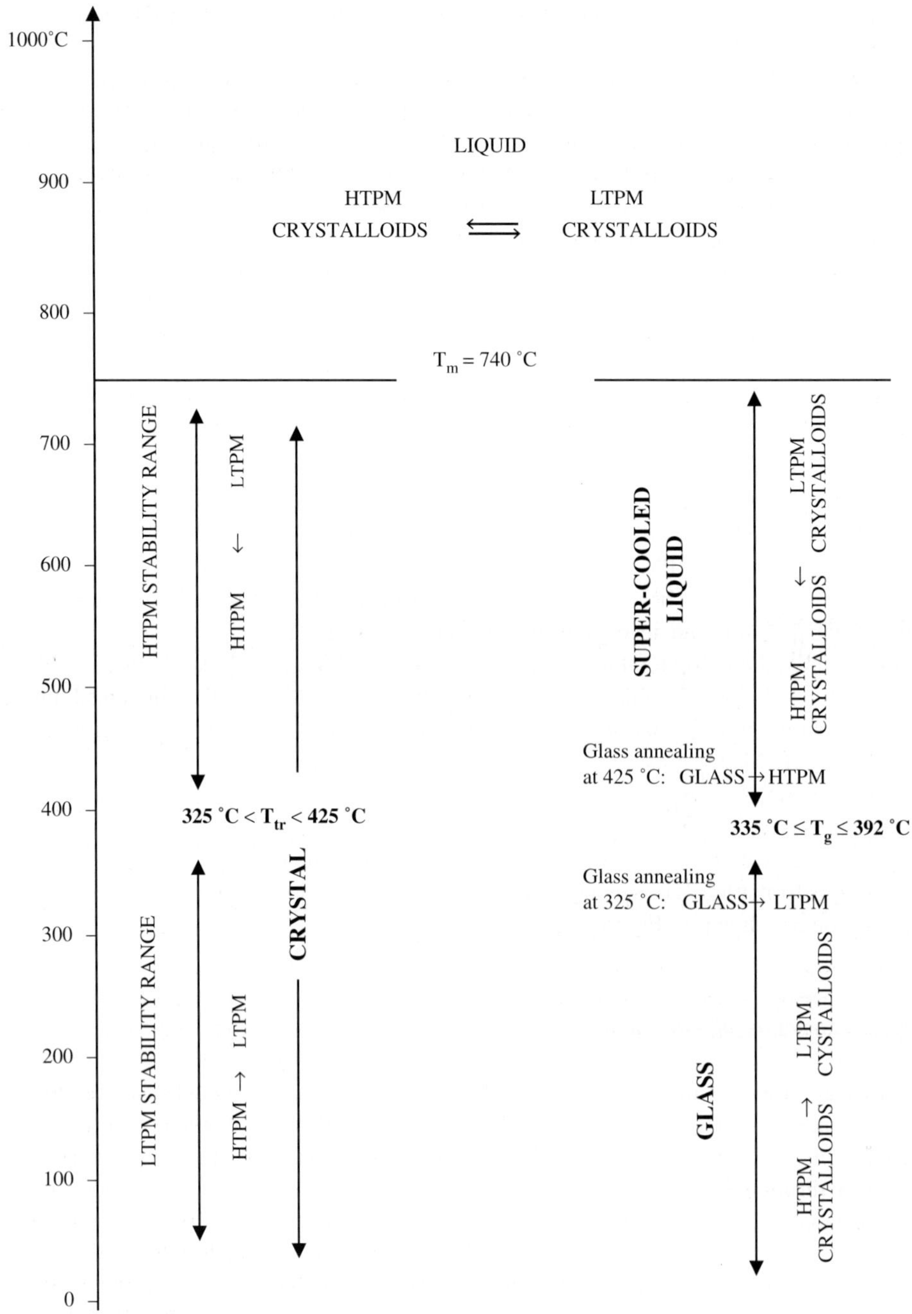

FIG. 1. Varieties of condensed state of $GeSe_2$ (see description in the text).

tetrahedrons, connected along edges, and tetrahedrons, connected in vertexes, which are the model of a COCRN.

Sugai (1986, 1987) proposed a new stochastic random network model of germanium and silicon chalcogenides. The model is based on the author's experiments that discovered the existence of two different micro-crystalline states (phases)—HTPM and LTPM obtained by photo-irradiation of vitreous $GeSe_2$ samples, and also obtained by vacuum deposition of amorphous films, by the beam of He–Ne laser (6328 Å) of various intensity—higher and lower than the threshold (≈ 0.7 kW cm^{-2}). Sugai has constructed his model from a stochastically joint three-dimensional network of two types of 'molecules': tetrahedrons $GeSe_{4/2}$ and ESTs $Se_{2/2}GeSe_2GeSe_{2/2}$ leaving aside the problem of short-range and intermediate-range orderings. The stochastic random network model is characterized by one parameter *P* representing the probability of the ratio of chemical bonds of ESTs and chemical bonds of corner-shared tetrahedrons. *P* can be changed: it decreases at irradiation with intensity lower than the threshold and increases at irradiation with intensity higher than the threshold (≈ 0.7 kW cm^{-2}). The stochastic model gives, in the author's opinion, equal possibilities for photo-induced crystallization in two different crystalline phases, while the 'outrigger-raft' model provides only one possibility.

Change of the ratio of chemical bonds of ESTs and corner-shared tetrahedrons (*P*) that was concluded based on the comparison of intensities of corresponding modes of Raman scattering A_I^c (219 cm^{-1}) and A_I (202 cm^{-1}) was confirmed by other investigators as well (Inoue, Matsuda and Murase, 1991; Uemura, Sagara, Muno and Satow, 1978; Wang et al., 1996; Wang, Matsuda, Inoue and Murase, 1998).

In many of the above works the problem of short-range and intermediate-range orderings has been mentioned directly or indirectly, but only the question of similarity of the intermediate range order in glass or liquid to the HTPM structure has been discussed. For example, in the work of Wang et al. (1998), it has been stated that at the lesser cooling rate of the liquid the medium-range structure is topologically more similar to a crystal structure, specifically a layered crystal, i.e., HTPM structure. The short-range ordering, if mentioned, has been considered in passing, without any detailed analysis.

According to the CPPCSGL concept (Minaev, 1996, 1998a), the feature of the short-range ordering in vitreous $GeSe_2$ is primarily the presence of two types of SRO, with two different atoms taken as a reference point. In the first case, it is a germanium atom surrounded by four selenium atoms $GeSe_{4/2}$, and in the second case, it is a selenium atom surrounded by two germanium atoms $SeGe_{2/4}$. As we can see, both SROs are in agreement with the principle of the chemically ordered network (Lucovsky et al., 1983; Nemanich et al., 1983). It must be noted that each of these SROs has its own variations: differences in bond lengths Ge–Se and interbond angles Ge–Se–Ge and Se–Ge–Se, in corner-shared and edge-shared $GeSe_{4/2}$.

The intermediate-range ordering in v-$GeSe_2$ is characterized mainly by two alternating IROs: IRO inherent to HTPM (the two-dimensional layered structure) and IRO adopted from LTPM (the three-dimensional network structure). Each of these IROs includes variations of the SRO corresponding to one or the other PM.

On the basis of the above-mentioned works and the concept of polymeric polymorphous-crystalloid structure, the conclusion has been made (Minaev, 1998a,c, 2000c,d) that the structure of vitreous $GeSe_2$ (and glass-forming liquid) is formed by co-polymerized (de-polymerized) crystalloids (structural fragments without an LRO) of layered (two-dimensional) HTPM and three-dimensional LTPM, which are joined in glass in a single three-dimensional network characterized by alternation of mainly two intermediate orders inherent to HTPM and LTPM.

6.2.3. *Structure of Amorphous Films and Liquid $GeSe_2$*

According to Tool's (1946) kinetic theory of glass formation, the structure of glass-forming liquid is a 'precursor,' and taking into account the fictitious (or structural) temperature T_f, it is the analog of glass structure.

Data on the viscosity of Ge–Se liquid (Glazov and Situlina, 1969) show that $GeSe_2$ in the liquid state starting from melt temperature ≈ 740 to ≈ 1000 °C behaves as a chemical compound, i.e., its short-range ordering is similar to that of a crystal substance. Data of Uemura et al. (1978) on neutron diffractometry of melt also evidence the conservation of strong covalent bonds Ge–Se in liquid $GeSe_2$. According to Sugai's (1987) data, Raman spectra of amorphous $GeSe_2$ films obtained by vacuum deposition show the same structure that glass obtained by tempering from melt.

Wang et al. (1996) assumed that similarity of liquid and glass structures is confirmed by similarity of Raman scattering spectra of vitreous samples at 420 °C and melt at 730 °C.

Thus, there is no principal difference between the structures of glass, deposited films, and glass-forming liquid. So, the conclusion can be made about the similarity of structures of glass, obtained by tempering from melt, films, obtained by vacuum deposition, and liquid. Differences of their structures are determined by the extent of co-polymerization of crystalloids, which is the most developed in solid glass, less so in a deposited film, and even lesser in liquid. As for structures of solid glass obtained in different conditions, or for liquids at different temperatures, in accordance with the CPPCSGL concept, they differ in the ratio of crystalloids of different PMs, and seemingly in their dimensions—the higher the temperature of tempering, the smaller the dimension Minaev, 2001a, 2002; Minaev and Timoshenkov, 2002; Minaev et al., 2002.

6.2.4. *Relaxation Processes in Vitreous $GeSe_2$ at Heating. Physical–Chemical Essence of Glass-Transition Temperature T_g*

In the temperature range of 250–350 °C, which directly precedes the $GeSe_2$ glass-transition temperature T_g (according to some authors, it lies in the range of 335–392 °C (Feltz, 1983; Afify, 1993; Wang et al., 1996), a smooth exothermal rise is observed on the thermogram of the differential scanning calorimetry (DSC) obtained at 10 K min^{-1} heating rate (Afify, 1993). Raman-spectra at temperatures 250, 300, and 350 °C show some increase of the relative intensity of the A_1 band inherent to the LTPM (201 cm^{-1}) and decrease of the intensity of the HTPM band of $GeSe_2$ (216 cm^{-1}) (Wang et al., 1996).

It is known that for individual substances the transition from HTPM to LTPM is followed by the exothermal effect (Lihtman, 1965). Therefore, data of DSC and Raman-spectra evidence that heating $GeSe_2$ glass, in the temperature interval that precedes T_g, a partial transformation of HTPM crystalloids in LTPM crystalloids takes place with the concentration increase of the latter that is confirmed by annealing of non-crystalline $GeSe_2$ films at 325 °C leading to LTPM crystallization (Inoue et al., 1991). DSC and Raman-spectra in described experiments show transformation of a part containing glass HTPM crystalloids in LTPM crystalloids in the range '250 °C–T_g'—the process that can be identified as the latent period preceding LTPM crystallization.

Above the T_g temperature, the endothermal effect on DSC thermogram (Afify, 1993) is observed. The Raman-spectrum (Wang et al., 1996) shows a decrease of the A_1 (201 cm^{-1}) band intensity that characterizes vibrations of tetrahedrons, connected in vertexes and forming LTPM, and increase of A_1^c (216 cm^{-1}) band intensity that characterizes tetrahedrons, connected along edges (HTPM). The ratio of band intensities $[A_1^c$(HTPM)]:$[A_1$(LTPM)] changes from 1:3 (below T_g) to 1:1 in the range T_g–T_c (T_c is the crystallization temperature).

Softening of $GeSe_2$ glass upon heating is nothing but the process of structural transformation of LTPM crystalloids in HTPM crystalloids. T_g is the most active stage of this transformation, the most active stage of disintegration of LTPM crystalloids, the most active stage of disruption of chemical bonds Ge–Se, which surpasses at the point T_g the process of unification of these bonds in crystalloids of another polymorphous modification, HTPM.

The ratio of Raman-spectra band intensities $I[A_1^c$(HTPM)]:$I[A_1$(LTPM)], equal to 1:1, characterizes overcooled liquid just before the beginning of the HTPM crystallization that coincides with the ratio of corner-shared and ESTs, also equal to 1:1, for crystallized HTPM. This fact allows quantity evaluation of corner-shared and ESTs from the ratio of Raman-spectra band intensities, and in this way the share of structural fragments (crystalloids) of HTPM and LTPM in glass and glass-forming liquid. In the interval 'T_g–T_c' the Raman-spectra, therefore, show disintegration of LTPM and growth of HTPM crystalloids concentration.

The polymorphous transition of LTPM in HTPM is followed, as it is known, by the endothermal effect (Lihtman, 1985) that is just observed in DSC thermograms at temperatures above T_g. The endothermal effect related with the softening temperature of glass, formed from ICS with the enantiotropic transformation in crystal state, represents the energy (heat) absorbed by glass at the structural transformation of LTPM crystalloids in HTPM crystalloids that is a part of the polymorphous transformation heat at the enantiotropic transition LTPM → HTPM in the crystal substance. If the latter is known, it is possible to evaluate the HTPM and LTPM crystalloids concentration ratio in glass, having measured the energy absorbed at the glass softening. It would be useful to compare such evaluation with the evaluation of the HTPM:LTPM concentration ratio based on intensity ratios of Raman-spectra. Thus, both DSC and Raman-spectra evidence that in the temperature range of T_g observable instability of LTPM structure fragments is observed upon annealing of non-crystalline $GeSe_2$ films at 425 °C (some lower than T_c in DSC thermograms) leading to HTPM crystallization (Inoue et al., 1991).

If during crystallization of non-crystalline $GeSe_2$ films at 325 °C the LTPM is formed, and at 425 °C the HTPM is formed, it means that the equilibrium temperature between these crystal modifications is located in the interval of 325–425 °C, i.e., the temperature of their enantiotropic transformation T_{tr}.

Therefore, upon heating glass in the range that directly precedes T_g, the transformation of HTPM structure fragments (crystalloids) in LTPM crystalloids takes place. Above T_g, in the T_g–T_c interval, the transformation process goes in the reverse direction, and significantly more intensively, due to higher temperature. T_g expresses itself as the temperature point where the inversion of inter-transformation of structural fragments of HTPM and LTPM takes place, as the temperature point separating stability regions of LTPM and HTPM crystalloids of $GeSe_2$, regions in which latent processes in glass take place preceding the LTPM crystallization (lower T_g) and the HTPM crystallization (above T_g), and processes of glass softening, its de-polymerization and transformation in overcooled liquid followed by an increase of the HTPM concentration and their consequent crystallization. T_g is genetically related with the temperature of enantiotropic transformation T_{tr} in crystal substance; T_g in glass is the analog of T_{tr} in crystal substance.

If glass is heated with the rate at which the HTPM:LTPM crystalloids ratio does not reach a critical level for the beginning of crystallization during the time of passing the T_g–T_m interval (T_m is the melting temperature), the substance will pass the temperature interval of overcooled liquid existence without crystallization and come to liquid state at T_m. At lesser heating rates, the substance crystallizes and then this crystalline HTPM phase melts.

6.2.5. *Relaxation of Glass-Forming Liquid*

Liquid state at a certain temperature is characterized by a certain concentration ratio of crystalloids of different PMs. The inverse crystalloid transformation reaction takes place at the temperature alterations: HTPM $\leftrightarrow$ LTPM.

During cooling of liquid below the melting temperature there are two main processes that take place. The first is the process of co-polymerization of crystalloids of different PMs which lead to viscosity increase and obstruction of melt crystallization.

The second process is conditioned by the transition in the interval T_m–T_{tr} from the HTPM stability range (Fig. 1) and the LTPM instability range. It is natural to expect that in these conditions the crystalloid inter-transformation will be shifted in the direction of increasing HTPM fragment concentration, and the LTPM fragment disintegration becomes apparent as the process of de-polymerization of overcooled liquid coincides with the main process of polymerization.

At high cooling rates the first process (co-polymerization) prevails, super-cooled liquid passes the LTPM instability range, having conserved, along with an increase of the HTPM crystalloids concentration, the LTPM crystalloids concentration which is sufficient for glass formation. Then the substance passes through the glass transition temperature T_g and is cooled being in the glass state.

The glass transition temperature T_g of ICS with enantiotropic transformation is the temperature of completion of the active stage of co-polymerization of HTPM and LTPM crystalloids, the temperature of the inversion of the process of transformation of LTPM crystalloids in HTPM crystalloids.

If cooling rates are less than the critical one, all LTPM crystalloids have enough time to transform to HTPM crystalloids before reaching the temperature range ($T < T_{tr}$) where LTPM is stable and HTPM is unstable. In this case overcooled liquid crystallizes as HTPM.

Fast cooling and long aging of a cooled substance at temperature directly under T_g in the high-temperature part of the LTPM stability range can lead to glass crystallization as LTPM, as it was shown for amorphous $GeSe_2$ films annealed at 325 °C (Inoue, 1991).

At some decrease of the cooling rate (still remaining higher than the critical cooling rate (CCR) causing crystallization) the HTPM crystalloid concentration in overcooled liquid and also in glass obtained from it increases and the LTPM crystalloid concentration decreases.

Wang's (1998) experiment illustrates this conclusion that compares threshold crystallization temperatures of $GeSe_2$ glass samples obtained by cooling in water with ice and by cooling in air (lesser cooling rate). The threshold crystallization temperature at irradiation of samples by the laser on argon ions (488 nm) was 430 °C in the former case and 130 °C in the latter case. Wang states that in the sample cooled at a lesser cooling rate its medium-range structure (according to Raman-spectroscopy data) is more similar to crystalline nuclei, or easier to transform into crystalline nuclei. At that, the author meant just the high-temperature modification (HTPM) the concentration of which, according to Minaev's (2001a) interpretation, becomes significantly higher at slow cooling.

Thus, depending on the change of the temperature factor, relaxation of $GeSe_2$ glass-forming liquid at cooling can result in the formation of glass, HTPM, or LTPM.

6.2.6. *Relaxation Processes in Vitreous $GeSe_2$ Caused by Photo-Irradiation*

Let us now consider the relaxation processes taking place in the result of the other type of external influence—photo-irradiation, in particular He–Ne laser irradiation ($\lambda = 6328$ Å).

In fresh vacuum-deposited non-crystalline $GeSe_2$ films, at He–Ne laser irradiation a shift of optical absorption edge is observed in the direction of shorter (the yellow region) wave lengths (photo-enlightenment) at 300 K, and in the direction of longer (the red region) wave lengths (photo-darkening) at 77 K (Kolobov, Kolomiets, Lubin, Sebastian and Tagirjanov, 1982).

In Sugai's (1986) experiment, the luminous flux of the He–Ne laser (<0.7 kW cm^{-2}) changed the intensities ratio $I(A_1^c){:}I(A_1)$ bands (217 and 201 cm^{-1}, characterizing edge-shared (A_1^c) and corner-shared (A_1) tetrahedrons) from 0.43 to 0.33 (the shift in the direction of LPTM concentration increase). At the flux more than 0.7 kW cm^{-2} the intensities ratio $I(A_1^c){:}I(A_1)$ changed from 0.43 to 1.0. (the shift in the direction of HTPM, to the yellow region of the spectrum). In this experiment it was possible either to crystallize LTPM of red color (in the first case) or to crystallize HTPM of yellow color (in the second case).

The visual shift in color (yellow-red) of the adsorption edge in the process of photo-irradiation (with corresponding change of the band intensities ratio responsible for LTPM and HTPM), demonstrates the physical–chemical nature of relaxation processes—mutual transformation of structural fragments of different PMs preceding photo-induced crystallization.

Photo (thermo)-darkening and photo (thermo)-enlightenment are photo (thermo)-indications of the latent pre-crystallization processes—the processes of transformation of HTPM crystalloids in LTPM crystalloids at darkening and LTPM crystalloids in

HTPM crystalloids at enlightenment take place during the latent periods which directly precede crystallization of LTPM and HTPM.

6.2.7. *High Pressure and Relaxation Processes in $GeSe_2$*

The relaxation process of crystalline HTPM of $GeSe_2$ under the influence of high pressure with phase identification by Raman-spectroscopy occurs in the following (Popovic et al., 1998). Single crystal HTPM at pressure increased to 6.2 GPa loses its crystal structure completely and transforms to a disordered non-crystalline substance, the color of the sample being changed from yellow to red. At further pressure increase to 7 GPa the transition from the disordered phase to the crystalline low-temperature polymorphous modification of $GeSe_2$ is observed. Further pressure increase (8–12.5GPa) leads to transformation of LTPM in the disordered phase that becomes dark and non-transparent. If pressure was not higher than 10 GPa, the process turned out to be reversible: pressure removal caused $GeSe_2$ relaxation leading to the recovery of HTPM. Correspondingly, the color reversal takes place back to yellow color.

The character of Raman-spectra change, given in this work, as well as change of the sample color both at increase and removal of pressure allows on the basis of CPPCSGL to make the conclusion that relaxation of $GeSe_2$ at pressure increase presents gradual transformation of HTPM in LTPM with the formation of an intermediate phase consisting of crystalloids of both PMs in which concentration ratio changes with pressure change: at its increase to 7 GPa LTPM crystalloids concentration increases and HTPM crystalloids concentration decreases. The reverse process takes place at pressure decrease (from 7 GPa).

Pressure increase above 8 GPa leads to LTPM destruction and the formation of crystalloids of, in our opinion, some new polymorphous modification ('dark and non-transparent'). As a result of that, the intermediate disordered phase is formed again. At further pressure increase higher than 12.5 GPa, the crystalloids concentration of the new PM can finally reach 100% and then a new crystalline phase of high pressure will appear, possibly one of that mentioned in the beginning of Section 2. We suggest this prognosis based on the main theses of CPPCSGL.

6.3. Chalcogenides GeS_2, $SiSe_2$, SiS_2. Relaxation Processes in Glass under Influence of Photo-Irradiation

GeS_2, like $GeSe_2$, exists at normal pressure as high-temperature and low-temperature polymorphous modifications (HTPM and LTPM). The former forms a layered structure from tetrahedrons $GeSe_{4/2}$, one half of which is corner-shared and the other edge-shared. The latter presents a three-dimensional network of corner-shared tetrahedrons (Dittmar and Schafer, 1975, 1976a,b). GeS_2 glass, according to interpretation of Raman-spectra and diffractometry data, is built of corner-shared and ESTs, and the average distance between atoms Ge–S corresponds with good approximation to bond lengths of both crystal PMs (Feltz, 1983) that evidence in favor of the concept of polymorphous-crystalloid structure of glass according to which glass consists of co-polymerized structural fragments of different PMs without an LRO (crystalloids) (Minaev, 2000b,c).

T_g of v-GeS_2 is equal to 495 °C, which practically coincides with the phase equilibrium temperature of HTPM and LTPM (with S excess) in the Ge–S phase diagram (Feltz, 1983). Here also, like the case of $GeSe_2$, a genetic relation is observed between T_g and the temperature of enantiotropic polymorphous transformation (Minaev, 1998a,b, 2000d).

Depending on the radiation intensity, photo-induced crystallization of vitreous GeS_2 causes crystallization of LTPM or HTPM that shows complete analogy of relaxation processes taking place in GeS_2 and $GeSe_2$ at photo-irradiation, according to Sugai's (1987) data.

Unlike germanium chalcogenides, SiS_2 and $SiSe_2$ have the HTPM where all tetrahedrons are edge-shared (Sugai, 1986).

Glass structures SiS_2, $SiSe_2$, GeS_2, $GeSe_2$, according to Sugai (1987), can be characterized by one parameter P representing a ratio of probabilities of existence of edge-shared and corner-shared bonds between tetrahedrons. Value of P can be changed by photo-irradiation. P decreases after irradiation with intensity lesser than the threshold and it increases if intensity is higher than the threshold. For example, at irradiation of $SiSe_2$ glass by Ar-ion laser (4579 A) with 15 mW power during 310 min the intensity ratio $I(A_I^c):I(A_I)$ decreases from 2.3 ($P = 1.05$) to 1.8 ($P = 0.8$) while irradiation with 30 mW power during just 10 min increases $I(A_I^c):I(A_I)$ to 3.0 ($P = 1.41$). Sugai's experiments are in complete compliance with our polymorphous-crystalloid concept of structure of these glasses, and show that physical–chemical essence of processes taking place in vitreous disulfides and diselenides of germanium and silicon effected by photo-irradiation is the mutual transformation of structural fragments of edge-shared and corner-shared tetrahedrons $AX_{4/2}$, i.e., mutual transformation of HTPM and LTPM crystalloids: HTPM $\leftrightarrow$ LTPM. If the intensity of the external excitation changes and exceeds a threshold, the direction of crystalloids transformation changes to the opposite direction and it reflects the ability of given substances to undergo enantiotropic polymorphous transformation in crystal state.

6.4. Arsenic Selenide $As_{50}Se_{50}$. Relaxation Processes

Crystalline arsenic selenide $As_{50}Se_{50}$ exists in two PMs. LTPM is isomorphic to molecular modification α-As_4S_4 (realgar). HTPM is not identified comprehensively by X-ray analysis. In works of Kotkata, Shamah, El-Den and El-Mously (1983) and Kolobov and Shimakava (1996) diffractograms of both PMs are presented.

Annealing of amorphous films $As_{50}Se_{50}$ obtained by vacuum deposition at temperatures less than T_g (~ 180 °C) leads to crystallization of LTPM, and at temperatures higher than T_g leads to crystallization of HTPM. This situation is similar to situation with $GeSe_2$ films behavior described earlier and evidences that amorphous $As_{50}Se_{50}$ is apparently formed by crystalloids of both PMs. At $T < T_g$ the process of transformation of crystalloids HTPM $\rightarrow$ LTPM and crystallization of the latter take place. At $T > T_g$ the relaxation takes place in direction of HTPM concentration increase. Like the case of v-$GeSe_2$ the softening temperature is in this case a point where the reverse of mutual transformation direction HTPM $\leftrightarrow$ LTPM takes place that is the analog of the enantiotropic transformation in crystalline substance. Thus, the relaxation of amorphous $As_{50}Se_{50}$ also goes by two ways determined by heating temperature of films and results in two-variant crystallization depending on annealing temperature: as LTPM or as HTPM. The additional confirmation of this relaxation model are data of DSC of $As_{50}Se_{50}$ films (Kolobov and Shimakava, 1996) that show two exothermal peaks one of which is lower T_g and another is higher T_g. Between these peaks the endothermic 'pit' of softening is located. The exothermal effect preceding T_g is the thermal effect of transformation of HTPM structural fragments of the non-crystalline film in LTPM structural fragments, the effect similar to that of heat release at the polymorphous transformation HTPM ↔ LTPM in the crystalline state (Lihtman, 1965). The endothermal slope after T_g is the result of the reverse transformation: LTPM crystalloids are transformed into HTPM crystalloids. The effect of polymorphous transformation in the crystalline state is also endothermal. And finally, the next exothermal effect is the effect of crystallization of the HTPM $As_{50}Se_{50}$. This model is well blended with the experiment on photo-amorphization of $As_{50}Se_{50}$ films crystallized as LTPM (Kolobov and Elliott, 1995; Kolobov and Shimakava, 1996). From the CPPCSGL point of view the process of transformation of LTPM into HTPM takes place here, but is not finished. The result is the formation of a structure containing crystalloids of both PMs with the concentration ratio depending on time and intensity of radiation. Increase of duration or intensity of these two factors must lead to crystallization of amorphous films as HTPM. Kolobov and Shimakava (1996) have interpreted described relaxation processes differently in some respects, although they are close enough to our point of view. They consider that $As_{50}Se_{50}$ glass has two structural modifications related correspondingly to low- and high-temperature forms of crystal. Irradiation and annealing can move glass from one modification to another.

Various assumptions regarding the existence of different modifications of non-crystalline, including vitreous, substances have been proposed during the past 20 years (see Minaev, 1991). It has been also shown, based on the concept of polymorphous-crystalloid structure of non-crystalline substance, that 'poly-amorphism' does not exist in the nature, and 'polyforms' of non-crystalline substance represent a composition of structural fragments of different PMs without IRO, having different concentration ratios which are able to change from 0 to 1 depending on time and intensity of external influence.

During the last 20 years, neither proofs of existence of amorphous polyforms nor a strict definition of this notion have appeared. The 'relationship' sign (Kolobov and Shimakava, 1996) is evidently not fully developed enough to be considered as such.

6.5. Selenium

6.5.1. Structure of Crystalline and Vitreous Selenium

Selenium forms three types of monoclinic PM (α, β and γ) whose structures are constructed from molecules Se_8. For Se_8 molecules, the *cis*-configuration in arrangement of four consecutive linked atoms is inherent; α-monoclinic Se melts at 144 °C and spontaneously transforms in hexagonal Se; β-modification behaves similarly but at temperatures above 100 °C (Feltz, 1983) (Fig. 2).

Rhombohedral selenium constructed from Se_6 molecules is also known. This modification melts at 120 °C and spontaneously transforms in hexagonal Se (Miyamoto,

1980). The thermodynamically stable Se modification is the hexagonal PM, consisting of spiral chains of atoms (Feltz, 1983) with *trans*-configuration atom arrangement in 4-fold fragments of the chain.

Analysis of IR- and Raman-spectroscopy, X-ray diffractometry, UV-photoelectron spectroscopy data as well as structural models proposed by Lucovsky (1979) and other authors (Feltz, 1983) allows us to make the conclusion based on CPPCSGL that vitreous selenium is constructed from a tangle of chains consisting of fragments of 8-fold rings of monoclinic PM, fragments of 6-fold rings of rhombohedral PM (in smaller amount), and fragments of chains of hexagonal selenium (Minaev, 1998a).

6.5.2. *Structural Relaxation in Crystalline and Vitreous Selenium at Heating*

Relaxation processes in metastable PMs (MSPMs) at temperature influence manifest themselves on the phenomenological level as polymorphous transformations 'monoclinic PM → hexagonal PM' (in single crystal samples the beginning of transformation is fixed after heating for 630 min at 70 °C (Murphy, Altman and Winderlich, 1977) and 'rhombohedral PM → hexagonal PM' at temperatures above 105 °C (Miyamoto, 1980). Both transformations are characterized by irreversibility that shows the monotropic character of the performing phase transition (Fig. 2).

Investigations of viscosity–elastic properties and strain relaxation in vitreous selenium, carried out by Bohmer and Angel (1993), have shown that during heating of glass in the temperature range of 27–42 °C, the fraction of ring-like conformations sharply decreases and the fraction of chain-like conformations sharply increases in softening glass structure ($T_g \approx 37 \pm 10$ °C). Such a change of the 'ring-chain' ratio is identified by authors as a phase transition between ring and chain structures. From CPPCSGL positions the presented facts are considered as structural transformation in glassy Se phase of crystalloids of monoclinic MSPM in crystalloids of stable hexagonal PM (SPM), as the latent pre-crystallization period preceding crystallization of hexagonal SPM. The beginning of this crystallization is fixed at 50 °C (Rawson, 1967).

As can be seen from the presented data, relaxation processes in vitreous selenium that take place upon heating are similar to those in the crystal substance (the transformation of MSPM crystalloids into SPM crystalloids—'MSPM → SPM'), but they begin at lower temperatures that are related, in our opinion, with the greater energy of the disordered glass network (Zachariasen, 1932), with the heterogeneity of the glass structure and the activation role of presented structural fragments of the hexagonal SPM. In monoclinic and rhombohedric crystalline PM, atoms are embedded at regularly arranged rings, and glass consists of fragments of such rings alternating in chains with fragments of the hexagonal PM (Minaev, 1998a, 2001a). For the structural transformation MSPM → SPM, in this which case there is no need to destroy chemical bonds in rings and to reconstruct the ring structure into the chain structure. It is sufficient to 'straighten' ring fragments into spiral chains that require significantly less energy consumption. Moreover, a part of glass structural fragments is already present as hexagonal chain links.

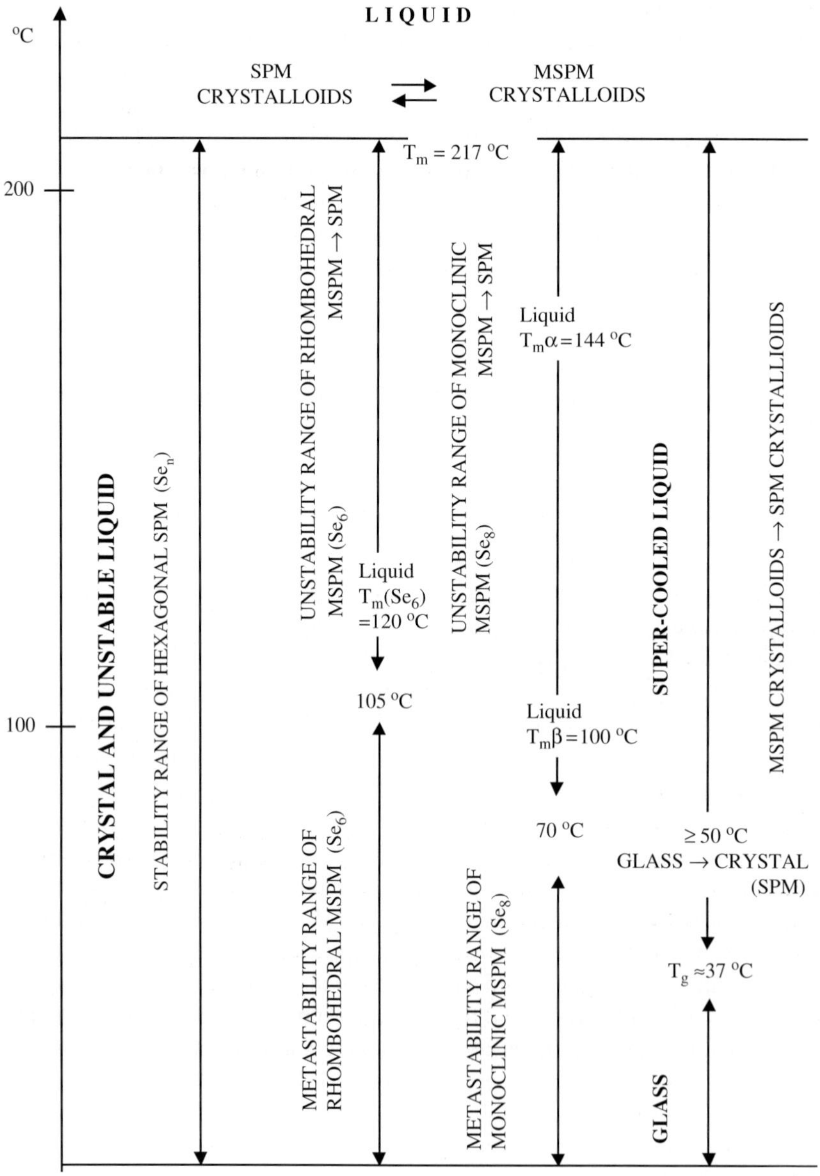

FIG. 2. Varieties of condensed states of Se (see description in the text).

6.5.3. Phenomenological Aspects of Liquid Selenium Relaxation at Cooling

Heating of the crystallized hexagonal PM of selenium leads to its melting at 217 °C. Diffractometry and IR-spectrometry data, as well as data on the viscosity temperature dependence (Rawson, 1967; Feltz, 1983), provide evidence that the melt structure is very similar to the glass structure, and there is also a statistical distribution of *cis*- and *trans*-configuration of selenium characteristic for ring and chain structures. Formation in melt of fragments of SPM and MSPM changes their ratio with temperature change (Rawson, 1967), evidences that the main physical–chemical feature of liquid selenium relaxation is mutual transformation of SPM and MSPM crystalloids 'MSPM $\leftrightarrow$ SPM,' unlike the one-directional transformation 'MSPM $\rightarrow$ SPM,'—in crystal and glass states.

Mutual transformation of crystalloids 'SPM $\leftrightarrow$ MSPM' and their growing polymorphous co-polymerization at melt cooling is the main factor that increases their viscosity and hinders melt crystallization up to solid glass state. However, if the melt cooling rate is less than the CCR, equal to 20 K min^{-1} (Feltz, 1983), all crystalloids of the MSPM will have enough time during cooling to transform into crystalloids of the SPM, and super-cooled liquid will crystallize. This observation demonstrates that CCR is a function of the concentration and the crystalloids transformation rate MSPM $\rightarrow$ SPM.

6.5.4. Photo-irradiation and Relaxation Processes in Vitreous Selenium

Sugai (1987) investigated the process of photo-crystallization in vitreous selenium at irradiation by the He–Ne laser ($\lambda = 6328$ Å) of different effective incident power (4, 8 W cm^{-2} and 8 kW cm^{-2}). Raman spectra of monoclinic selenium is characterized by the peak at 251.5 cm^{-1} attributed to the A_1 mode of Se rings, hexagonal selenium chains are characterized by the 234 cm^{-1} peak. In glass spectra, bands corresponding to modes of both PMs are observed. At irradiation by laser beam with increasing effective power, the intensity of 251.5 cm^{-1} band decreases to almost complete disappearance at 0.8 kW cm^{-2} power, and the intensity of 234 cm^{-1} band increases, providing evidence of transformation of monoclinic modification crystalloids (*cis*-configuration of atom arrangement) in hexagonal modification crystalloids (*trans*-configuration) (Minaev, 2001a). The same structural changes are observed under thermal annealing.

As these observations demonstrate, the relaxation processes in vitreous Se and v-$GeSe_2$ under thermal influence and by He–Ne laser irradiation are principally different. The relaxation in $GeSe_2$ follows two-variant LTPM- and HTPM-crystallization that is characteristic for substances with enantiotropic polymorphous transformation. For selenium, with its monotropic polymorphous transformations in crystal state, the influence of temperature and photo-irradiation leads to the single way of structural relaxation in glass—an increase of concentration of crystalloids of thermodynamically SPM at the expense of decrease of concentration of structural fragments of metastable monoclinic, and possibly rhombohedral, modifications. As the result, only single-variant crystallization is characteristic for v-Se (Minaev, 2000c, 2001a).

7. Nanoheteromorphism in Ge–Se and S–Se Glass-Forming Systems

The main cause and physical–chemical essence of glass formation is co-polymerization of topologically different fragments of substance structures without translation symmetry (LRO), fragments in which atomic positions are connected between themselves by different ways. Differences in fragment structures of the nanometric scale do not allow the co-polymerizing substance to organize translational symmetry (LRO), i.e., to be crystallized.

Nanoheteromorphism in non-crystalline substances is the co-existence of different structural fragments (structural inhomogeneities) having no LRO. Nanoheteromorphism and nanoheteromorphous co-polymerization are the necessary and sufficient conditions for glass formation in any glass-forming substance.

In the case of ICSs that are able to exist in non-crystalline states (and vitreous states in particular) as one-component glass, such structural fragments are crystalloids of different PMs. If glass contains more than one component, i.e., it is formed of two and more ICSs, the presence of structural fragments of these ICSs is sufficient for glass formation irrespective of existence or non-existence of their different PMs.

There is a possibility of the intermediate variant: the presence in glass of crystalloids of different PMs of one of the ICS and structural fragments of other ICS belonging to only one PM (Minaev, 2001a).

Taking this into account, let us consider the structure of multi-component vitreous substances by the example of two binary systems, $GeSe_2$–Se and S–Se.

7.1. Intermediate-Range and Short-Range Ordering in Glass-Forming System $GeSe_2$–Se

Glass formation in ICSs $GeSe_2$ and Se is caused by co-polymerization of crystalloids of different PMs of $GeSe_2$ and Se (Minaev, 1996, 1998a). The minimal crystalloid $GeSe_2$—the bearer of the IRO of one or another PM—contains two joined tetrahedrons. The IRO of LTPM is characterized by tetrahedrons joined by vertexes only. IRO of HTPM is formed by tetrahedrons joined by both vertexes and edges (Minaev, 2001a).

Besides there are different SROs in v-$GeSe_2$. Germanium atom surrounded by four selenium atoms forms two closely resembling but different SROs: $GeSe_{4/2}$. They are different in angles Se–Ge–Se and the length of the bond Ge–Se because in LTPM $GeSe_{4/2}$ tetrahedrons are joined by vertexes, and in HTPM both vertexes and edges that in the latter case lead to distortion of tetrahedrons, and consequently SRO. Two different SROs exist around Se atoms connected with two Ge atoms (these SROs ($SeGe_{2/4}$) are different in angles Ge–Se–Ge and the length of the bond Ge–Se due to the same reason as SRO inside $GeSe_{4/2}$ tetrahedron).

The above-mentioned in compliance with the glass model described by the COCRN and is based on the assumption that for all compositions, heteropolar bonds having larger energy are more preferable than homopolar bonds (Lucovsky and Hayes, 1979; Nemanich et al., 1983). In real glass $GeSe_2$ many authors (for example, Boolchand, Bresser, Georgiev, Wang and Wells (2001) have observed Raman-spectra peaks which are characteristics of chemical bonds Ge–Ge ($\sim 170\ cm^{-1}$) with concentration up to 2% and Se–Se bonds ($240–250\ cm^{-1}$).

It means that in real v-$GeSe_2$ two more SROs appear:

(1) Ge atom surrounded by three Se atoms and one Ge atom—$Se_{3/2}\ GeGe_{1/4}$ and
(2) Se atom connected with one Se atom and one Ge atom—$Se_{1/2}SeGe_{1/4}$.

These SROs are formed from fragments of SROs of initial ICSs: $GeSe_2$ ($GeSe_{4/2}$) and Se ($SeSe_{2/2}$—chains of three Se atoms). Therefore, these SROs can be considered to be compound.

The minimum number of atoms necessary for the creation of different IROs of selenium, in *cis-* and *trans*-configurations, typical for monoclinic and hexagonal PMs of selenium, is equal to 4 (Minaev, 1996). The SRO of selenium is described above.

Now let us consider intermediate compositions of the glass-forming system $GeSe_2$—Se from the point of view of intermediate-range and short-range ordering.

When Se is added to $GeSe_2$, selenium atoms embed between $GeSe_{4/2}$ tetrahedrons and first form 2-fold chains, then 3-fold, 4-fold, and so on. When selenium content of 80 at.% is achieved[2)] (at ideal mixing of $GeSe_2$ and Se), all tetrahedrons (more exactly, the germanium atoms within them) are connected through 2-atom bridges –Se–Se–, and as a result all crystalloids of high-temperature and LTPMs, from which vitreous $GeSe_2$ (v-$GeSe_2$) is formed, and in which Ge atoms are connected with each other by the 1-atom bridge –Se–, are destroyed (Minaev, 2002). This means that in the composition $Ge_{20}Se_{80}$, IROs typical for both PMs of $GeSe_2$ disappear (Figs. 3 and 4). All germanium atoms at this composition are connected with each other by 2-fold selenium chains, which are also not typical for any selenium PM; they cannot form IRO of any selenium PM, as it is necessary to have no less than four Se atoms in the chain. The first chain of four selenium atoms appears in the composition when the ratio of selenium and germanium is greater than 6:1 ($Ge_{14.29}Se_{85.71}$).

Whereas in ICS $GeSe_2$ and Se, the glass formation in the 'ideal' case is provided for by different structural, heteromorphous crystalloids, bearers of IRO of the nanometric scale inherent to different PMs of corresponding ICS, in the composition range of $80 > Se > 66.6(6)$ at.% there are, besides $GeSe_2$ crystalloids, nano-structural selenium fragments composed of 2-fold chains which are not IRO bearers of any selenium PM. In this composition range, nanoheteromorphous structural fragments which form two IROs and all SROs of $GeSe_2$ are typical (Fig. 3), as well as nanoheteromorphous structural fragments composed of 2-fold chains of Se which unite $GeSe_2$ tetrahedrons, and as a result, organize the compound SRO $Se_{1/2}SeGe_{1/4}$.

Therefore, the nanoheteromorphism here—the assembly of nanoheteromorphous structural fragments—is presented by both structural nanofragments of different PMs and nanofragments which are not typical for any PM.

In the composition range of $80.00 \leq Se \leq 85.71$ at.%, there are neither structural fragments with IRO of $GeSe_2$ nor Se (Fig. 3).

In this range, nanoheteromorphous structural fragments are presented which are typical only for three different SROs: (1) the SRO reproducing $GeSe_2$ tetrahedrons, (2) the SRO organizing 3-fold selenium chains, and (3) the compound SRO $Se_{1/2}SeGe_{1/4}$.

For the composition range of $85.71 < Se < 100$ at.%, IROs of two selenium PMs with inherent selenium SRO, SRO $GeSe_{4/2}$ and the compound SRO $Se_{1/2}SeGe_{1/4}$, are typical.

Therefore, at ideally homogeneous mixing of $GeSe_2$ and Se in the system $GeSe_2$–Se, which glass formation region is characterized by concentration alteration of each element

---

[2] Hereinafter, percentage of Ge and Se is indicated not in regard to the particular system $GeSe_2$–Se but it regards the general system Ge–Se. For example, the composition $GeSe_2$ contains 33.3(3) at.% Ge and 66.6(6) at.% Se.

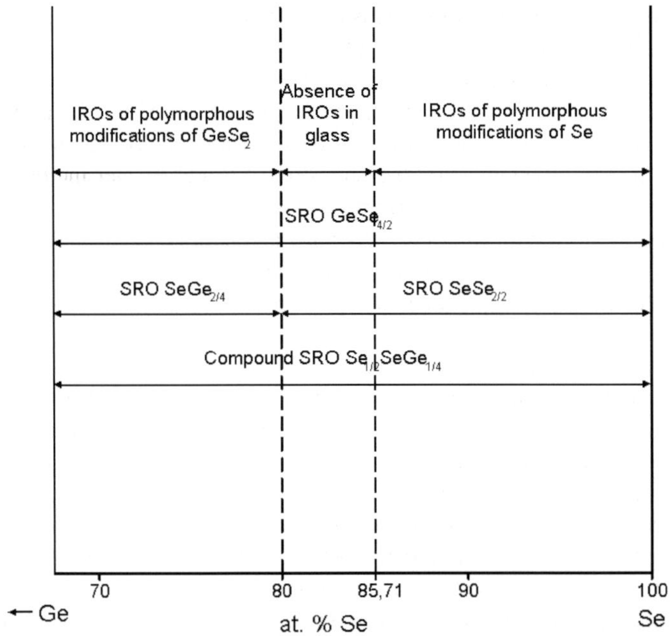

FIG. 3. Location of intermediate-range and SROs in the glass-forming system $GeSe_2$–Se at ideally homogenous distribution of components.

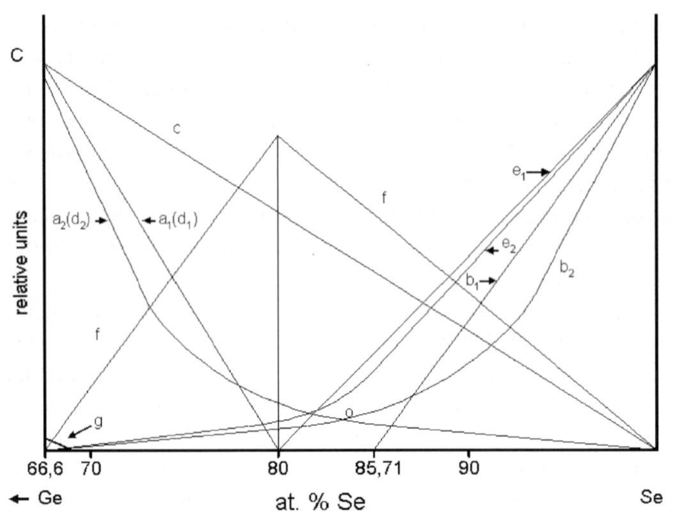

FIG. 4. The dependence of the concentration C (relative units) of crystalloids—bearers of IROs of different polymorphous modifications of $GeSe_2$ (a_1—for the case of ideally homogenous distribution of components in glass, a_2—for real glass) and Se (b_1—'ideal' glass, b_2—real glass), concentration of structural fragments—bearers of the SRO in $GeSe_2$ (c—$GeSe_{4/2}$; d_1, d_2—$SeGe_{2/4}$ in 'ideal' and real glasses), in Se (e_1, e_2—$SeSe_{2/2}$ in 'ideal' and real glasses), the concentration of structural fragments—bearers of compound SRO (f—$Se_{1/2}SeGe_{1/4}$ in 'ideal' glasses, g—$Ge_{1/4}GeSe_{3/2}$ in real glasses) in the glass-forming system $GeSe_2$–Se vs. composition. ○—the point of intersection of curves a_2 and b_2.

by 33.3(3) at.%, the nanoheteromorphism is not related with structural fragments of different PMs in the range of 5.71 at.% only.

It follows from above that the concentration of crystalloids—bearers of IROs of PMs of $GeSe_2$—decreases from the maximum at 66.6(6) at.% Se to zero at 80.00 at.%. At selenium concentrations of 80.00–85.71 at.%, in glass at ideal homogenous distribution of components, crystalloids of any PMs are absent, and at further increase of selenium concentration from 85.71 at.% to 100% concentration of selenium crystalloids—bearers of its IRO—increases from zero to maximum. So, on the IRO bearers' concentration vs. composition dependence, we have two rectilinear sections: 66.6(6)–80 and 85.71–100 at.% Se divided by the section of 80.00–85.71 at.% Se with zero concentration of IRO bearers (Fig. 3).

Naturally, we considered the ideally homogenous distribution of components $GeSe_2$ and Se. But it is known that in almost all real glass, and also in glass-forming liquids, which contain two or more components, a structural feature appears, the so-called 'concentration fluctuation' (Mazurin, 1986). Therefore, in real glass in the whole glass formation region of $GeSe_2$–Se system, some amount of nanoheteromorphous formations characterizing IROs of different PMs of $GeSe_2$ and Se will be present, and in this case the alteration of concentration of crystalloids—bearers of IRO of PMs of $GeSe_2$ and Se—from maximum to minimum will deviate from the polyline composed of straight lines and will follow concave curves above points 80 and 85.71 at.% Se and asymptotically tending to zero in points 100 and 66.6(6) at.% Se, respectively. The summarized curve of the analyzed dependence will present a smooth concave curve with maximums at compositions of 66.6(6) and 100 at.% Se and the minimum above the zero line in the region of 80–85.71 at.% Se (Fig. 4).

It is interesting that the very same shape has been obtained for the dependence of endothermic non-reversing heat flow at heating of glasses of the $GeSe_2$–Se system in the softening region (T_g) by temperature-modulated DSC in the work of Boolchand et al. (2001). From the point of view of the concept of polymeric polymorphous-crystalloid glass structure (Minaev, 1991, 1996, 1998a), such dependence is quite natural. It is known (Lihtman, 1965) that the polymorphous transformation of LTPMs in HTPMs occurs with an endothermic effect. Both v-$GeSe_2$ and v-Se are formed of co-polymerized crystalloids of these PMs. And in the region T_g, where the transformation LTPM → HTPM is the most intensive, the endothermic effect is observed (Minaev, 2001a). It has its maximum where concentration of crystalloids of these PMs is highest, i.e., for ICS $GeSe_2$ and Se. The minimum of the effect coincides with the minimum of concentration of crystalloids—bearers of IRO of PMs. Boolchand's experiment is one more evidence of correctness of the concept of polymeric polymorphous-crystalloid glass structure.

It is necessary to note, however, that in Boolchand's (2001) experiment the minimum of the curve 'the value of the endothermic effect vs. composition' corresponds to the glass in which the average coordination number r is equal to 2.46; i.e., the composition $Ge_{23}Se_{77}$. According to our calculations, this minimum should be located in the interval of compositions containing 20–14.29 at.% of germanium and 80–85.71 at.% of selenium, i.e., in the interval $r = 2.4 \div 2.29$.

Such shift of the minimum of the endothermic effect is likely determined by three factors.

(1) The presence of a number of bonds Ge–Ge ($\approx$2% for $GeSe_2$ compound) decreasing the concentration of crystalloids of different PMs of $GeSe_2$ and, correspondingly, increasing the concentration of Se crystalloids;
(2) Appearance of 4-fold atomic chains of selenium; i.e., crystalloids of its PMs, as the result of the non-ideal mixing of Ge and Se atoms in compositions containing more than 20 at.% of Ge;
(3) The higher value of the endothermic effect of the polymorphous transformation in selenium when compared to $GeSe_2$, as is seen in Boolchand's (2001) data.

It is interesting that the minimal endothermic effect in the Boolchand's experiment is quite distant from the zero level and equals to $\approx$0.1 cal per s.g. In our opinion, this means that in the minimum point of the effect ($r = 2.46$), i.e., in the composition $Ge_{23}Se_{77}$ both $GeSe_2$ and Se crystalloids are present.

7.2. Intermediate-Range and Short-Range Ordering in Glass-Forming System S–Se

The two-component system S–Se is presented by the glass formation region expanding on all compositions from sulfur to selenium (Shilo, 1967).

Selenium is characterized by three monoclinic and one hexagonal modifications, sulfur—19 PMs (Feltz, 1983). At the present time, there are no sufficient experimental data for a detailed analysis of glasses based on structures of all PMs of chalcogens. Therefore, we will not go beyond reliably confirmed data on the presence in vitreous sulfur and selenium of *cis*- and *trans*-fragments in arrangements of four chalcogen atoms characterizing different PMs. We take that chalcogen crystalloids of the minimum size contain four atoms and are in one of the mentioned configurations (*cis* and *trans*) that in crystalline selenium form monoclinic and hexagonal modifications constructed from 8-fold rings and 'endless' chains, correspondingly, in sulfur—the group of *cis*-modifications consisting of rings with different number of atoms (6, 7, 9, 10, 11, 12, 18, 20, etc.) (Feltz, 1983) and the *trans*-modification consisting of 'endless' chains, correspondingly.

Vitreous selenium is constructed mainly from co-polymerized 4-atom structural fragments consisting of *cis*- and *trans*-configurations (Lucovsky, 1979), i.e., crystalloids monoclinic and hexagonal PMs (Minaev, 1996). When sulfur is added to selenium, its atoms embed between selenium atoms. At 10% sulfur in the alloy, as in the case of its 'ideal' dissolving and distribution, sulfur atoms will separate fragments of chains consisting of nine selenium atoms. In a 20% sulfur alloy, sulfur atoms will separate fragments of chains of 4 selenium atoms, fragments which can still form both *cis*- and *trans*-configurations, i.e., represent crystalloids which are bearers of IROs of two different PMs. At 25 at.% S in Se, all selenium chains consist of three atoms united by sulfur atoms in 'endless' compound chains. In this case, there are no selenium crystalloids, neither monoclinic nor hexagonal modifications, and there are no IROs characterizing these polyforms. And naturally, there is no IRO characterizing sulfur poly-forms. They are also absent in sulfur-enriched compositions including $Se_{25}S_{75}$ composition (Figs. 5 and 6).

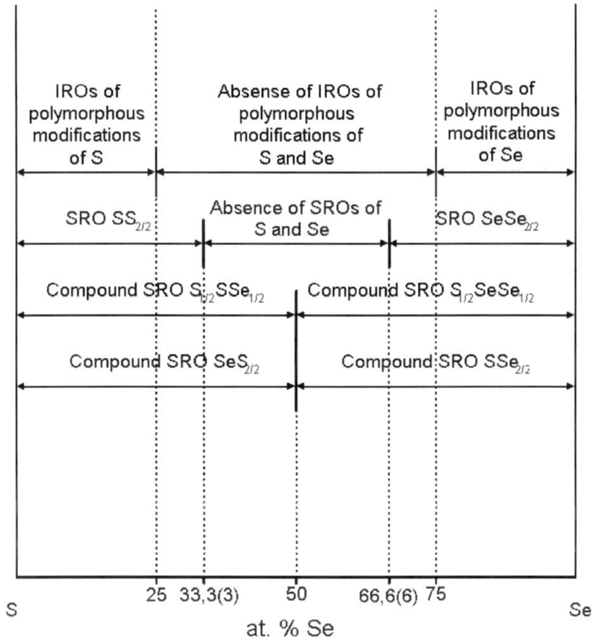

FIG. 5. Location of intermediate-range and SROs in the system S–Se at ideally homogenous distribution of components.

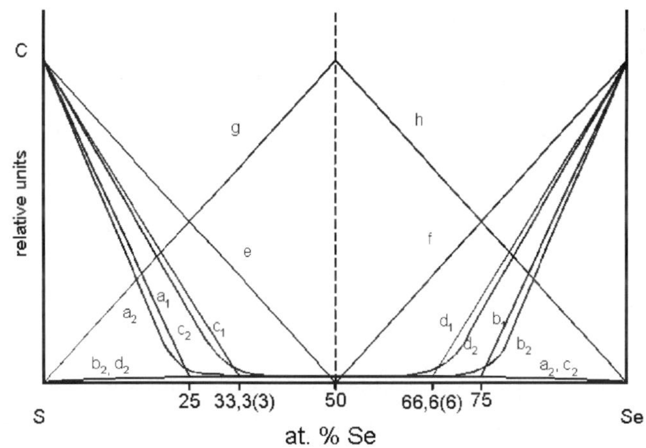

FIG. 6. The dependence of the concentration C (relative units) of crystalloids – bearers of IROs of different polymorphous modifications of S (a_1 – for the case of ideally homogenous distribution of components in glass, a_2 – for real glass) and Se (b_1 – 'ideal' glass, b_2 – real glass), the concentration of structural fragments – bearers of the SRO in sulfur $SS_{2/2}$ (c_1 and c_2 for 'ideal' and real glass, respectively), selenium SRO $SeSe_{2/2}$ (d_1 and d_2 – for 'ideal' and real glass, respectively), the concentration of structural fragments – bearers of compound SRO $S_{1/2}SSe_{1/2}$ (e), $S_{1/2}SeSe_{1/2}$ (f), $SeS_{2/2}$ (g), $SSe_{2/2}$ (h) – for the case of ideally homogenous distribution of components in glass vs. composition.

At further increasing sulfur content ($S > 75$ at.%) in alloys (again, in the 'ideal' case), first 4-fold sulfur chains appear which are able to organize *cis*- or *trans*-configuration of atoms and corresponding IRO characterizing ring-like or chain-like modifications.

So, in the range of $75 \geq Se(S) \geq 25$ at.% 'ideal' glasses in the S–Se system do not have IRO of any PM. The intermediate-range ordering in this system is proper only to $100 > Se(S) > 75$ at.% compositions (Fig. 5). It means that on the dependence 'concentration of crystalloids—bearers of IRO of selenium and sulfur vs. composition,' like the $GeSe_2$–Se system, there will be two rectilinear section (100–75 and 25–0 at.% Se) divided by the section of 75–25 at.% Se with zero concentration of IRO bearers (Fig. 6). Due to concentration fluctuation in real glass, the summarized curve of the dependence will present, like the $GeSe_2$–Se system, a smooth concave curve with maximums at 100% S and Se and the minimum in the area of 50 at.% of each component above the zero level of the crystalloids concentration of S and Se. The dependence 'endothermic effect vs. glass composition' must have the similar shape in the Te-region.

The short-range ordering in sulfur and selenium is characterized by the chain of three atoms in which the central atom is connected with two atoms of the same type: sulfur's SRO –S–S–S– and selenium's SRO –Se–Se–Se–, i.e., $SSe_{2/2}$ and $SeS_{2/2}$.

Short-range ordering in glasses of the S–Se system is characterized by nanoheteromorphous fragments consisting of three atoms of one element or both elements of the S–Se system.

In the 'ideally' mixed alloy of sulfur and selenium chains –S–S–S– and –Se–Se–Se– can exist only at the sulfur or selenium concentration higher than 66.6(6) at.% of the given element. At the concentration of this element equal to 66.6(6) at.% all its chains consist of two atoms united in the 'common' chain by individual atoms of another element.

It means that the SRO of initial components of the S–Se system at ideal mixing of S and Se takes place only at concentrations of $100 \geq Se(S) > 66.6(6)$ at.%. The range of 33.3(3)–66.6(6) at.% Se(S) is characterized by the absence of SROs of initial components of the binary system S–Se.

In the system S–Se there are, however, other SROs as well: SRO around the sulfur atom characterizing the chain –S–S–Se– ($S_{1/2}SSe_{1/2}$) and SRO around the selenium atom characterizing the chain Se–Se–S–($Se_{1/2}SeS_{1/2}$), and also SROs characterizing chains – S–Se–S–($SeS_{2/2}$) and –Se–S–Se–($SSe_{2/2}$). The former two SROs are in the composition range of $100 > Se(S) > 50$ at.%, the latter two SROs are in the composition range of $100 > Se(S) \geq 50$ at.% (Figs. 5 and 6).

Naturally, in real alloys the borders of the existence ranges of different intermediate-range and short-range orders get diffused, being greater at lower temperature and shorter time of the glass synthesis. Apparently, any of the orders described can be revealed in the whole composition range $100 >$ at.% $S(Se) > 0$, but the main part of structural fragments corresponding to all orders listed is in the indicated above ranges (Fig. 5).

7.3. Some General Regularities of Glass Structure in Binary Glass-Forming Systems

The analysis of structure of binary glass-forming systems demonstrates that in a non-single component vitreous substance, nanoheteromorphism, the concurrent existence of topologically different structural fragments of the nanometric scale without an LRO, is typical.

These fragments can be divided into three types:

(1) crystalloids—bearers of IROs of PMs of initial components of glass-forming systems;
(2) structural fragment—bearers of SROs of initial components; and
(3) structural fragments—bearers of compound SROs formed from atoms of both initial components.

Glass formation is realized due to co-polymerization of fragments of different structure—nanoheteromorphous fragments without an LRO which are not able to form a crystalline substance because of their heteromorphous character. For one-component substances (ICSs), such fragments are fragments of the first type: crystalloids—bearers of IROs of different PMs.

Depending on their composition, two-component glasses can be formed: first, from fragments of all the three above-mentioned types, for example, compositions $Ge_{30}Se_{70}$, $S_{20}Se_{80}$, and others; second, from fragments of second and third types (for example, compositions $Ge_{17}Se_{83}$, $S_{30}Se_{70}$; third, from fragments of the third type (for example, the composition $S_{50}Se_{50}$).

Thus, in binary systems there are some glass compositions in which intermediate-range ordering is absent, in which there are no IROs characterizing initial components. And in the system S–Se we meet the case where in some glass compositions both intermediate-range and SROs typical for initial components are absent. In this case, there are only compound SROs.

All above-mentioned are related with glasses having ideally homogenous distribution of components. In real glasses there are always 'concentration fluctuations' of components which lead to the situation where all three analyzed structural fragments are present in any two-component glass. But in the second case (glass formation from structural fragments of 2nd and 3rd types—see above), there is a very insignificant amount (fractions of percent, or several percents, in the case of poorly synthesized glass) of fragments—bearers of IRO. In the third case, there are only trace amounts of bearers of IRO and SRO of initial components.

Any concept of structure of multi-component glass-forming substances (binary, in particular) can be considered if it includes a particular concept of the structure of individual vitreous chemical substances, a particular concept of the structure of one-component glass-forming systems.

The systems analyzed above show a genetic relationship between the concept of polymeric polymorphous-crystalloid glass structure that is applicable to ICSs (Minaev, 1996, 1998a, 2000b), and the more general concept of polymeric nanoheteromorphous glass structure (PNHGS) that is applicable to multi-component systems as well.

Some features of these concepts, as well as differences between them are shown above, recognizing the fact that the former is the special case of the latter. Other differences as well as generalized theses of the PNHGS concept will be stated later, after analysis of other binary and multi-component systems. But it appears that the main thesis of the new concept can already be stated now: *nanoheteromorphism and nanoheteromorphous co-polymerization are necessary and sufficient conditions of glass formation in any substance.*

8. Conclusions

The comparative analysis of structure, properties, and relaxation processes taking place in liquid, glass, and various crystalline PMs of ICSs has shown that neither the concept of a COCRN, nor the 'crystallite' concept, nor any other concept of glass structure is in complete compliance with experimental data obtained.

The first-mentioned concept is not able to explain evidence of simultaneous influence of different PMs on properties and structure of glass and its crystallization as different PMs.

The second concept contradicts diffractometric data (X-ray, electron-, and neutronography) that demonstrates the absence of even smallest crystals—crystallites—in glass and liquid.

The latter contradiction was the basis for the refusal by most researchers to consider the 'crystallite' concept, and to continue to accept COCRN as the model.

A paradoxical situation has arisen: glasses without crystals manifest properties of different crystalline polymorphous modifications.

The variance between the concept of the chemically ordered random network with a significantly large set of experimental data on glass structure and properties suggests the COCRN paradigm is incomplete, and casts doubt upon its basic principles.

There is no doubt that the principle of a chemically ordered network is applicable to glass, when considered with the acknowledgment of fluctuations of the chemical composition.

The principle of disorder and randomness within a chemically ordered network, compared with the 'long-range ordering' in crystal structure, also does not raise doubts. The intermediate-range (average) order, introduced in the random network, remains ambiguous in interpretation and has no generally accepted definition. At that, it is implied that there is only one IRO in glass.

And, finally, accepting the idea of the short-range ordering, the COCRN concept has in mind the SRO, just the same as in the crystal; in the best case, meaning one of the concrete SROs—the SRO of the HTPM after which melting the glass-forming liquid is obtained. Here, existence of a single SRO in glass is also implied.

The principal formation of the existence of a single SRO in glass, 'the same as in crystal,' has hindered development of structural–chemical approaches to glass formation for decades.

The concept of polymeric nanoheteromorphous structure of glass and its particular case—the concept of polymorphous-crystalloid structure of one-component glass—proposes a new paradigm of structure of glass and glass-forming liquid to resolve the paradox of glass formation manifesting the properties of crystalline polymorphous modifications without crystals.

According to the new concept, the main cause of glass formation is the co-polymerization of fragments of a substance without an LRO and with a different structure that does not allow this co-polymerizing substance to organize a translation symmetry (a LRO), i.e., to crystallize.

Such structural difference on the nanometric level, or nanoheteromorphism and nanoheteromorphous (polymorphous) co-polymerization is the necessary and sufficient condition of glass formation—formation of the continuous net or chain structure.

In the case of individual (one-component) chemical substance, such structural fragments (without an LRO) are crystalloids of different polymorphous modifications. Crystalloids of each PM possess the IRO inherent to this PM only, that is, a certain combination (topology) of SROs typical for this PM and not having a LRO. Therefore, co-polymerizing one-component substance is characterized as a combination of different IROs and SROs whose co-existence is just a basis of nanoheteromorphism of one-component substance and its ability to nanoheteromorphous co-polymerization, in this case to polymorphous co-polymerization, i.e., to glass formation.

At melting of any PM of substance, along with crystalloids of this melting PM, crystalloids of other PMs appear in liquid. When thermodynamical equilibrium in liquid is reached, a certain concentration ratio of crystalloids of different PMs corresponds to each temperature, pressure, or field impact. Alterations of intensity of external influences, or a qualitatively new impact cause alteration of the concentration ratio of crystalloids of different PMs both in glass-forming liquid and in glass, formed from it. This alteration is the physical–chemical essence of all relaxation processes going in liquid and glass in the direction of the thermodynamical equilibrium corresponding to the given impact, i.e., new conditions of existence of the substance.

At alteration of intensity of temperature influence on one-component glass-forming liquid or glass, the following relaxation processes take place in them:

(1) Glass formation as a two-in-one process taking place at cooling of liquid:
  (a) co-polymerization of crystalloids of different PMs;
  (b) de-polymerization of the forming co-polymer due to disintegration of crystalloids of low-temperature (enantiotropic substance) or unstable (monotropic substance) PMs in the glass formation region.

(2) Crystallization of overcooled liquid at cooling rates lesser than the CCR where crystalloids of different PMs have enough time to transform into crystalloids of the single PM which crystallizes in the temperature range $T_m - T_g$.

(3) Annealing of glass leading to obtaining of a certain concentration ratio of crystalloids of different PMs after which glass possesses necessary properties (for example, close to properties of LTPM or HTPM).

(4) Coursing of the latent pre-crystallization process preceding crystallization of HTPM or LTPM depending on the annealing temperature or the intensity of another influence (for example, photo-irradiation).

(5) Crystallization of glass in high- or low-temperature modification (depending on the annealing temperature) for enantiotropic substances or in stable PM for monotropic substances.

(6) Softening of glass (at heating), the process of de-polymerization of the vitreous polymorphous co-polymer due to disintegration of crystalloids of low-temperature (for enantiotropic substances) or metastable (for monotropic

substances) PM (LTPM or MSTM) in the temperature range of their instability where formation of new crystalloids of high-temperature or stable (for monotropic substances) polymorphous modifications (HTPM or SPM) take place simultaneously.

The softening temperature is the temperature of the most active stage of transformation of LTPM or MSPM crystalloids into HTPM or SPM crystalloids. Correspondingly, it is the temperature of the most active stage of disintegration of LTPM or MSPM crystalloids where the process of disintegration surpasses the process taking place simultaneously of the formation of new crystalloids of HTPM or SPM, with the result of the amount of broken bonds increases and glass gets soft.

(7) Formation of overcooled liquid as the result of softening of glass at heating whose further behavior depends on the heating rate:
  (a) if the cooling rate is lesser than the critical heating rate (CHR) all LTPM or MSPM have time to transform into HTPM or SPM, correspondingly, and overcooled liquid crystallizes in the form of these PMs (see item 2);
  (b) if the cooling rate is higher than the CHR not all LTPM or MSPM have time to transform into HTPM or SPM, correspondingly, before the melting temperature is reached and overcooled liquid at T_m turns into glass-forming liquid that is characterized (above T_m) by a certain (depending on temperature) concentration ratio of different PMs (see above).

Thus, based on the idea of polymorphous (nanoheteromorphous) co-polymerization of structural fragments without an LRO (crystalloids) of different PMs, the new non-contradictory CPPCSGL, describing features of their structure and physical–chemical processes taking place under influence of external impacts and leading to alteration of their properties or both properties and state, has been constructed.

This concept was successfully implemented for the analyses of structure, properties, and relaxation processes in some one-component oxide and halogenide substances as well as in halcogenides $GeSe_2$, GeS_2, SiS_2, $SiSe_2$, $As_{50}Se_{50}$, and Se. Unexamined remaining chalcogenide compounds As_2S_3, P_4Se_{10}, P_4Se_3, P_4Se_4, where obviously expressed polymorphism is observed at the pressure of 1 atm, and some other chalcogenide glasses, for example, As_2Se_3, data on polymorphism of which were out of our sight. These glasses will be examined in future works.

Nanoheteromorphism is also typical for multi-component glass-forming systems, binary in particular—simultaneous presence of different structural fragments of the nanometric scale without LROs in a vitreous polymer. These fragments can be divided in three types:

(a) crystalloids—bearers of IROs of different polymorphous modifications of source components of glass-forming systems;
(b) structural fragments—bearers of SROs of source components;
(c) structural fragments—bearers of compound SROs formed from atoms of different source components.

Two-component glasses can be formed, depending on composition, from:

- fragments of all three types;
- fragments of the second and third types;
- fragments of the third type only.

In the first case, both short-range and IROs of source components are present in glass. In the second and third cases, IROs inherent to source components are absent in glass, and in the third case SROs inherent to source components are also absent in glass. Here, only compound SROs are present. All the above-mentioned concern glasses with ideally homogenous distribution of source components. In real glasses 'fluctuations of concentration' of components are present that cause in the second and third considered cases, the presence of some amount (usually one-digit percent) of crystalloids—bearers of IROs, and fragments—bearers of SROs of source components.

A fundamental relation exists between the concept of polymeric polymorphous-crystalloid structure of glass, applied to ICSs, and the more general concept of polymeric nanoheteromorphous structure of glass, which can also be applied to multi-component systems, allowing to consider the former as a fundamental basis of the latter, general concept.

While the concept of polymeric polymorphous-crystalloid structure of glass has been developed for about 15 years, the general concept of polymeric nanoheteromorphous structure of multi-component glass has been 'tested' only with three binary systems: chalcogene system S–Se, chalcogenide system $GeSe_2$–Se and oxide system SiO_2–GeO_2 (Minaev, 2002). Analysis from its position of other binary as well as ternary glass-forming systems allows apparently to reveal additional physical–chemical features of glass formation processes, to interpret in a new fashion the liquation processes in glass-forming systems as well as to disclose other features which cannot be forecasted. But even now it is possible to state that the main thesis of the general concept of glass formation is: *nanoheteromorphism and nanoheteromorphous co-polymerization are necessary and sufficient conditions for glass formation in any substance.*

To conclude, the author hopes that the proposed concept of PNHGS will stimulate the further development of thermodynamic and kinetic approaches to glass formation, and to promote future unification of these approaches with structural–chemical approach in a joint general theory of glass formation.

References

Afify, N. (1993) Structural relaxation of $GeSe_2$ chalcogenide glass studied with use of the radial distribution function, *Phys. Rev. B*, **48**, 16304–16309.

Azoulay, R., Thibergen, H. and Brenac, A. (1975) Devitrification characteristics of Ge_xSe_{1-x} glasses, *J. Non-Cryst. Solids*, **18**, 33–53.

Bates, J.B. (1972) Raman spectra of α and β cristobalite, *J. Chem. Phys.*, **57**, 4042–4047.

Bernal, J.D. and Fowler, R.H. (1933) A theory of water and ionic solution, with particular reference to hydrogen and hydroxyl, *J. Chem. Phys.*, **1**, 515–548.

Blinov, L.N. (1985) On correlation between glass formation and polymorphism in chalcogenide systems, *Abstracts of All-Union Conference "Vitreous Semiconductors"*, Leningrad, USSR, pp. 176–177.

Blinov, L.N., Balmakov, M.D. and Pochentsova, N.S. (1988) Structural modifications of substance and glassformation, *Lett. J. Tech. Phys.*, **14**(1), 86–89.

Bohmer, R. and Angel, C.A. (1993) Elastic and viscoelastic properties of amorphous selenium and identification of the phase transition between ring and chain structures, *Phys. Rev. B*, **48**, 5857–5864.

Boolchand, P., Bresser, W.J., Georgiev, D.G., Wang, Y. and Wells, J. (2001) Evidence for the intermediate phase in chalcogenide glasses. In *Phase Transitions and Self-Organization in Electronic and Molecular Networks* (Eds, Phillips, J.C. and Thorpe, M.F.), Kluwer, New York, pp. 65–84.

Bridenbaugh, P.M., Espinosa, G.P., Griffits, J.E., Phillips, J.C. and Remeika, J.P. (1979) Microscopic origin of the companion A_1 Raman line in glassy $Ge(S,Se)_2$, *Phys. Rev. B*, **20**, 4140–4144.

Brückner, R. (1970) Properties and structure of vitreous silica, *J. Non-Cryst. Solids*, **5**, 123–175.

Brückner, R. (1971) Properties and structure of vitreous silica, *J. Non-Cryst. Solids*, **5**, 177–216.

Deeg, E. (1957) Uber die Zeitabhangigkeit der inneren Dampfung von Spiegelglas and uber Anomalien im Temperaturgang der inneren Dampflung von Quartz-glass, *Naturwiss*, **44**, 303.

Dittmar, G. and Schafer, H. (1975) Die Kristallstrutur von H.T.-GeS_2, *Acta Cryst. B*, **31**, 2060–2064.

Dittmar, G. and Schafer, H. (1976a) Die Kristallstrutur von Germanium diselenid, *Acta Cryst. B*, **32**, 2726–2728.

Dittmar, G. and Schafer, H. (1976b) Die Kristallstrutur von L.T.-GeS_2, *Acta Cryst. B*, **32**, 1188–1192.

Elliott, S.R. (1984) Physics of Amorphous Materials, Longman, London.

Encyclopedic Dictionary of Physics (1961) Vol. 2, Pergamon, Oxford, London-New York-Paris, p. 219.

Feltz, A. (1983) Amorphe und glasartige anorganische Festkorper, Akademie-Verlag, Berlin.

Frankenheim, M.L. (1835) Die Lehre von der Kohäsion, umfassend die Elastizitat der Gase, die Elasizitat und Kohärenz der flussigen und festen Korper, und die Kristallkunde. A. Schultz und Komp, Breslau.

Frankenheim, M.L. (1851) Kristallization und Amorphie, *J. Prakt. Chem.*, **54**(7–8), 430–476.

Gerber, Th., Himmel, B., Lorenz, H. and Stachel, D. (1988) Phase transitions in vitreous and amorphous SiO_2, *Cryst. Res. Technol.*, **23**, 1293–1302.

Glazov, V.M., Chizhevskaya, S.I. and Glagoleva, N.N. (1962) Liquid Semiconductors, Nauka, Moscow.

Glazov, V.M. and Situlina, O.V. (1969) Physical–chemical analysis of binary liquid systems A^{IV}–Se, *Proc. AS USSR Ser.: Chem.*, **187**, 799–802.

Golubkov, V.V. (1992) The structure and structural transition in vitreous SiO_2, *Fiz. Khim. Stekla. Sov. J. Glass Phys. Chem. (English transl.)*, **18**, 28–34.

Goodman, C.H.L. (1975) Strained mixed-cluster model for glass structure, *Nature*, **241**, 370–372.

Hagg, G. (1935) The vitreous state, *J. Chem. Phys.*, **3**, 42–49.

Inoue, K., Matsuda, O. and Murase, K. (1991) Raman spectra of tetrahedral vibrations in crystalline germanium dichalcogenides. GeS_2 and $GeSe_2$, in high and low temperature forms, *Solid State Commun.*, **79**, 905–910.

Ioffe, A.P. and Regel, A.R. (1960) *Non-Crystalline Amorphous and Liquid Electronic Semiconductors*, Progress in Semiconductors, Vol. 4, Heywood Company Ltd., London, pp. 237–291.

Kolobov, A.V. and Elliott, S.R. (1995) Reversible photo-amorphization of a crystallized $As_{50}Se_{50}$ alloy, *Philos. Mag. B*, **71**, 1–10.

Kolobov, A.V., Kolomiets, B.T., Lubin, V.M., Sebastian, N. and Tagirjanov, M.A. (1982) *Solid State Phys. (Russian)*, **24**, 1062–1067.

Kolobov, A.V. and Shimakava, K. (1996) Photo-induced metastability in CGS. In *Electronic Phenomena in Chalcogenide Vitreous Semiconductors* (Ed., Tsendin, K.D.), Nauka, Sankt-Petersburg, pp. 365–395.

Kotkata, M.F., Shamah, A.M., El-Den, M.B. and El-Mously, M.K. (1983) An X-ray study of AS–Se–Te compounds, *Acta Phys. Hungarica*, **54**, 49–63.

Landa, L.M., Landa, K.A. and Tananaev, I.V. (1984) Short-order polymorphism, *Proc. AS USSR Non-Org. Mater.*, **20**, 1681–1683.

Landa, L.M. and Nikolaea, I.N. (1979) Polymorphism in glasses, *Proc. AS USSR*, **244**, 1407–1410.

Lebedev, A.A. (1921) On the polymorphism of glass and glass annealing, *Proc. State Opt. Inst. (Russian)*, **2**(10), 1–20.

Lebedev, A.A. (1924) On the annealing of optical glass, *Proc. State Opt. Inst.*, **3**(4), 1–24.

Leko, V.K. (1993) The structure of vitreous silica, *Fiz. Khim. Stekla*, **19**, 673–715 (*Glass Phys. Chem.* (*English Transl.*), **19**, 351–374).

Lihtman, V.I. (1965) In *Polymorphism*, Brief Chemical Encyclopedia, Vol. 4 (Ed, Knuniants, I.L.) Soviet Encyclopedia Publishers, Moscow, pp. 194–196.

Lucovsky, G. (1979) In *Selenium, the Amorphous and Liquid States*, The Physics of Selenium and Tellurium (Eds, Gerlach, E. and Grosse, P.) Springer, Berlin, Heidelberg, New York, pp. 178–192.

Lucovsky, G. and Hayes, T.M. (1979) In *Short-Range Order in Amorphous Semiconductors*, Amorphous Semiconductors (Ed, Brodsky, M.H.) Springer, Berlin, Heidelberg, New York, pp. 268–310 (Russian Translation).

Lucovsky, G., Wong, C.K. and Pollard, W.B. (1983) Vibrational properties of glasses: intermediate range order, *J. Non-Cryst. Solids*, **59**, 839–846.

Mackenzie, J.D. (1960) Fusion of quartz and cristobalite, *J. Am. Ceram. Soc.*, **43**, 615–620.

Maeno, N. (1988) Science of Ice, Mir, Moscow, (Translation from Japanese).

Mazurin, O.V. (1978) Glasstransition and Stabilization of Non-organic Glasses, Nauka, Leningrad.

Mazurin, O.V. (1986) Glasstransition, Nauka, Leningrad.

Mendeleev, D.I. (1864) Glass Production, Sankt, Petersburg.

Minaev, V.S. (1989) Polymorphism and varieties of solid substance state. *Non-Crystalline Semiconductors-89*, Proc. Int. Conf., Vol. 1, pp. 190–192, Uzhgorod, USSR.

Minaev, V.S. (1991) Stekloobraznye Poluprovodnikovye Splavy (Vitreous Semiconductor Alloys), Metallurgy Publishers, Moscow.

Minaev, V.S. (1992) Definition of non-crystalline solids and its varieties, *Fiz. Khim. Stekla*, **18**, 175–179 (*Sov. J. Glass Phys. Chem. (Engl. Transl.)*, **18**, 95–97).

Minaev, V.S. (1994) Presentation on seminar-discussion, *Fiz. Khim. Stekla*, **20**, 661–663 (*Sov. J. Glass Phys. Chem. (Engl. Transl.)*, **20**, 455–456).

Minaev, V.S. (1996) Polymorphic-crystalloid structure of glass, *Fiz. Khim. Stekla*, **22**, 314–325 (*Sov. J. Glass Phys. Chem. (Engl. Transl.)*, **22**, 235–242).

Minaev, V.S. (1998a) Integration of basic hypotheses of glass structure in the single concept of the polymer polymorphous-crystalloid structure, *Glastech. Ber. Glass Sci. Technol*, **71C**, 416–419.

Minaev, V.S. (1998b). "Polymorphous-crystalloid nature of glass transition temperature." In: Proc. XVIII Int. Congress on Glass, San Francisco ICG 402.D009-005. CD-ROM.

Minaev, V.S. (1998c) Polymorphism and polymer-crystallite concept of glass structure, *Fiz. Khim. Stekla*, **24**, 597–603 (*Sov. J. Glass Phys. Chem. (Engl. Transl.)* **24**, 422–426).

Minaev, V.S. (2000a) *Properties of Vitreous Silicon Dioxide and Their Relation with Polymorphous-Crystalloid Glass Structure*, Abstracts of II All-Russian Conf. on Material Science and Physical–Chemical Foundations of Technology of Doped Silicon Crystal Production ("Silicon-2000"), Moscow State Institute of Steel and Alloys (Technical University), Moscow.

Minaev, V.S. (2000b). Glass in the new millennium: Conceptual breakthrough in view of glass structure. In: "Proc. of ICG Annual Meeting 2000." CD-ROM. Amsterdam.

Minaev, V.S. (2000c) Bivariant (high- and low-temperature) crystallization of some simple glass-forming substances and polymorphous-crystalloid structure of glass, *Glastech. Ber. Glass Sci. Technol.*, **73C1**, 186–193.

Minaev, V.S. (2000d). Polymorphous-crystalloid structure and properties of glassforming and liquid $GeSe_2$. In: "Proc. of ICG Annual Meeting 2000." CD-ROM. Amsterdam.

Minaev, V.S. (2001a) Polymorphous-crystalloid structure and relaxation processes in some chalcogenide glass-forming substances, *J. Optoelectron. Adv. Mater.*, **3**, 233–248.

Minaev, V.S. (2001b) *Polymorphous-Crystalloid Nature of Glassforming Process and Diversiform Crystallization of Some Simple Glass-Forming Substances*, Proc. XIX Int. Cong. Glass, Vol. 2, pp. 72–73, Extended Abstracts.

Minaev, V.S. (2002) On the polymeric nano-heteromorphous glass structure, *J. Optoelectron. Adv. Mater.*, **4**, 843–855.

Minaev, V.S. and Timoshenkov, S.P. (2002) The concept of polymeric nano-heteromorphous glass structure. In: "Proc. 6th ESG Conference "Glass Odyssey-2002" CD-ROM.' Montpellier, France.

Minaev, V.S., Timoshenkov, S.P. and Tchernykh, S.P. (2002). Formation and relaxation of glass-forming liquid of some individual chalcogenides, halogenides and oxides. In: "Extended Abstracts of "XIII Int. Symposium on non-oxide glasses and new optical glasses." Part I, 190–192.

Miyamoto, Y. (1980) Structure and phase transformation of rombohedral selenium composed of Se_6 molecules, *Jpn. J. Appl. Phys.*, **19**, 1813–1819.

Mott, N.F. and Davis, E.A. (1979) *Electron Processes in Non-Crystalline Materials*, Clarendon Press, Oxford.

Mozzi, R.L. and Warren, B.E. (1969) Structure of vitreous silica, *J. Appl. Cryst.*, **2**, 164–172.

Murphy, K.E., Altman, M.B. and Winderlich, B. (1977) The monoclinic-to-trigonal transformation in selenium, *J. Appl. Phys.*, **48**, 4122–4131.

Nemanich, R.J., Galeener, F.L., Mikkelsen, J.C., Jr., Connell, G.A.N., Etherington, G., Wright, A.C. and Sinclair, B.N. (1983) Configurations of chemically ordered continuous random network to describe the structure of $GeSe_2$ glass, *Physica*, **117B–118B**, 959–961.

Nemilov, S.V. (1995). Moritz Ludvig Frankenheim (1801–1869)—the author of the first scientific hypothesis on glass structure, *Fiz. Khim. Stekla* **21**, 221–235. (*Sov. J. Glass Phys. Chem. (Engl. Transl.)*).

Nemilov, S.V. (2001) Physical ageing of silicate glasses at room temperature: selection of quantitative characteristics of the process and construction of sequence of glasses in accordance with their disposition to ageing, *Fiz. Khim. Stekla*, **27**, 326–346 (*Sov. J. Glass Phys. Chem. (Engl. Transl.)*).

Pauling, L. (1970) *General Chemistry*, W.H. Freeman and Co., San Francisco.

Pavlotou, E.A. and Papatheodorou, G.N. (2000) Raman spectroscopic study of $BeCl_2$ in the crystalline, glassy and liquid states and a molten $BeCl_2$–CsCl mixtures, *Phys. Chem. Chem. Phys.*, **2**, 1035–1043.

Poltavtsev, Yu.G. and Pozdnyakova, V.M. (1973) Structure of vitreous $GeSe_2$ and GeSI, *Proc. AS USSR Non-org. Mater.*, **9**, 853–854.

Popovic, Z.V., Raptis, Y.S., Astassakis, E. and Jakis, Z. (1998) Pressure-induced amorphization of germanium diselenide, *J. Non-Cryst. Solids*, **227–230**, 794–798.

Porai-Koshits, E.A. (1942). "On the vitreous state (X-ray diffraction study)." Cand. Sci. (Phys.–Math.) Dissertation, Kazan, USSR.

Porai-Koshits, E.A. (1959). Glass structure. In: Bezborodov, M.A., Bobkova, N.N., Brekhovskikh, S.M., Ermolenko, N.I., Mazo, E.E., Porai-Koshits, E.A., "Diagrammy stekloobrazuyushchikh sistem (Phase Diagrams of Glass-forming systems)." pp. 13–29. Belarus. Politekh. Inst., Minsk.

Porai-Koshits, E.A. (1977) On structure of one-component glasses, *Fiz. Khim. Stekla*, **3**, 292–305.

Porai-Koshits, E.A. (1992) The structure of glass: problems and outlook, *Fiz. Khim. Stekla*, **18**(6), 3–9 (*Sov. J. Glass Phys. Chem. (Engl. Transl.)*, **18**, 408–411).

Prokhorov, A.M. (1982) *Soviet Encyclopedia Vocabulary*, Soviet Encyclopedia Publishers, Moscow, p. 1129.

Rawson, H. (1967) *Inorganic Glass-Forming Systems*, Academic Press, London.

Sharma, S.K., Mammone, J.F. and Nicol, M.F. (1981) Raman investigation of ring configuration in vitreous silica, *Nature (London)*, **292**(5819), 140–141.

Shilo, V. (1967). "Regions of vitreous state in chalcogenide alloys". Cand. Sci. (Chemistry) Dissertation. Institute of General and Inorganic Chemistry AS USSR, Moscow.

Shimada, M. and Dachille, F. (1977) Crystallization of amorphous GeS_2 and $GeSe_2$ under pressure, *Inorg. Chem.*, **16**, 2094–2097.

Sosman, R.D. (1927) *Properties of Silica*, American Chemical Society Monograph Series No. 37, 37, Chemical Catalog Co. Inc., New York.

Sugai, S. (1986) Two-directional photo-induced crystallization in $GeSe_2$ and $SiSe_2$ glasses, *Phys. Rev. Lett.*, **57**, 456–459.

Sugai, S. (1987) Stochastic random network model in Ge and Si chalcogenide glasses, *Phys. Rev.* B, **35**, 1345–1361.

Tamman, G.A. (1933) Der Glaszustand, Leonard Voss, Leipzig.

Tananaev, I.V. (1972) Scientific foundations of new materials creation for new technologies, *Bull. AS USSR*, **2**, 21–29.

Tarasov, V.V. (1959) *New Problems of Glass Physics*, State Publishers on construction industry, Moscow.

Tarasov, V.V. (1979) *Problems of Glass Physics*, Stroiizdat Publishers, Moscow.

Tool, A.Q. (1946) Relation between inelastic deformability and thermal expansion of glass in its annealing range, *J. Am. Ceram. Soc.*, **29**(9), 240–253.

Tudorovskaya, N.A. (1938) Variations in the refractive index of glass below 300 °C, *Izv. Akad. Nauk SSSR*, **1**, 107–124.

Uemura, O., Sagara, Y., Muno, D. and Satow, T. (1978) The structure of liquid As_2Se_3 and $GeSe_2$ by neutron diffraction, *J. Non-Cryst. Solids*, **30**, 155–162.

Uhlman, D.R. (1977) Glass formation, *J. Non-Cryst. Solids*, **25**, 43–85.

Vukcevich, M.R. (1972) A new interpretation of the anomalous properties of vitreous silica, *J. Non-Cryst. Solids*, **11**, 25–63.

Wang, R. and Merz, M.D. (1977) Polymorphic bonding and thermal stability of elemental non-crystalline solids, *Phys. Status Solidi (a)*, **39**, 697–703.

Wang, Y., Matsuda, O., Inoue, K. and Murase, K. (1998) Photo-induced crystallization in amorphous $GeSe_2$ studied by Raman scattering, *J. Non-Cryst. Solids*, **227–230**, 728–731.

Wang, Y., Nakamura, M., Matsuda, O., Inoue, K. and Murase, K. (1996) Temperature dependence of Raman-spectra in amorphous, crystalline and liquid $GeSe_2$, *J. Non-Cryst. Solids*, **198-200**, 753–757.
Warren, B.E. (1937) X-ray determination of the structure of liquidus and glass, *J. Appl. Phys.*, **8**, 645.
Winter, A. (1943) Structural homogeneity of glass, *J. Am. Ceram. Soc.*, **26**(8), 277–284.
Wright, A.C., Etherington, G., Desa, E., Sinclair, R.N., Connell, G.A.N. and Mikkelsen, J.C., Jr. (1982) Neutron amorphography, *J. Non-Cryst. Solids*, **49**, 63–102.
Zachariasen, W.H. (1932) The atomic arrangement in glass, *J. Am. Chem. Soc.*, **54**, 3841–3851.

CHAPTER 5

PHOTO-INDUCED TRANSFORMATIONS IN GLASS

Mihai Popescu

NATIONAL INSTITUTE OF MATERIALS PHYSICS, STR ATOMISTILOR, 105 BIS, P O BOX MG7, BUCHAREST-MAGURELE (ILFOV), ROMANIA

In comparison to the crystalline chalcogenides that exhibit photoconductivity and photoluminescence during light irradiation but revert to their initial properties after light fluency is stopped, in non-crystalline chalcogenides the energy released during the non-radiative recombination may be used to introduce metastable structural modifications that change the properties of the material. The chalcogenide glasses are susceptible to light-induced changes because they are characterized by intrinsic structural flexibility. The chalcogen elements are twofold co-ordinated. The atoms possess lone pair electrons, which are normally non-bonding, but they undergo light-induced reactions and give rise to structural defects of three- and onefold co-ordinated chalcogen. The states associated with the non-bonding electrons lie at the top of the valence band and hence are preferentially excited by illumination.

Evidences of significant changes in the physical properties and structural modifications induced by light have been observed in many chalcogenide glasses and amorphous chalcogenide films. The first reports on the changes induced by light in chalcogenide glasses were published by Berkes, Ing and Hillegas (1971), Keneman (1971) and Pearson and Bagley (1971). Recently, several reviews on the photo-induced metastable effects have been published (Tanaka, 1990; Pfeiffer, Paesler and Aggarwal, 1991; Shimakawa, Kolobov and Elliott, 1995; Kolobov and Tanaka, 1999). The photo-induced modifications in chalcogenide glasses include changes in density (Tanaka, 1980a), hardness (Kolomiets, Lantratova, Lyubin and Puh, 1976), rheological properties (Berkes, 1971), chemical reactivity (Elliott, 1986), dissolution rate (Kolomiets, Lyubin and Shilo, 1978; Owen, Firth and Ewen, 1985), electrical (Agarwal and Fritzsche, 1974; Shimakawa, Hattori and Elliott, 1987; Shimakawa and Elliott, 1988) and optical properties (Tanaka and Ohtsuka, 1978; de Neufville, Moss and Ovshinsky, 1973/1974; Tanaka, 1980a; Elliott, 1986) as well as decomposition (Keneman, Bordogna and Zemel, 1978; Tanaka, 1980a; Owen et al., 1985) and crystallization (Balkanski, 1987).

As a function of the experimental conditions and of the material composition, the modifications induced by light can be reversible, partially reversible and irreversible.

The photo-induced modifications have been observed in elemental chalcogens (Ch) sulfur and selenium, in binary systems (As–Ch, Ge–Ch), in ternary systems (As–Ch–Ge) and in more complex compositions.

ISBN 0-12-752187-9
ISSN 0080-8784

1. Irreversible Modifications

1.1. Photo-physical Transformations

1.1.1. Photo-vaporization

Firstly, Janai and Rudman (1974) and Janai (1982) observed the photo-vaporization in As_2S_3. They concluded that this phenomenon is controlled by the photo-oxidation reaction followed by thermal evaporation of the volatile product.

Under the action of high-power laser pulses, the amorphous chalcogenides exhibit local evaporations. Feinleib has shown that the modifications of the optical properties of the chalcogenide semiconductors are closely related to the formation of bubbles induced by rapid vaporization during the impact of the light pulse.

In the chalcogenide films, the softening temperature is several tens of degrees above the ambient temperature and the evaporation temperature is situated above the softening temperature by about several hundreds of degrees. As a consequence, the intense illumination of the amorphous material can induce melting and vaporization. The vapor pressure increases during light absorption and some cavities (bubbles) filled by vapors will gradually appear. If the light is switched off, then a rapid cooling occurs and the bubble wall strengthens while the internal space becomes empty due to the vapor condensation. The bubbles migrate from the film surface towards the film center. If the substrate is transparent, then it will preserve its temperature and the amorphous layer in contact will be cooled. As a consequence, the bubbles will be formed within the film at some distance from the interface. If the sample is heated up to T_g, then the material will flow and the bubbles will disappear. Such behavior is specific to amorphous alloys, e.g., the selenium-based alloys, which soften by heating, without decomposition. The selenium alloys exhibit T_g down to ~ 50 °C. In order to facilitate the formation of the bubbles, it is important to choose an amorphous film with a narrow temperature interval for the transition in the glassy state. As a function of the rheological properties of the amorphous film, the bubbles form clusters, disappear or remain as distinct spheres.

In the case of the amorphous selenium films, which exhibit high plasticity at room temperature, the bubbles take a discoid shape due to the decrease in the vapor pressure as a consequence of the condensation of selenium vapors on the bubble walls.

In the thin film of $Se_{97.5}S_{2.5}$ alloy, it was possible to inscribe bubbles of 1–3 μm diameter by irradiation with pulses of 200 mW and 4 μs duration (800 nJ per pulse) generated by a krypton laser. Other alloys based on Ge, As and Te show higher softening temperatures (150–200 °C) and as a consequence, the bubble shell will be more rigid and the bubbles will be more stable at room temperature.

1.1.2. Photo-crystallization and Photo-amorphization

The crystallization of the amorphous films under illumination is a challenging physical effect. The photo-crystallization was observed firstly in amorphous selenium by Dresner and Stringfellow (1968). Some scientists suggest that the crystallization is the result of the direct action of the light and not a thermal effect. Other scientists believe that the thermal effect plays an essential role. Chaudhari (1972) has shown that the irradiation of the amorphous films of composition $Te_{81}Ge_{15}As_4$ with He–Ne or Ar laser beam induces structural modifications. In the illuminated regions crystallites are formed that disappear after further illumination in a different regime of irradiation. The illuminated zones show optical properties at variance from the material in its initial state. This fact can be explained both by the appearance of small crystallites and by the morphological change of the amorphous film subjected to the laser beam.

After Feinleib, de Neufville, Moss and Ovshinsky (1971) and Adler and Feinleib (1971), during the absorption of light of a given wavelength takes place a step-wise increase in the free-carrier concentration, related to the breaking of the covalent bonds. The breaking of the bonds weakens the metastable amorphous state and therefore, the rate of crystallization increases. The transition from the amorphous to the crystalline state under the action of the light is the direct consequence of the electron–hole formation by the absorption of light quanta. The reversible crystallization takes place during illumination. The formation of the electron–hole pairs is a very efficient process in the case of absorption of the high-energy photons. As a consequence, the following effects must be considered:

(a) the recombination of the charge carriers thus formed followed by increased temperature and of the mobility of the atoms and
(b) the weakening of the bonds between the atoms as a consequence of the formation of the charge carriers.

Paribok-Alexandrovitch (1969) found that the crystallization rate of the amorphous selenium increases under light irradiation. A significant effect on the crystallization rate was observed for the light of wavelength less than 560 nm. The energy of these quanta determines the breaking of the selenium chains, thus facilitating the reordering of the shorter chains. A crystallization under light irradiation was also observed in Se–Te (Feinleib et al., 1971). The crystallization rate of a film with the composition $Te_{81}Ge_{15}Sb_2S_2$, under the laser pulses ($\lambda = 5145$ Å, $\tau = 1-16\ \mu s$, film thickness: 0.1 μm) has been analyzed. The rate of advancement of the crystallization front was found $v = 4 \times 10^{-3}$ cm s^{-1}. The crystallization is accompanied by a strong change in the reflection and transmission of the film. It has been concluded that the mechanism of crystallization is optical in nature and not thermal. The peculiarities of the re-amorphization process suggest the optical mechanism of crystallization as an explanation. In the photo-crystallized and then amorphized zones under the action of a different light pulse, the material becomes completely amorphous. Therefore, the heat dissipation is not sufficient for the crystallization of the surrounding amorphous material. This conclusion is based on the fact that the thermal crystallization cannot be produced in a very short time (less than a few milliseconds). Nevertheless, no definite proof for the purely optical character of the photo-crystallization effect does exist (von Gutfeld, 1973; Weiser, Gambino and Reinold, 1973).

Haro, Xu, Morhange, Balkanski, Espinosa and Phillips (1985) studied the photo-crystallization induced by a laser beam in amorphous $GeSe_2$. They demonstrated that the initiation of the process needs a minimum density of energy. Above this threshold value, the energy transferred to the system is spent either for electronic transitions (with the consequence of bond breaking) or for mechanical transitions (vibrations and rotations of the clusters). This energy transfer to the system leads to structural changes in the material. Three stages in the photo-transformations as a function of light intensity have been observed. The first one consists in the formation of nuclei and their growth due to the appearance around them, of free volumes. The growth continues up to the limit of a void-free system with crystallites embedded in the glassy matrix. In the second stage, when the irradiation power increases, the crystallites coalesce and a new dense crystalline material is formed. This transformation is partially reversible. If the laser power is diminished, the system will relax towards an equilibrium state between micro-crystallites and crystallites. The third stage appears for long-time irradiation with light beams of very high energy. The system becomes completely fixed in the crystalline state (the transformation is irreversible).

Dresner and Stringfellow (1968) observed an effect of amplification with a factor of 20 of the crystallization rate of selenium during illumination, this corresponding to the creation of 3×10^{18} electron–hole pairs per μm^2 s. The crystallization rate is 3×10^{-5} cm s^{-1}. From the spectral dependency of the crystallization rate, it was concluded that the process is controlled by the concentration of the charge carriers and not by the absorbed energy.

Jecu (1986) and Jecu, Zamfira and Popescu (1989) observed the crystallization effects in glassy compositions from the system Se–S exposed to ruby laser pulses. The glass $Se_{0.42}S_{0.58}$ corresponds in the binary diagram to the eutectic with minimum melting temperature of 105 °C. The absorption edge of this material is situated in the neighborhood of the photon energy of the laser pulse (the photon wavelength in the laser pulse is $\lambda = 0.6943$ μm and the wavelength corresponding to band gap in the material is 0.606 μm). The darkening of the material appears immediately after irradiation if the energy of the pulse is more than 980 mJ ($t = 500$ μs) or after 1 h (for energies of ~750 mJ) or after 24 h at energies of ~200 nJ, or does not appear anymore for energies less than 100 mJ. The focal spot was maintained constant with ~0.5 mm in diameter. It was concluded that as a function of the power of the electromagnetic radiation, sooner or later, ordered nuclei do appear and they induce the crystallization. It is quite remarkable that in all the samples, the crystallization induced by light appears after some weeks of storage. The crystalline phase Se_3S_5 was revealed.

Oriented crystallization of amorphous selenium induced by linearly polarized light was observed by Tikhomirov, Hertogen, Glorieux and Adriaenssens (1997). Illumination with polarized (laser) light above T_g causes the formation of crystalline nuclei with a specific preferred orientation. Prolonged illumination causes anisotropic crystal growth from those nuclei.

In general, the crystallization of glassy and amorphous chalcogenide semiconductors is induced by near or above band-gap light. The photo-induced crystallization process occurs at lower temperatures than the thermal crystallization temperature in the dark. This implies that the most expensive part of the thermal crystallization process is substituted by the photo process. The photo-induced crystallization implies the long-range ordering as well as the local or mesoscopic ordering. In chalcogenide semiconductors, the electronic excitation by photon absorption is supposed to be able to achieving the bond switching, which is the origin of the other photo-induced structural changes such as the photo-induced anisotropy (Tikhomirov and Adriaenssens, 1997).

The same kind of bond-switching mechanism should also be involved in the photo-induced crystallization process. On the other hand there should also be required a spatially widespread excitation to produce the long-range order. This excitation is achieved by thermal process and requires lower temperature than the thermal crystallization temperature in the dark. During the photo-induced crystallization process, the high-temperature crystalline phase seems to grow much more favorable than the low-temperature crystalline phase does, though the crystallization temperature is well below thermal crystallization temperature in the dark, as shown by Matsuda, Takeuchi, Wang, Inoue and Murase (1997) in the case of $GeSe_2$ amorphous evaporated films.

Gazsó, Hajtó and Zénai (1976) observed that the illumination of the amorphous chalcogenide by high-intensity light can produce a transformation to a new amorphous state after melting and quenching of the material. Due to high light intensity involved in this process ($\sim 10^6$ W cm^{-2}), the mechanism is probably of thermal nature. The illumination causes the film to melt and the subsequent rapid quenching in air gives rise to an amorphous film of a somewhat different structure. If the photo-crystallization is the intermediary step, then this effect can be considered as photo-amorphization.

The transition from the amorphous material to the ordered crystalline phase is a long-time process. Much less time is necessary for the amorphization of the crystalline phase and this feature is exploited in the system for optical recording of information.

Elliott and Kolobov (1991) have discovered a new type of photo-induced structural change in amorphous chalcogenide materials in AsSe molecular films. This is the *athermal photo-induced transformation* in the amorphous state, i.e., a process that is not caused by local melting/quenching. The AsSe films were firstly crystallized by annealing at ~ 140 °C for time intervals up to 24 h. Thereafter, the crystallized films were exposed to broad band white light (2 W cm^{-2}) for ~ 150 min at room temperature and fully amorphous films were obtained. Although, the mechanism of photo-induced transformation to the amorphous state is not known, two possibilities were suggested: either photon-induced intramolecular bond breaking, which leads by cross linking to a continuous random network (CRN) or intermolecular bond-breaking resulting in an orientationally disordered molecular glass.

It was suggested in the case of photo-induced amorphization of crystallized $As_{50}Se_{50}$ glass that the presence of the amorphous phase and strain were the driving forces for the amorphization (Kolobov, 1995).

An interesting observation was that annealing of the photo-amorphized film at temperatures above T_g led to crystallization in a new structure (Kotkata, Shamah, El-Den and El-Mously, 1983). This structure is stable and cannot be amorphized. The crystallization of the films carried out by annealing for 1 h at 180 °C leads to the formation of As_4Se_4 phase and the photo-amorphization can be repeated.

Photo-induced amorphization was also observed in As_2S_3 (Frumar, Firth and Owen, 1995). The starting material was natural orpiment crystal. Investigation of the temperature dependence of the process has led the authors to the conclusion that amorphization has essentially an electronic and not a thermal origin.

Roy, Kolobov and Tanaka (1998) have discovered laser-induced suppression of photo-crystallization in amorphous selenium films. A film of amorphous selenium exposed to the simultaneous action of two different (Kr^+ and Ar^+) lasers, whose photon energies are on different sides of the optical band gap, crystallizes more slowly than the one exposed to only one of the above laser beams. A decisive role of the polarization of the two laser beams has been demonstrated, namely, the suppression of the crystallization rate is observed only for the polarization of the two light sources being parallel to each other. The crystallization suppression is probably due to the fact that while the sub-band gap light creates nuclei with a certain optical axis, the cross-band gap light breaks them, and vice versa.

The role played by the silicon substrate in the light-induced vitrification of $As_{50}Se_{50}$ thin films was investigated by Prieto-Alcón, Márquez and Gonzales-Leal (2000). It was shown that the films crystallize in different structures depending on the substrate they are attached to. Silicon plays an active role during illumination. The spectral composition of the radiation emitted by the light source influences the photo-amorphization phenomenon, because the existence of a larger proportion of photons with higher energy will probably cause a larger degree of disorder in the films.

1.1.3. Photo-contraction and Photo-expansion

Some amorphous chalcogenide films show significant modifications of thickness (contraction) when exposed to light. The magnitude of the effect depends on the film composition. Bhanwar Singh, Rajagopalan, Bhat, Pandhya and Chopra (1980), have studied this effect in Rajagopalan, Bhat, Pandhya and Chopra, Ge–Se films. They found that the photo-contraction is produced only in obliquely deposited films and increases appreciably for depositions at incident angles of more than 40°. No photo-contraction was found in amorphous selenium films, or in amorphous germanium films, or for normal incidence in Ge–Se films. Exposure to band-gap illumination results in a maximum photo-contraction of 12% in $GeSe_2$ and 19% in GeS_2 (Rajagopalan, Harshavardhan, Malhotraa and Chopra, 1982).

The analysis of the experimental data allows for the conclusion that the photo-contraction determines the modification of the topography and the inclination of the columns of material specific to the oblique depositions. After exposure to light, the inclination of the columns to substrate decreases. Therefore, the columnar structure seems to be essential in the photo-contraction process. The light finally determines the collapse of the columns. If the amorphous film is annealed before exposure to light or if the film is deposited at high temperatures, the photo-contraction phenomenon is strongly diminished and can be even inhibited. The phenomenon seems to be more complex because, in the above conditions, the columns do not disappear. Bhanwar Singh et al. (1980) explained the photo-contraction by a volume change induced electronically. In this process, the volume density of the dangling bonds plays a main role, which controls the magnitude of the effect. If the re-ordering of the network takes place at a large scale, then, even the collapse of the columns can be induced.

It was demonstrated that glassy As_2S_3 films expand by $\sim 0.5\%$ when illuminated with band-gap light (Hamanaka, Tanaka, Matsuda and Iijima, 1976; Kimura, Nakata, Murayama and Ninomya, 1981; Mikhailov, Karpova, Cimpl and Kosek, 1990), that is, with radiation of $\hbar\omega \geq E_g$, where E_g is the optical band-gap energy. The expansion can be recovered by annealing at the glass-transition temperature $T_g \sim 470$ K and the phenomena can be repeated by exposure and annealing. However, X-ray structural studies have not been able to provide reliable results accounting for the macroscopic expansion phenomena (Hamanaka, Minomura and Tsuji, 1991; Zhou, Sayers and Paesler, 1993) and the mechanism of the photo-expansion is still speculative (Elliott, 1986; Tanaka, 1990). It is also known that stress accumulated in chalcogenide glasses can be released during illumination (Matsuda and Yoshimoto, 1975; Koseki and Odajima, 1982; Tanaka, 1984; Teteris and Manika, 1990).

Hisakuni and Tanaka (1994) have shown that when illuminated with a focused beam from He–Ne lasers ($\hbar\omega = 2.0$ eV) for a duration around 10 s, the As_2S_3 films ($E_g = 2.4$ eV) with thickness of ~ 50 μm exhibit a thickness expansion up to 3 μm, which is approximately 10 times as great as that expected from the conventional photo-expansion phenomenon. The photon energy is located in the Urbach tail ($2.0 \leq \hbar\omega \leq 2.4$ eV). This effect was called *giant photo-expansion*. The expansion enhancement was explained by the photo-relaxation of strains generated by photo-expansion. The phenomenon was found in GeS_2 also.

In addition, it was found that when As_2S_3 is illuminated with focused Ar laser light ($\hbar\omega = 2.41$ eV) of 10 mW, the sample surface takes a concave shape in contrast to the expansion induced by illumination of the He–Ne laser light. The convex structure with a typical dimension of 2–4 μm in height and 10–100 μm in diameter can be formed during exposure to the Ne laser light.

1.1.4. Photo-induced Softening and Hardening and Photo-induced Deformation

Both softening and hardening of the amorphous films in Ge–As–S system, subjected to ultraviolet irradiation have been observed (Popescu, Sava, Lorinczi, Skordeva, Koch and Bradaczek, 1998). Amorphous films with the thickness of ~ 4 μm were prepared by vacuum thermal evaporation. The films were irradiated by UV light ($\lambda = 300$–400 nm) and power density (0.05 W cm^{-2}). Long-time irradiation (1–4 days) was used in order to ensure the saturation of the transformations induced in films. The sample heating did not exceed 40 °C during irradiation in flowing air.

In the $Ge_xAs_{40-x}S_{60}$ ($0 \leq x \leq 40$) system, the hardness increases with the germanium content and exhibits a maximum at $x = 27$, which corresponds to the threshold of the topological transition from 2D to 3D amorphous network in these alloys (Skordeva, 1995). For $x < 19$, strong softening was obtained by UV irradiation, while for $x > 19$, a hardening effect was revealed. Both effects tend to saturation for 90–100 h of irradiation. The explanation for the hardness results is based on the sulfur loss during UV irradiation, and by specific modifications of the non-crystalline network that accompany the chemical changes.

Hisakuni and Tanaka (1995) demonstrated that light or electron beam exposures give rise to deformations in the chalcogenide glasses (e.g., As_2S_3). This phenomenon may be referred to as photo-induced softening (or photo-induced glass transition). The As_2S_3 glass becomes viscous in the illuminated area. This unique phenomenon is due to the fact that the non-crystalline chalcogenide behaves as a soft semiconductor. The softness is due to the twofold co-ordinated chalcogen atoms, which are susceptible to exhibit electro-atomic responses. This chalcogenide glass exemplifies a flexible electron–lattice coupling system.

1.2. Photo-chemical Modifications

1.2.1. Photo-decomposition and Photo-amplified Oxidation

These are processes characterized by the modification of the chemical composition of the material. Electromagnetic radiation approximately equal to band-gap energy has been established as responsible for the dissociation of amorphous As_2S_3 and As_2Se_3. The dissociation is accompanied by an optical identification observable as a 'photographic' effect in thin films of these materials. The densification is reversible by thermal cycling to higher temperatures. After Berkes et al. (1971), the photolysis of arsenic trisulphide follows the reaction:

$$As_2S_3 \xrightarrow{h\nu} 2As + 3S \tag{1}$$

A photo-dissociation in sulfur-rich regions and arsenic regions takes place. The amorphous sulfur produced during photo-decomposition transforms into rhombic sulfur. For the photolysis process, the photon energy should be larger than the semiconductor band gap, and this is the case of the argon laser used for irradiation. In the next step the oxidation of free arsenic in the presence of oxygen and moisture is produced:

$$4As \xrightarrow{H_2O} 2As_2O_3 \tag{2}$$

The electron microscopy studies assessed the formation of the As_2O_3 crystallites on the surface of As_2S_3 film exposed to light.

In the case of As_2Se_3, the photolysis process exhibits peculiar features (Berkes et al., 1971). Initially, the decomposition follows the scheme:

$$As_2Se_3 \xrightarrow{h\nu} xAs + As_{2-x}Se_3 \tag{3}$$

where $0 < x < 2$. Firstly, a non-stoichiometric composition is formed. Then the process proceeds as in the case of arsenic trisulphide.

In the case of arsenic trisulphide, we are in fact dealing with a photo-catalyzed oxidation of arsenic, which creates a depletion of arsenic in glass and determines the release of sulfur rings (S_8) that crystallize. The photo-chemical reaction produces the optical darkening of the glass, in direct connection with the presence of the arsenic phase (Terao et al., 1972). By thermal annealing, a decrease in the precipitate arsenic results and, consequently, a partial recovery of the optical transmission coefficient is obtained. Some authors (Tanaka and Kikuchi, 1973) consider that in the case of compounds, e.g., As_2S_2, the energy received during exposure is enough for the release of the arsenic atoms, which will form finally small crystallites. The enrichment in sulfur of the material will give rise to a bleaching effect due to the fact that the optical absorption edge of sulfur is situated at smaller wavelengths than the absorption edge of As–S.

Matsuda and Kikuchi (1973) have analyzed the photobleaching (PB) effect in the crystalline films of As_2S_2 and concluded that this effect is conditioned by the formation of amorphous regions with an excess of sulfur as a result of the decomposition of the As_2S_2

crystallites:

$$As_2S_2 \xrightarrow{h\nu} xAs + \underset{\text{crystal with defects}}{As_{2-x}S_2} \quad (4)$$

for $0 < x < 2$.

The photo-chemical transformations have been observed in amorphous Ge–S (Tichú, Tycha and Bandler, 1987) and Ge–Se (Tanaka, Kasanuki and Odajima, 1984; Tichú, Tríska, Tichá and Frumar, 1986) films. The occurrence of such transformations is confirmed by the presence of the Ge–O bonds evidenced in infrared spectra. These bonds accompany the irreversible PB of the material. Therefore, in ambient conditions, the photo-oxidation process is triggered during exposure to light. The composition $Ge_{35}S_{65}$ (Wágner and Frumar, 1990) is one of the most sensible. The *photo-amplified oxidation* is very important in the applications of chalcogenide thin films.

Exposure to UV radiation determines the effusion of chalcogen from chalcogenide films or bulk glasses. Popescu et al. (1998) have shown that exposure to UV radiation of thin films in the system Ge–As–S induces the decomposition of the films with the release of sulfur. The process seems to reach saturation after ~3 days of irradiation at the power density of 0.05 W cm^{-2}. Recently, Popescu has found that As_2Se_3 bulk glass is strongly transformed when it is UV irradiated at temperatures near T_g. Selenium is released and AsSe crystallites appear at the surface of the sample. The irradiation in air determines the appearance of large-size As_2O_3 (arsenolite) crystallites.

For the chalcogenide films ($As_{50}Se_{50}$) deposited on silicon wafers, the photo-oxidation can be enhanced due to a chemical reaction of the chalcogenide material with silicon (Prieto-Alcón et al., 2000).

1.2.2. Photo-dissolution, Photo-diffusion and Photo-doping

The photo-dissolution of the metals into the chalcogenide alloys, mainly the diffusion of a metal layer deposited on the surface of a chalcogenide film, has been observed as early as 1966 (Kostishin, Mihailovskaia and Romanenko, 1966). Later, a great variety of thin metallic films were discovered (Ag, Cu, In, Zn) as well as metallic alloys that can be dissolved in the amorphous chalcogenides (As_2S_x, As_2Se_x, GeS_x, $GeSe_x$) under the light exposure. The dissolution is followed by the diffusion of the metal ions through the chalcogenide layers along the direction of the incident light. The diffusion inside the unexposed regions is very weak. It is worthwhile to mention that the diffusion of the metals in chalcogenides occurs in the absence of the light too, but the diffusion rate is considerably diminished and no preferential direction of diffusion was observed. Intense research on photo-doping is under way (Wagner, Kasap, Vlcek, Frumar and Nesladek, 2001).

Among the metals with the largest diffusion rates in chalcogenide glasses is silver. The silver concentration in chalcogenide films can reach 29.1% (Kluge, 1987). The first explanation of the phenomenon of photo-diffusion of silver in As_2S_3 was given by Kokado, Shimizu and Inoue (1976). The incident light excites the silver atoms:

$$Ag \longrightarrow Ag^* \quad (5a)$$

$$\mathrm{Ag}^{*} \xrightarrow{h\nu} \mathrm{Ag}^{+} + \mathrm{e}^{-} \tag{5b}$$

and the excited atoms (ions) are dissolved in the amorphous chalcogenide film thus forming an amorphous photo-doped solid solution:

$$\mathrm{Ag}^{*} + \mathrm{As_2S_3} \longrightarrow \mathrm{As_2S}_x\ :\ \mathrm{Ag} \tag{6}$$

A possibility exists for a non-specific excitation of As_2S_3 (Goldsmidt and Rudman, 1976):

$$\mathrm{As_2S_3} \xrightarrow{h\nu} (\mathrm{As_2S_3})^{*} \tag{7}$$

followed by the reaction with silver of the excited chalcogenide:

$$\mathrm{As_2S_3} + \mathrm{Ag} \longrightarrow \mathrm{As_2S_3}\ :\ \mathrm{Ag} \tag{8}$$

Kluge (1987) considered that the photo-dissolution might be described as a solid-state reaction, which requires the intercalation of the silver ions in the amorphous chalcogenide. Accordingly, he supposed two diffusion coefficients: D_{I} and D_{II} (D_{I} for the concentration of silver $C < C^{*}$ and $\mathrm{D_{II}}$ for $C > C^{*}$ with $D_{\mathrm{II}} > D_{\mathrm{I}}$; C^{*} is a certain threshold concentration of the silver in the doped chalcogenide).

Wágner and Frumar (1990) interpreted the metal-dissolution process as a photo-enhanced diffusion in a diphasic system with immiscibility gap. This interpretation is based on the existence of two regions of glass formation in the system Ag–As–S (Kawamoto, Agata and Tsuchihashi, 1974) with different silver contents ($C_{\mathrm{I}} < 1.6$ at.% and $C_{\mathrm{II}} > 17.4$ at.%). On this basis, they succeeded to explain the steepness of the diffusion edge and supported the existence of two diffusion coefficients.

The lateral diffusion rate (diffusion rate in the plane of the substrate) is much higher for the chalcogenide films deposited on conducting substrates than for those deposited on insulators (Frumar, Firth and Owen, 1984; Firth, Ewen, Owen and Huntley, 1986; Wágner, Frumar and Benes, 1990). The time dependence of the position of the diffusion edge follows a parabolic law, $x \sim t^{1/2}$ (Frumar et al., 1984; Wágner et al., 1990) or varies linearly with the time, $x \sim t$ (Firth et al., 1986). These experimental results have been interpreted in Oldale and Elliott (1997) by the diminishing of the spatial charge created during the diffusion of the more mobile silver ions (Ag^{+}).

de Neufville et al. (1973/1974) observed that photo-doping does not lead to the formation of crystalline Ag_2S, as required if the light should produce a segregated sulfur-rich phase. The photo-enhanced oxidation and the silver diffusion, can be described by the following reactions:

$$(\mathrm{As_2S_3})_{\mathrm{amorph}} \xrightarrow{h\nu+\mathrm{O_2}} (\mathrm{As_2O_3})_{\mathrm{cryst}} + \mathrm{S_{cryst}} \tag{9}$$

$$\mathrm{Ag_{cryst}} + (\mathrm{As_2S_3})_{\mathrm{amorph}} \xrightarrow{h\nu} (\mathrm{Ag{-}As{-}S})_{\mathrm{amorph}} \tag{10}$$

These reactions are thermodynamically favorable (they are accelerated at high temperatures) and are strongly inhibited in the absence of light of energy near E_{g}. Therefore, the light can be regarded as a catalyst that diminishes the activation energy, which prevents a thermodynamically possible but kinetically inhibited chemical reaction. The As_2S_3 is not in fact a true catalyst because it participates in both reactions.

Nevertheless, its photo-enhanced reactivity might indicate enhanced catalytic activity with respect to some chemical reaction in which As_2S_3 cannot participate.

Elliott (1991) described a detailed model of the photo-induced dissolution of silver, exemplified by Ge–S. The general reaction between Ag and Ch during photo-dissolution can be described as a redox reaction:

$$Ag(0) + Ch(II) \rightarrow Ag(I) + Ch(I) \tag{11}$$

where the metal atoms oxidize and become positive ions and the chalcogen atoms become negative ions with non-bridged configurations. For the germanium chalcogenides, one can write the following reaction:

$$Ag^0 + Ch_3Ge{-}Ch{-}GeCh_3 \rightarrow Ch_3Ge{-}Ch{-}Ag^+ + Ge^*Ch_3 \tag{12}$$

where the breaking of a Ge–Ch bond leads to a negatively charged non-bridged center and a half-filled orbital on the germanium atom (Ge^*).

Although reaction (11) shows that the concentration limit for the silver dissolution under light exposure is controlled by the amount of chalcogen in the glass, in fact, this conclusion can be invalidated if the steric effect is important. Thus, it would be more favorable from the energetic point of view to transform the dangling bonds of the incompletely co-ordinated germanium atom in a new configuration, eventually with the participation of some homopolar Ge–Ge bonds:

$$Ch_3Ge^* + Ge^*Ch_3 \rightarrow Ch_3Ge + GeCh_3 \tag{13}$$

This reconstructive transformation will be facilitated by the decrease in the connectivity of the network that gives rise to layer flexibility due to the appearance of the non-bridged chalcogens.

If the microphase separation during photo-dissolution is only bounded to the formation of homopolar Ge–Ge bonds, then the limit structure of the ternary photo-doped material saturated by silver will correspond to the case of two silver atoms for every new Ge–Ge bond. Thus, the structural mechanism of photo-doping predicts that the maximum amount of Ag dissolved in glass is equal to that of germanium and this was experimentally confirmed (Rennie, Elliott and Jeynes, 1986; Oldale et al., 1997).

In the case of the photo-dissolution of silver in arsenic chalcogenides, one supposes a mechanism similar to that in the germanium chalcogenides. With the complete segregation of chalcogen, the maximum silver amount in the glassy network As_xCh_{1-x} corresponds to the total amount of chalcogen necessary for the formation of Ag_2S, as observed experimentally. Therefore, the maximum content of photo-dissolved silver (e.g., in As_2S_3) corresponds to the chemical formula of the glass: $Ag_6As_2S_3$. The experimental values found for the limit composition are situated between $Ag_{2.6}As_2S_3$ and $Ag_{4.6}As_2S_3$ and they prove that only partial phase segregation occurs during photo-dissolution. The segregation of arsenic during photo-doping will lead finally to the formation of amorphous domains based on arsenic. Amorphous arsenic has been evidenced in the photo-doped Ag/As_xS_{1-x} materials by Raman scattering experiment.

In conclusion, the redox mechanism described by Eq. (11) seems to lie at the basis of all the silver photo-dissolution processes in amorphous chalcogenides. Nevertheless, the type of the reaction products and therefore, the maximum amount of silver able to be dissolved will be dependent on the type of chalcogen. In the case of the germanium chalcogenides, the tetrahedral co-ordination will reduce the flexibility of the local structure and the phase segregation will be bounded by the formation of homopolar Ge–Ge bonds. In the case of the arsenic chalcogenides, the low arsenic co-ordination as compared to germanium will facilitate the mobility of the arsenic atoms and the dominant factor for controlling the maximum amount of silver able to enter into the glassy matrix will be the chalcogen concentration.

A novel photo-effect related to photo-doping has been observed in bilayers and alternately deposited films consisting of glassy chalcogenides. The phenomenon is remarkable, in As_2Se_3/Se films. Light exposure induces an increase in the photocurrent (photoconductivity response) in the bilayer and disordered multilayer structures. The mechanism seems to be due to the photo-induced diffusion of selenium atoms at the heterojunctions.

It is useful to compare the photo-diffusion at heterojunctions with the photo-doping phenomenon observed, typically, in Ag/As–S structures. A remarkable difference is that the motion of silver is much more dramatic than that of selenium. In fact, it is not difficult to introduce silver atoms into a depth of 1 μm, but for selenium this depth has been estimated to be less than 100 Å. This difference seems to originate from different removal mechanisms. Although the photo-doping mechanism has not been completely elucidated, a common idea is that ionized silver atoms migrate electrically. However, since selenium has a mixture of chain-like and ring-like molecular structures, photo-induced breaking of the molecules into isolated single atoms may occur with difficulty. The ionization of the selenium fragments may not occur. Then one speculates that the neutral Se fragments, which could be produced by illumination, thermally diffuse into the neighboring amorphous regions, with much faster rate than in conventional thermal diffusion.

The phenomenon of photo-induced surface deposition of metallic silver is a photo-chemical reaction in which a large number of Ag particles deposit on the surface of Ag-rich chalcogenide glasses or films following illumination with band-gap light. Possible application to optical recording devices has aroused much interest, since this allows direct positive patterning to be achieved with high contrast. The mechanism, especially the force inducing the migration of silver towards the illuminated surface, has been discussed from the viewpoint of the electrical properties of the Ag-rich chalcogenide glasses being mixed ion–electron (hole) conductors. Furthermore, from a thermo-dynamical consideration, this phenomenon has been explained to be a photo-chemical reaction toward a thermodynamically stable state, with segregation of excess Ag^+ ions. After Kawaguchi, Maruno and Elliott (1997), the origin of the compositional dependence of the photo-induced surface deposition of metallic silver in Ag–As–S glasses can be accounted for by the combined effects of diffusivity and insolubility of Ag^+ ions in the Ag-rich phase.

The photo-migration of Ag^+ ions in Ag–As–S glasses was observed by Tanaka and Itoh (1994). Migration over distances larger than 1 μm was induced by light illumination and dramatic changes in atomic compositions were revealed.

Arsh, Froumin, Klebanov and Lyubin (2002) have recently demonstrated the photo-induced dissolution of zinc in As_2S_3, As_2Se_3 and $As_{50}Se_{50}$, only at a certain temperatures. The dissolution of Zn in As_2S_3 film, in darkness, starts at 115–120 °C while under the action of light the beginning of dissolution starts at 70–75 °C.

The photo-dissolution, photo-diffusion and photo-doping are important processes of relevance in applications like holography, memory cells, diffraction gratings, submicron lithography. Utsugi (1990) has shown that by applying a scanning tunneling microscope, it was possible to develop a lithographic process with the resolution of several tens of angstroms.

1.2.3. Photo-polymerization

The polymerization of the glass during exposure to light, accompanied by a red shift of the optical absorption edge, is a typical irreversible phenomenon, which was observed both in evaporated As–Ch films (de Neufville, 1975) and simple chalcogen films (Tanaka, 1986).

In the case of the amorphous evaporated films of composition As_2S_3 and As_2Se_3, de Neufville et al. (1973/1974) have shown that both annealing and exposure to light of photon energy corresponding to E_g determine structural transformations that are essentially identical, excepting the appearance of a reversible component for illumination. In both cases, the films were irradiated in vacuum in order to protect them against oxidation.

The film structure and its relaxation after illumination are undoubtedly related to the molecular nature of the vapor phase from which the film is born.

By evaporation, mainly As_4S_4 (or As_4Se_6) molecular species are formed. These molecules knock the deposition substrate, migrate, coalesce and form a molecular glass. Other molecular species are also present in the vapor phase in significant concentrations, e.g., As_4S_4, As_2S_4, As_4S_5 and S_2 (Buzdugan, Vataman, Dolghier, Indricean and Popescu, 1989) or selenium homologues. Therefore, the amorphous film must consist initially of distorted packing of molecules. During relaxation of the film (by heating or by illumination) the molecules polymerize, i.e., form large molecules by bond breaking and interconnection.

Because the photo-chemical activity of the arsenic sulfide is accompanied by structural modifications, the two effects are closely related. The response of the material to the light consists, on one hand, in the photo-polymerization and on the other hand in the creation of trapped non-equilibrium carriers (holes and electrons). Occurring in both cases are a perturbation of the equilibrium concentration of the holes, of the trapped electrons and of the dangling bonds (in fact the free uncharged radicals). An excess of concentration of broken bonds is necessary in a given stage of the polymerization process for the triggering of the reconstructive transformation of the network. The catalytic effects of the semiconductors are also associated with the free charge carriers on the surface, which are influenced by light and impurities (Schwabb, 1957). The enhanced chemical reactivity of As_2S_3, which accompanies the exposure to light of band-gap energy (e.g., the oxidation in a 10^{-6} Torr vacuum or the reaction with a thin silver layer) is related to the structural transformations by the intermediary of the localized electron defects that include trapped electrons and broken bonds.

Onari, Asai and Arai (1985) have studied the photo-induced modifications in the amorphous evaporated films of $(As_2S_3)_{1-x}(As_2Se_3)_x$ by irradiation with a Hg-lamp (38 mW cm^{-2}). The chopped infrared light with $\lambda > 1$ μm was used. The observed polymerization process was explained by the formation of defect pairs D^+ and D^- during the absorption of phonons and subsequent re-arrangement of the local configurations by switching the As–As bonds and the D^+D^- pairs, which lead, to As–S or As–Se bonds (Street, 1977).

In the model of configurational co-ordinate (Mott and Stoneham, 1977), the polymerization is explained as follows. Firstly, electron–hole pairs are created by the excitation of the non-bonding electrons of the chalcogen atoms. After the relaxation of the network by electron–lattice interactions, trapped pairs D^+D^- do appear and the final configuration of the network is obtained by switching the interatomic bonds. During the polymerization process, the rate of variation of the optical absorption coefficient of As_2S_3 grows with the temperature and this fact suggests an activated process. The variation in the number of polymerized molecules, m, can be expressed by the equation:

$$\frac{dm}{dt} = -Pm \tag{14}$$

where P is the probability of the polymerization process that can be related to the activation energy E_a and the temperature T by the equation:

$$P = P_0 \exp(-E_a/kT) \tag{15}$$

In the hypothesis of a linear dependence between m and $(\Delta\alpha_0 - \Delta\alpha)/\Delta\alpha_0$, where $\Delta\alpha_0$ is the modification of the light-absorption coefficient, after a very long irradiation time and $\Delta\alpha$ is the modification of the absorption coefficient after the time t, it is possible to calculate the probability P for every temperature. From an Arrhenius plot $\ln P \sim f(1/kT)$ was obtained for As_2S_3 an activation energy of 0.008 eV (Onari et al., 1985).

Polymerization causes densification and in general, after prolonged exposure, the density and structure of the thin film become virtually identical with those of melt-quenched glasses and well-annealed films (Chang and Chen, 1978; Kolwicz and Chang, 1980).

Photo-polymerization of As_4S_4 (Porter and Sheldrick, 1972) has been reported in crystals of α- and β-As_4S_4 (Matsuda and Kikuchi, 1973), so that this particular photo-induced effect is not unique to the amorphous state.

The photo-polymerization of the molecular fraction was shown to have the major contribution to photodarkening (PD) during the first stages of illumination of As_2S_3 by CW and pulsed laser radiation (Bertolotti, Michelotti, Chumash, Cherbari, Popescu and Zamfira, 1995).

Under the action of light, the depolymerization can occur in certain conditions. It is suggested that by simultaneous annealing below T_g and illumination a partial depolymerization is possible.

1.2.4. Photo-induced Phase-Changes

The order–disorder phase-change induced by light has been discovered by Ovshinsky (1968) and applied for optical memory under the name of ‘ovonic memory’. The enthalpy of the amorphous state is higher than in the crystalline state. Under the action of a laser spot, the structure changes from the amorphous state to the crystalline state. Then, the refractive index and the reflectivity of the film change. This represents the optical memory effect. The information is inscribed by changes in reflectivity. When a high-power laser spot locally irradiates the crystalline film, the temperature of that portion of material goes over the melting temperature. After stopping the irradiation, the temperature is rapidly decreased through the supercooling state and the material becomes amorphous. When a low-power laser spot irradiates the amorphous material, the temperature is raised enough to determine its crystallization. Thus, the 'information' is erased, that is, the reflectivity of the film in the irradiated region is strongly reduced. The materials for optical memory based on phase-change must exhibit rather high-speed crystallization characteristics. Examples are In–Sb–Te, Ge–Te–Sb and Ag–In–Sb–Te (Ohta, 2001).

1.2.5. The Oxygen-Assisted Photo-induced Phenomena

The effect of oxygen on the photo-induced changes in non-crystalline chalcogenides has been recently demonstrated. In general, the photo-induced effects are believed to arise from the excitation of electrons above the bandgap. Nevertheless, it was suggested by Marquez, Gonzalez-Leal, Prieto-Alcon, Jimenez-Garay and Vlcek (1999) that photo-induced oxidation may be responsible for the photoeffects. Khrishnawami, Jain and Miller (2001) have demonstrated that the presence of oxygen is catalytic, if not a requirement, for creating photo-induced changes in the electronic structure of arsenic selenide glasses. When illuminated with a laser under vacuum, no significant compositional or structural changes occur on the surface of the $As_{50}Se_{50}$ film. On the other hand when irradiation takes place in ambient atmosphere, the surface is depleted of As and enriched in Se, a fact that can be understood in terms of the difference in the energy of formation of As and Se point defects. Also, prolonged illumination in air causes conversion from Se–As bonds to Se–Se bonds. The results confirm that oxygen from ambient diffuses into the surface, with a concurrent depletion of As at the surface.

2. Reversible Modifications

The reversible photo-structural transformations, which will be discussed in this section, are modifications induced by light in thin amorphous films annealed below T_g (de Neufville et al., 1973/1974) and in bulk glass (Hamanaka, Tanaka and Iizima, 1977a,b), which can be eliminated by an appropriate thermal treatment. The reversible modifications are independent of the sample history and therefore, they are of interest for understanding the fundamental properties of the amorphous solids.

The class of the reversible phenomena comprises the small but significant decrease in the optical gap, the PD. The reversible PD is accompanied by reversible modifications of hardness, softening temperature, density, rate of dissolution in various solvents and elastic, dielectric, photo-electric and acoustic constants (Tanaka, 1980).

The reversible modifications of the hardening during exposure to light have been observed in many chalcogenide compositions. The effect was called *photo-hardening*. The photo-hardening is typical for stoichiometric As_2S_3. The light irradiation determines the decrease while the annealing determines the increase in the microhardness.

The influence of the substrate seems to be very important (Terao et al., 1972). As a function of substrate, the photo-hardening effect can change its magnitude and sign. For arsenic sulfide films deposited on quartz, glass, or CaF_2, the annealing of the sample exposed to light leads to a decrease in microhardness and the next light exposure determines a new increase in the hardness. For KCl, NaCl or even crystalline As_2S_3-supported films, for free-standing films and for bulk glasses, the annealing after exposure leads to the increase in microhardness, and the next exposure determines the decrease in the hardness. It is supposed that the effect of the substrate is related to the magnitude of the expansion coefficient of the material. For the first type of substrate materials, the expansion coefficients are lower than those of the film, while for the second type the situation is reversed. After annealing near T_g and cooling at room temperature, it seems that in the chalcogenide film mechanical strains are developed, and these strains are dependent on the substrate properties.

As opposed to As_2S_3 films, the light exposure of As_2Se_3 films leads to the increase in microhardness and density. The maximum effect in As–Se system is observed for the compositions with the maximum disordered structure (Tanaka and Kikuchi, 1973). The reversible phenomena seems to be caused by the excitation of the As–As and As–Se bonds, and their switching with the consequence of the formation of the charged defects As^{2-} and Se^{5+}. The optimum condition for bond switching is the presence of one As–As bond for every Se atom in the system (Tanaka and Kikuchi, 1973).

The change in the rate of dissolution of amorphous chalcogenides (in particular As_2S_3) in solvents under light exposure and by thermal annealing is practically the same. By light irradiation of the annealed As_2S_3 films, i.e., in the reversible cycles, the change in the rate of dissolution in NH_3, CH_3NH_2 and $(CH_3)_2NH$ solutions is no more than two times while in the first cycle starting from the virgin films this change is by a factor of 10–1000. The change in the dissolution rate after photo-structural transformations and by annealing seems to be related to the variation of surface absorption ability of the solvents. The solvent molecules are activated by S_n fragments and by oxidation, which create As–As bonds and finally leads to the formation of thio-arsenates (Zenkin, Khirianov, Lobanov, Mihailov, Iusupov and Iakovuk, 1989).

Evidences of reversible optical anisotropy effects induced by the polarized light were observed (Tanaka, 1980a,b) with energy identical to that used for triggering the PD.

In spite of the close relation between the optical edge shift and the modifications of the macroscopic properties of the chalcogens, many papers point out the fact that the mechanism of some photo-structural changes may be different from that which produces the shift of the optical absorption edge.

2.1. Photodarkening and Photobleaching

2.1.1. *Photodarkening*

The reversible PD phenomenon was reported by de Neufville et al. (1973/1974). If an amorphous chalcogenide film (e.g., As_2S_3) is irradiated by light with the photon energy near E_g (2.4 eV), then one observes a shift of the absorption edge towards lower energies down to a saturation limit. The new state called darkened state, due to the lower transparency of the film when regarded in white light, and it can be erased by annealing the glass near the softening temperature. Because the photoconductivity spectrum shifts also towards lower energies (red shift) (Babacheva, Baranovski, Lyubin, Taghirdjanov and Feodorov, 1984), it was assumed that the PD phenomenon is a photo-induced decrease in the optical gap, which can be recovered by annealing. The recovery was called PB. The PB is the reverse process of PD.

The PD is a common phenomenon in many glassy materials especially in chalcogenides ones. The chalcogens themselves show such phenomena (amorphous selenium is a typical example) but they exhibit specific features (Bogomolov, Poborcii, Holodkevici and Shagin, 1983; Katayama, Yao, Ajiro and Inui, 1989). The hydrogenated As–S and As–Se films show also PD (Fritzsche, Said, Ugur and Gaczi, 1981). The amorphous tetrahedral materials, e.g., a-Si : H, do not exhibit PD. As regards the pnictide materials, the experimental results are controversial: the arsenic does not show darkening (Mytilineou, Taylor and Davis, 1980) but phosphorus exhibits such an effect although with different characteristics when compared to the amorphous chalcogens (Kawashima, Hosono and Abe, 1987).

The oxygen belongs to the same chalcogen group of the periodic table. The studies of As_2O_3 and GeO_2 glasses have not issued a definite answer concerning the existence of the PD in these materials (Pontushka and Taylor, 1981; Schwartz and Blair, 1989).

The PD can be practically detected only in amorphous films thinner than 10–20 μm because this effect enhances the absorption coefficient for the light of wavelength corresponding to optical gap, and consequently, the effective depth of light penetration in the sample is diminished. Thus, although the intense, long-time illumination can, in principle, lead to the PD of the thick samples, the limited exposure time (practically no more than 1 week) puts a limit to the sample thickness (Tanaka and Ohtsuka, 1977). Most experiments have been carried out on thin films (or very thin plates) which were previously annealed below T_g for stabilization. It is remarkable that the bulk glasses are preferred for the investigation of the mechanism of the photo-induced phenomena because the films deposited by evaporation, sputtering, etc., contain a high amount of defects that are not completely eliminated by annealing (Tanaka, 1987a,b).

The structural nature of the reversible PD phenomena is demonstrated by the observation that only the disordered materials show reversible PD (Hamanaka et al., 1977a,b) and also by the observation that the PD is sensible to hydrostatic pressure (Tanaka, 1987). This is because the photon energy necessary to induce structural changes in crystals is much higher than those needed to induce PD. The non-crystalline state exists in multiple structural configurations and is characterized by local minima of the distortion energy, not very different from another, so that the photon energy can be enough for triggering a transition to a metastable neighboring structural configuration.

A fundamental feature of the materials with PD properties is the presence of a significant concentration of at least one element with non-tetrahedral bonds. This gives more freedom for steric arrangements as a consequence of the change of the atomic-scale interactions.

For $As_{50}Se_{50}$ layers deposited on silicon wafers, an interaction with the substrate was demonstrated during illumination. In this case, too, the reversible structural changes induced by light are accompanied by reversible PD (Prieto-Alcón et al., 2000).

Finally, we must remark that recently, Tanaka and Nakayama (2000) have shown that the fundamental photoconductive edge in As_2Se_3 glass is located at nearly the same photon energy with the absorption edge in the corresponding crystals. This correspondence implies that the mobility edges in the glass are located at the band-edge positions in the crystals. Such an electronic similarity must reflect the structural similarity in amorphous and crystalline chalcogenides. Therefore, the crystalline features should be important for the explanation of the PD effects.

Kuzukawa, Ganjoo and Shimakawa (1998) have studied, recently, the effect of band-gap illumination and annealing below the glass transition temperature on the thickness and the optical band gap of As-based (As_2S_3, As_3Se_3) and Ge-based (GeS_2, $GeSe_2$) obliquely deposited chalcogenide films. It was observed that in the case of arsenic-based glasses, illumination increases the thickness (expansion effect) and the band gap decreases (darkening effect), while for germanium-based glasses, both thickness and band-gap show an opposite behavior to that of arsenic-based glasses. By annealing the samples, before and/or after illumination, the trends of the changes in thickness and band-gap are reversed. These changes have been explained on the basis of ordering of the structure by annealing, and repulsion and slip motion by illumination, the latter processing being due to the negative charging of layers by electron accumulation in conduction band tails (Kuzukawa, Ganjoo and Shimakawa, 1999). An important observation is that the changes in thickness and gap are higher for the obliquely deposited amorphous films than in the case of normally deposited films.

2.1.2. *Photobleaching*

In most chalcogenide films, there was observed a bleaching phenomenon under the influence of light: *photobleaching*. The PB is the reverse effect of PD. In order to get PB, it is important to illuminate the sample at temperatures a little but larger than those that favor the PD (Averyanov, Kolobov, Kolomiets and Lyubin, 1980). The effect is illustrated in Figure 1 (Lyubin, 1985). The line AB corresponds to the PD at room temperature. The lines AC and BD describe the modifications in the transmission of

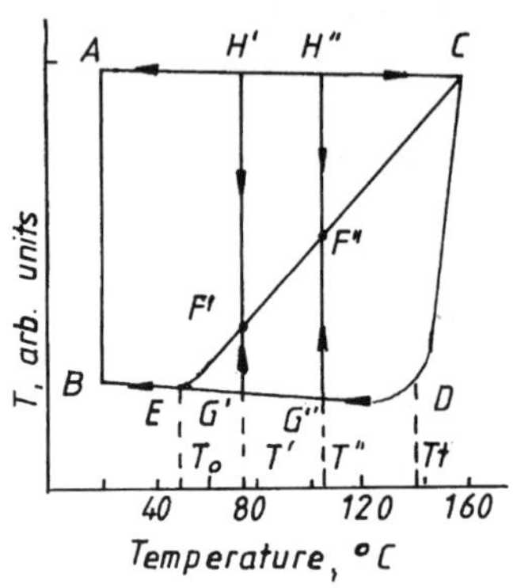

FIG. 1. The modification of the transmission in an As_3Se_2 film during irradiation by a He–Ne laser light ($\lambda = 0.633\ \mu m$) at various temperatures.

the non-illuminated and photodarkened films with temperature, respectively. The light irradiation of a photodarkened film at the temperatures T' and T'' produces an increase of the transmission (lines G′F″ and G″F″).

It is remarkable that the illumination of the virgin films at the same temperatures $T'$and T'' gives rise to a *PD effect* (lines H′F′ and H″F″) with the same resultant transmission. If a film photodarkened at room temperature is heated under light irradiation, its transmission coefficient changes following the line BEF′F″C. Thus, the final value of the transmission is determined only by the temperature corresponding to the last irradiation but not on the irradiation sequence or by the heating processes.

The above-described phenomena are typical for a great number of glasses in the systems As–S and As–Se. In some cases, e.g., in the As_3Se_2 films, the PB is characterized by a thermal threshold and this means that the bleaching can be observed only above a given temperature called the *optical bleaching threshold*, T_0 which is lower than the thermal bleaching temperature, T_t.

It is worthwhile to mention that the PB process has a higher sensibility by a factor of 10–50 when compared to PD.

Many experiments (Lyubin, 1985) have shown that the value of the final transmission depends on the intensity of the incident light (Fig. 2).

The curves 1 and 2 in the figure correspond to the PD of an As_3Se_2 film during light irradiation at 75 °C, using a He–Ne laser for the intensities I_1 and $I_2 = 10I_1$. If a sample, previously illuminated by I_2, is irradiated with I_1 then the PD is produced (curve 3) up to the value $T = T_1$ of the transmission.

In the case of As_2S_3 films, the illumination at room temperature by red light (band–band energy) can bleach the material if a previous illumination by blue light (in-band energy) was performed.

In conclusion, we can affirm that after successive PD and photo-bleaching of an amorphous chalcogenide film, the value of the transmission and the refractive index is defined by the temperature at which the last irradiation is carried out, by the wavelength used and by the intensity of the light.

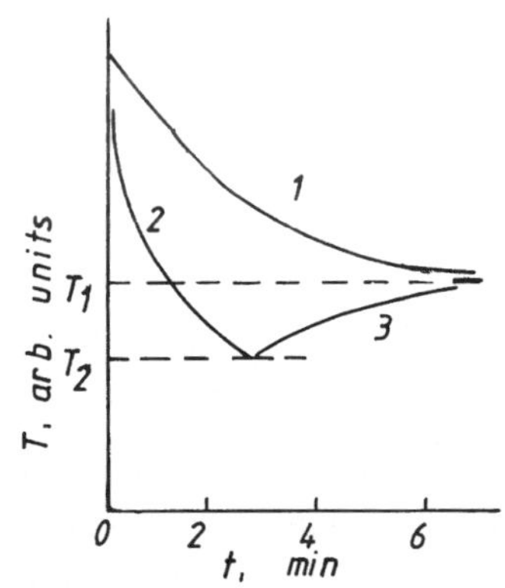

FIG. 2. The modification of the optical transmission in a film of As_3Se_2 for various illumination intensities.

2.1.3. Photo-induced Anisotropy

The amorphous materials are essentially isotropic while the crystals show anisotropy, i.e., most of the properties depend on the crystal orientation. The photo-induced anisotropy was originally discovered by Weigert (1920). Then, the phenomenon was observed in organic polymers including liquid crystals, in oxides and in phase-separated systems. Besides, the temporary anisotropy induced by electrical and magnetic fields or by mechanical forces which disappear when the inducing factor is switched off, the chalcogenide amorphous films (Jdanov, Kolomiets, Lyubin and Malinovski, 1979; Jdanov and Malinovski, 1980; Hajtó, Jánossy and Forgács, 1982a,b) and bulk glasses (Tikhomirov and Elliott, 1995a–c) show quasi-stable optical and electronic anisotropy when illuminated by polarized light. The anisotropy induced by the electric field of the light wave is maintained after switching off the illumination.

The main photo-induced anisotropy phenomena, also called vectoral phenomena, are dichroism and birefringence. The other phenomena are: difference in the intensity of the photo-luminescence and difference in the fine structure of the X-ray absorption edge for polarizations of the control light beam parallel and perpendicular to the direction defined by the polarization of the beam used in PD.

The optical anisotropy induced by exposure to polarized light is observed not only in annealed glasses but also, and essentially undiminished, in glasses photodarkened by light exposure. The photo-induced anisotropies can be reversibly reproduced after annealing, and the axes, which define the optical anisotropy can be rotated by turning the polarization direction of the exposing light. These anisotropic phenomena differ from the isotropic photo-structural changes in their induction and annealing kinetics, as well as in their dependence on temperature and on photon energy.

The dichroism induced by light is experimentally easily accessible and it is more directly related to the scalar PD (i.e., isotropic PD) than to birefringence.

The plot of the variation of the absorption coefficient in the Urbach tail as a function of the photon energy is given in Figure 3 for the case of As_2S_3 films.

The absorption coefficients in the direction parallel ($\alpha_{\parallel}$) and perpendicular ($\alpha_{\perp}$) to the polarization plane defined by the polarized beam, which gives rise to PD, are different. The absorption along the perpendicular direction is higher than along the parallel direction while both are lower than in the case of irradiation by unpolarized light (Murayama, 1987).

By studying the difference between $\alpha_{\parallel}$ and $\alpha_{\perp}$ during the PD under polarized light, there was observed an oscillation dependent on the light intensity and on temperature, between the states with positive dichroism ($\alpha_{\parallel} > \alpha_{\perp}$) and those with negative dichroism ($\alpha_{\parallel} > \alpha_{\perp}$) (Lee, Pfeiffer, Paessler, Sayers and Fontaine, 1989; Lyubin and Tikhomirov, 1989; Lee, 1990). The most interesting feature is the cycle negative–positive–negative dichroism produced when the annealed As_2S_3 sample is photodarkened sequentially to 80 and 300 K (Lee, 1990).

The dichroism phenomenon was also observed in the chalcogenide films that do not photodarken at room temperature (Lyubin and Tikhomirov, 1989).

The photo-induced dichroism can be re-oriented by changing the polarization of the light. The existing dichroism is destroyed within a time span much shorter than that

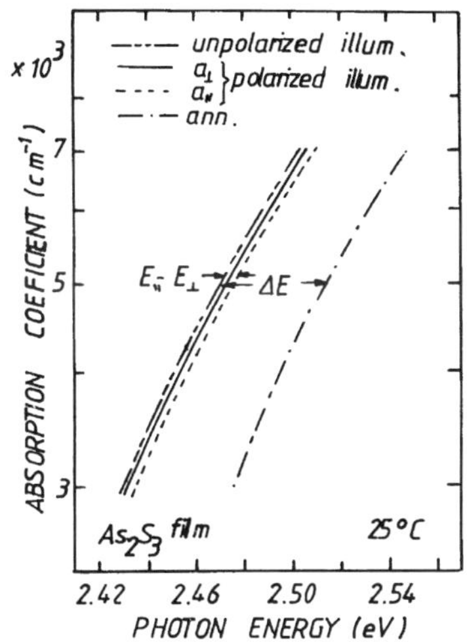

FIG. 3. The dichroism. The variation with the incident photon energy, of the optical absorption coefficient of an As_2S_3 film photodarkened in polarized light.

necessary for its creation. A similar dichroism is created afterwards in an orthogonal direction, i.e., reversal of dichroism is possible. Such re-orientation can be performed several hundred times without any sign of a decrease in the effect. This light-induced re-orientation of erasure is not associated with any change in the scalar PD.

The largest value of the photo-induced dichroism was observed in Sb_2S_3 and $Ge_{28.5}Pb_{15}S_{56.5}$ (Lyubin and Tikhomirov, 1990a,b), i.e., in materials, which do not exhibit reversible PD. A further proof that PD and photo-induced anisotropy are different phenomena is the fact that thermal destruction of the PD and photo-dichroism are also characterized by different behavior (Lyubin and Tikhomirov, 1990a,b).

The photo-induced anisotropy can not only be re-oriented by switching the polarization of the beam which produces the PD but it can also be eliminated by irradiation with unpolarized light, without influencing the PD (Hajtó, Janossy and Forgacs 1982a,b). The comparison between the magnitude of the effect as a function of the excitation energy of the two effects (dichroism and PD) shows that the dichroic effect reaches a maximum value in the region of the Urbach tail where the threshold of the efficiency of the PD is reached. Thus, the dichroism will be largest when the electrons are excited with the highest probability from the states situated at the top of the valence band originating from the orbitals of the lone pair.

The photons of higher energy excite the electrons from the lower states and give rise to PD without anisotropic characteristics, associated to the excitation of the lone pair orbitals. Based on this argument, Lee (1990) suggested that PD is a more general effect which implies global changes in the atomic network at the level of medium-range order while anisotropy is induced as a special case where the polarized light of appropriate energy excites the lone pair electrons leading to the re-orientation of the local anisotropic structures. Thus, the two effects take place simultaneously in certain conditions but some of their properties are different.

The polarization image can exhibit a very high contrast that is limited only by the homogeneity of the polarizer elements (Pontushka and Taylor, 1981). The comparison between polarizers must be done in the transparency range where dichroism is absent and the photo-induced birefringence is high.

The photo-induced anisotropy is not a property specific to the amorphous chalcogenide films. Similar phenomena have been observed in bulk As_2S_3 glasses too (Serbulenko, Iu, Tiscenko and Nenashev, 1974) and in photosensitive emulsions dispersed in gelatine (Jdanov, Malinovski, Nikolova and Todorov, 1979). The anisotropy has been revealed also in As–(S, Se), in $GeSe_2$ (Hajtó et al., 1982a,b) and in other amorphous materials. The maximum value of the anisotropy can reach $\sim 1/10$ of the intensity of the PD effect and of the modification of the refractive index related to this effect (Kimura, Murayama and Ninomiya, 1985; Lyubin, Yasuda, Kolobov, Tanaka, Klebanov and Boehm, 1998).

Lyubin et al. (1998) have revealed the possibility of re-orientation with the linearly polarized light not only of defects and scattering centers but also of interatomic covalent bonds in various chalcogenide glasses.

Tikhomirov and Elliott (1995a–c) remarked that ordered chirality is a property of crystalline analogues of chalcogenide and oxide glasses (e.g., spirals in c-As_2S_3 and c-As_2Se_3 or right- and left-hand modifications in c-SiO_2) in contrast to the disordered chirality in glasses. The glasses can be nevertheless ordered by irradiation with polarized light and this explains the photo-induced anisotropy.

Metastable photo-induced anisotropy generated by linearly polarized sub-gap light was also shown to appear in the undoped and Pr-doped Ge–S–I, Ge–Ga–S and As–Ga–S glasses with varying composition and Pr content (Tikhomirov, Hertogen, Adriaenssens, Krasteva, Sigel, Kirchhof, Kobelke and Scheffler, 1998). Compositional trends of the photo-induced anisotropy are related to the light-stimulated re-orientation of intrinsic anisotropic centers and their environments, which vary with composition. These centers can be modeled with pairs of over- and under-coordinated sulfur atoms: the valence alternation pairs (VAPs).

Kolobov, Lyubin, Yasuda, Klebanov and Tanaka (1997) have investigated the photo-induced anisotropy in a model of chalcogenide glass As_2S_3 using reflectance difference spectroscopy. They found that the anisotropy can be induced in the energy range much exceeding the energy of the photons of the exciting light and that not only defects but also main covalent bonds of the glass are re-oriented by linearly polarized light. The sign of the photo-induced anisotropy, especially at higher energies, strongly depends on the photon energy of the exciting light. This feature was explained by the photo-induced change in the bond topology involving a conversion between bonding and non-bonding electrons. Pre-irradiation of the glass by unpolarized light increases substantially the magnitude and creation rate of the photo-induced anisotropy indicating that both native and photo-induced defects play a role. The processes occurring in bulk glasses and thin films are essentially identical and the observed difference in reflectance difference spectroscopy is caused only by an interference phenomenon.

Tanaka, Gotoh and Nakayama (1999) have recently discovered that linearly polarized light can produce an anisotropic surface corrugation in amorphous chalcogenide films of Ag–As–S. The corrugation resembles a mouth whisker consisting of narrow fringes, which are parallel to the electric field of light, and streaks, which radiate from the illuminated spot to directions nearly perpendicular to the electric field. Optical birefringence of about 0.01 appears with this pattern.

Fritzsche (1995) has shown that optical isotropic materials such as chalcogenide glasses can become optically anisotropic because they consist of and contain entities that are anisotropic. The original macroscopic anisotropy originates from the random orientations of the microscopic anisotropic entities. A recombination event, which leads to a structural change of a microscopic anisotropic entity, will change the orientation or nature of this anisotropy. This constantly happens everywhere in the material during illumination without, however, necessarily producing a macroscopic anisotropy. For this to happen, it is necessary that the recombining electron–hole pair be excited in the same microscopic anisotropic entity, which undergoes the structural change. This means that macroscopic anisotropies result from geminate recombination of electron–hole pairs, which do not diffuse out of the microscopic entity in which they were created by absorbed photons. The lack of electron–hole pair diffusion and the geminate nature distinguish the recombination events leading to anisotropies from all the other events, which yield isotropic (or scalar) photo-induced changes.

In order to explain the optical-induced anisotropy in bulk glasses, it was supposed that the microscopic mechanism comprises two parts: the optical irreversible scalar component due to creation of randomly formed dipole moments and the reversible vectoral component caused by the re-orientation of intrinsic dipole moments (structural units) according to the electrical vector of the inducing light (Tikhomirov and Elliott, 1995).

Emelianova, Hertogen, Arkhipov and Adriaenssens (1999) have developed a model for explaining some basic characteristic features of photo-induced anisotropy in glassy semiconductors. The model assumes the occurrence of correlated pairs of localized states for electrons and holes, and relates photo-induced anisotropy to generation of geminate electron–hole pairs trapped by these localized states.

2.1.4. Photo-induced Defect Creation

The photo-induced defect creation occurs in amorphous chalcogenides during illumination and this induces changes in electron properties. Biegelsen and Street (1980) by studying electron spin density of chalcogenide glasses, observed that prolonged exposure to strongly absorbed light induces a large density of metastable defects. Shimakawa, Inami, Kato and Elliott (1992) observed a similar result for defect creation while studying the photoconductivity in amorphous chalcogenides. Meherun-Nessa, Shimakawa and Ganjoo (2002) have studied the effect of structural flexibility on photo-induced defect creation in obliquely and normally deposited films of amorphous As_2Se_3, at different temperatures by band-gap and sub-band gap illumination. It has been stated that the defect creation is larger for obliquely deposited films compared to normally deposited films, which may be due to a larger flexibility of obliquely deposited films with many voids. A large value of quantum efficiency was observed even for sub-gap illumination.

2.2. Other Reversible Photo-induced Effects

Lyubin and Tikhomirov (1990a,b, 1991) have shown that the photo-anisotropy in chalcogenide glasses can be induced not only by light of energy above the optical gap but also by light of energy below the gap. New phenomena were discovered: the anisotropy of the transmittance without PD, the photo-induced girotropy, the photo-induced scattering of light accompanied by depolarization. These phenomena were revealed in As_2S_3 glasses by irradiation with polarized light of sub-gap energy.

2.2.1. The Anisotropy of the Transmittance Without PD

If one measures the variation of the relative transmittance of the linearly polarized light during irradiation, one observes that the transmittance decreases with more than one order of magnitude in the case of a prolonged irradiation (Fig. 4). On the other hand if one follows the time evolution of the transmittance of the anisotropy defined by $2(I_{\|} - I_{\perp})/(I_{\|} + I_{\perp})$, then the kinetics of the process looks differently.

The curve of evolution is not monotonous (Fig. 4, curve 2). Firstly, the anisotropy of the transmittance increases then maximum follows and finally decreases and even changes the sign. The maximum value of this anisotropy is 0.4 and this exceeds the value observed in non-crystalline chalcogenide films (Lyubin and Tikhomirov, 1991). By heating the sample, the transmission anisotropy disappears around 120 °C, therefore, at a lower temperature than the T_g of the As_2S_3 (185 °C). The disappearance of the anisotropy by heating at temperatures substantially lower than T_g has been observed in amorphous $As_{50}Se_{50}$ films also (Lyubin and Tikhomirov, 1991).

The oscillatory character of the kinetic curves of the transmittance anisotropy with the alternative change of sign is typical. As there is no reason to consider a priori that the photo-induced dichroism can have alternated signs, then, in order to explain the non-monotonous character of the variation of the transmission anisotropy, Lyubin and Tikhomirov (1991) suggested that in fact we are dealing with the rotation of the polarization plane of the polarized light during sample penetration.

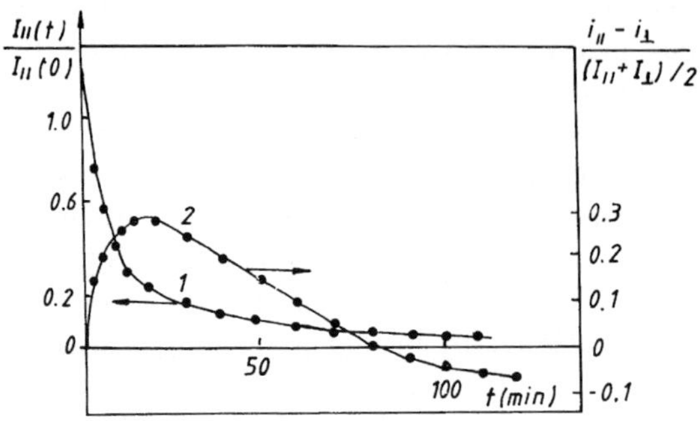

FIG. 4. The kinetics of the evolution of the optical transmittance (1) and of the transmittance anisotropy (2) for a control beam, induced in a sample of As_2S_3 of 2.5 cm thickness by a control light beam of power density 5 W cm^{-2}.

Hájto and Ewen (1979) have shown that the As_2S_3 glass rotates the polarization plane of a polarized ray before light irradiation (therefore, As_2S_3 exhibits natural optical activity) and after irradiation, the rotation angle changes (therefore photo-induced optical activity exists).

Lyubin and Tikhomirov (1991) have shown that simultaneously with the rotation of the polarization plane, an ellipticity of the transmitted light does appear. The ellipticity is low enough before the sample irradiation but during irradiation changes the sign and increases up to a saturation value situated well above the initial values. Moreover, it was shown that the transmitted light is completely polarized before irradiation and the depolarized component appears during irradiation and its weight increases up to a saturation value. A significant variation of the optical activity, of the ellipticity and of the degree of the depolarization of the transmitted light, as a function of the time of action of the incident beam, was observed when the incidence position of the laser beam on the sample was changed.

2.2.2. Photo-induced Girotropy

Starting from the results obtained from the investigation of the optical activity and of the photo-induced ellipticity, Lyubin and Tikhomirov (1991) have suggested that the appearance of the photo-induced girotropy in the chalcogenide glasses, i.e., the photo-induced circular birefringence, which leads to optical activity and photo-induced circular dichroism that leads to ellipticity, therefore the optical properties of the investigated chalcogenide glasses are essentially determined by the spatial dispersion.

The photo-induced circular dichroism was observed and studied (Lyubin and Tikhomirov, 1991). Figure 5 (curve 1) shows the kinetics of the transmittance girotropy induced by the linearly polarized light. The As_2S_3 sample exhibits initially a very small girotropy of the transmittance that differs in magnitude and sign for various regions of the sample. The transmittance girotropy changes its sign and increases up to large values. Figure 5 (curve 2) shows the kinetics of the transmittance girotropy induced by right-hand circularly polarized light. The circular dichroism reaches in this case, values 3 times greater than in the previous case.

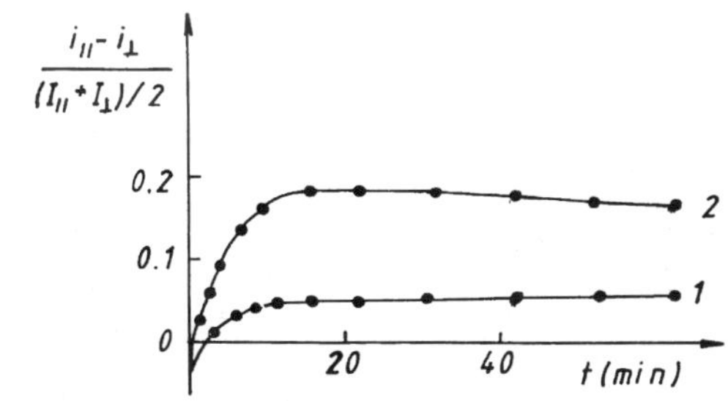

FIG. 5. The kinetics of the transmittance girotropy induced in an As_2Se_3 sample of thickness 2.5 μm by the linearly polarized light (1) and circular polarized light (2).

2.2.3. *The Photo-induced Scattering of the Light*

It was observed (Lyubin and Tikhomirov, 1991) that the light of energy below the gap exhibits a strong photo-induced scattering. This effect is revealed by the change of the shape of the cross-section of the transmitted laser beam. Before the irradiation, the image of the transmitted laser beam on a screen is circular, as for the initial beam. During irradiation a diffuse halo is formed around the original spot. The image is gradually eroded and finally is stabilized as a nebula covered by spots (speculae).

Evidences of the above-described photo-induced effects, in the As_2S_3 glass, have also been revealed in other chalcogenide compositions. It was found that for $As_{34}S_{52}I_{14}$, the values of the transmittance anisotropy and girotropy exceed those found in As_2S_3 and are reached in a shorter time. It is remarkable that the sign of these effects is negative and their kinetics has a monotonous character.

The amplification of these effects in the iodine glass can be ascribed to the following causes:

(i) The introduction of the single-valent iodine atoms leads to the formation of polymeric chain structures with enhanced micro-anisotropy as compared to the stoichiometric As_2S_3. This feature was stated in 1973 on the basis of the conductivity anisotropy in As–Se–I glasses (Kolomiets, Lyubin and Shilo, 1973).

(ii) The addition of iodine atoms determines the diminishing in the E_g. In this case, the radiation used in experiments ($\lambda = 633$ nm) will be situated in the energy range of the Urbach tail of the absorption spectrum where the effects can be very different.

The photo-induced scattering of the light is only partially thermo-reversible, as opposite to other photo-induced effects, which are completely reversible.

2.2.4. *Anisotropic Opto-mechanical Effect*

This effect consists in the appearance of optically controllable, reversible nano-contraction and nano-dilatation induced in chalcogenide glassy films by linearly polarized light (Krecmer, Moulin, Stephenson, Rayment, Weland and Elliott, 1997). Very good correlation of this effect with the photo-induced dichroism was observed. The effect seems to be electronic in nature and not thermal and is believed to be caused by the same photo-induced structural arrangements that are responsible for the optically induced anisotropy observed in chalcogenide glasses (Stuchlik, Krecmer and Elliott, 2001).

Krecmer et al. (1997) discovered the reversible anisotropic volume change induced by polarized light in a thin film of $As_{50}Se_{50}$. Contraction occurs along the direction of the electric field vector and dilatation perpendicular to that direction. Light from a He–Ne laser was used whose energy (~ 2 eV) falls into the Urbach absorption region. This experiment shows that the anisotropies produced extend to other material properties besides the optical tensor. The magnitude of the effect suggests that the anisotropic microvolumes of the whole material are involved and not only IVAP species. These new results imply that the elastic properties, sound propagation and probably many other material parameters of chalcogenide glasses become anisotropic with light exposure.

2.2.5. *Photo-induced Fluidity*

Photo-induced fluidity or photo-induced glass transition has been discovered by Hisakuni and Tanaka (1995). Sample of As_2S_3 (optically processed film: thickness of 50 μm, size 0.2×2 mm^2) was annealed at the glass-transition temperature ($\sim$450 K) to stabilize the glass structure. The free flake was obtained by peeling off from a substrate. Next, one end of the sample was pasted to a glass slide and the other end was bent with a stick. The bending was confirmed to be elastic. Then, the curved As_2S_3 flake was illuminated locally by a laser-focused light beam (the spot diameter: 50–100 μm) from a He–Ne laser (633 nm, 10 mW) for 2 h. A permanent deformation occurred. The irradiated part became fluid.

The phenomenon becomes more conspicuous if illumination is provided at lower temperatures. This indicates that the photo-structural fluidity occurs through athermal processes. If the temperature rise induced by light were responsible, the phenomenon would become more prominent at higher temperatures (in the case of As_2S_3 the estimated temperature rise is less than 0.1 K). The photon energy during illumination is 2.0 eV and is substantially smaller than the Tauc optical band-gap of $\sim$2.4 eV in As_2S_3 (Elliott, 1991a,b). In this sense, 2.0 eV photons can be regarded as sub-band gap light, or more precisely Urbach tail light, since at this photon energy, As_2S_3 exhibits the so-called Urbach tail.

It was suggested that the light from the Urbach tail is responsible for the photo-induced fluidity. Koseki and Odajima (1982) have demonstrated that photo-induced stress relaxation is observed in a-Se when subjected to band-gap illumination.

A photo-conductive measurement of As_2S_3 glass using the constant photo-current method showed that the photoresponse induced by 2.0 eV light is smaller by 10^{-2}–10^{-3} than that by 2.4 eV light. That is, the number of photo-excited carriers by the Urbach tail light intensity is considerably smaller. However, this ratio merely reflects the absorption coefficients. If the numbers of generated carriers normalized to an absorbed photon are evaluated, no appreciable difference exists between 2.0 and 2.4 eV photons. This fact implies that since holes are responsible for photo-currents, only free electrons are needed for the photo-induced change, irrespective as to whether electrons being trapped (for Urbach tail excitation) or free (for band-gap excitation). Microscopically, some electro-atomic processes follow the photo-excitation of holes, and then slipping of molecular clusters will occur, which phenomenon appears as photo-induced fluidity.

2.2.6. *Giant Photo-expansion*

Tanaka (1996) observed that when As_2S_3 glass is illuminated by tightly focused He–Ne laser light under unstressed condition, macroscopic surface expansion occurs. The thickness expansion for 50 μm thick annealed sample illuminated for 1000 s by 10 mW laser light is $\sim$4%. This is greater by an order of magnitude than that observed

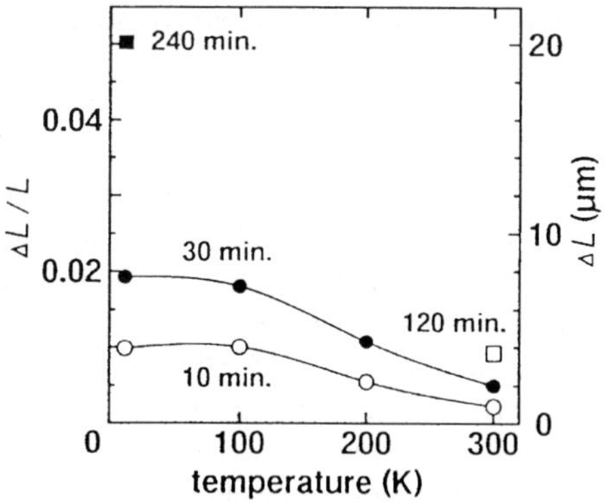

FIG. 6. Giant photo-expansion in a 0.4 mm thick As_2S_3 as a function of temperature at which the sample is illuminated. The exposure time is indicated. The light source is a 25 mW He–Ne laser, focused by a $\times 5$ objective lens. Illumination is performed at room temperature.

($\sim$0.4%) in the conventional photo-expansion phenomenon in As_2S_3 (Janai, 1982). The giant photo-expansion has been observed also in As_2Se_3 and GeS_2 when excited by 1.6 and 2.7 eV light, respectively. The phenomenon is specific to chalcogenide glasses subjected to Urbach tail illumination.

The expansion becomes greater if the sample is illuminated at lower temperatures. This means that the process is an athermal one. Figure 6 shows that the maximum expansion observed amounts to 20 μm ($\Delta L/L = 5\%$; $L = 400$ μm). When $L \geq 100$ μm, L is determined by the self-focused depth of light ($\sim$100–200 μm). Accordingly, $\Delta L/L \sim$ 10–20%.

As illustrated in Figure 7, we can assume that the giant photo-expansion is induced through a combination of the conventional photo-expansion and photo-induced fluidity. That is, the irradiated volume with a thickness of L and a diameter of $2r$ is able to expand with a ratio α ($\sim$0.4% at room temperature) of the conventional photo-expansion. However, the lateral photo-expansion is practically impossible due to the existence of non-illuminated region and accordingly the strain components will flow to the vertical direction through the photo-induced fluidity. The apparent expansion $\Delta L/L$ becomes $\Delta/L = \alpha(1 + L/r)$. That is, the conventional photo-expansion is seemingly amplified by L/r (Popescu et al., 1998).

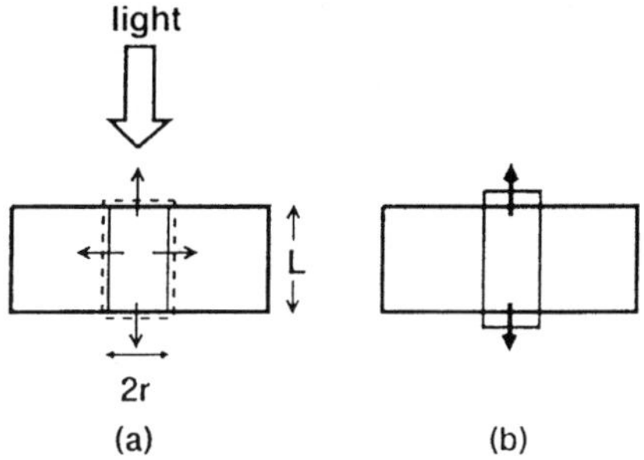

FIG. 7. A model for giant photo-expansion (cross-sectional view).

This model is consistent with the observations that the giant photo-expansion is prominent only when $L \gg r$.

As to the explanation of the photo-expansion effect, this is related to an increase in structural randomness of amorphous networks. It is known that a glass is less dense than the corresponding crystal by $\sim 10\%$. It is also known that the density of As_2S_3 can be modified by $\sim 1\%$ under some treatments such as annealing. Accordingly, illumination may also be capable to increase the specific volume by $\leq 1\%$, which is comparable to the magnitude observed in the giant photo-expansion.

Kuzakawa, Ganjoo and Shimakawa (1999) studied the effect of illumination in obliquely deposited thin films of As_2S_3 and As_2Se_3. It was observed that on annealing, the thickness decreases and the band gap increases while illumination is found to increase the thickness and decrease the band gap. Annealing the samples before or after illumination always shows an effect, which is opposite to that of illumination. The authors observed large changes in both thickness and band gap with illumination. It was also observed that post-illumination annealing causes the changes to revert to nearly the initial conditions. These giant changes have been explained on the basis of the presence of voids and an easy motion of layers in the films, resulting in an easier expansion and slip motion of the layers.

In the studies of interaction of polarized light with chalcogenide glasses, it was demonstrated in the last years that many new phenomena are possible. Examples of such phenomena are: polarization-dependent photo-crystallization, polarization-dependent metal photo-doping, photo-induced anisotropy of photoconductivity, anisotropic surface deformation and photo-induced anisotropy in the ionic conducting amorphous chalcogenide films (Lyubin and Klebanov, 2001).

An important discovery is the possibility of inducing optical anisotropy by non-polarized light (Tikhomirov and Elliott, 1994). In this case, the bonds oriented perpendicular to the **k**-vector and parallel to **E**-vector of the light will be most absorbing and anisotropy between the direction of the light propagation and the plane of **E** and **H** vectors of the light should be observed.

References

Adler, D. and Feinleib, J. (1971) In *The Physics of Optoelectronic Materials* (Ed, Albers, W.A. Jr.) Plenum Press, New York, p. 233.

Agarwal, S.C. and Fritzsche, H. (1974) *Phys. Rev.*, **B10**, 4351.

Arsh, A., Froumin, N., Klebanov, M. and Lyubin, V. (2002) *J. Optoelectron. Adv. Mater.*, **4**(1), 27.

Averyanov, V.L., Kolobov, A.V., Kolomiets, B.T. and Lyubin, V.M. (1980) *Phys. Status Solidi (a)*, **57**, 81.

Babacheva, M., Baranovski, S.D., Lyubin, V.M., Taghirdjanov, M.A. and Feodorov, V.A. (1984) *Fiz. Tverd. Tela (Russ.)*, **26**, 1331.

Balkanski, M. (1987) In *Physics of Disordered Materials* (Eds, Adler, D., Fritzsche, H. and Ovshinsky, S.R.) Institute for Amorphous Studies, Plenum Press, New York, p. 405.

Berkes, J.S., Ing, S.W. and Hillegas, W.J. (1971) *J. Appl. Phys.*, **42**, 4908.

Bertolotti, M., Michelotti, F., Chumash, V., Cherbari, P., Popescu, M. and Zamfira, S. (1995) *J. Non-Cryst. Solids*, **192/193**, 657.

Bhanwar Singh, Rajagopalan, S., Bhat, P.K., Pandhya, D.K. and Chopra, K.L. (1980) *J. Non-Cryst. Solids*, **35/36**, 1053.

Biegelsen, D.K. and Street, R.A. (1980) *Phys. Rev. Lett.*, **44**, 803.

Bogomolov, V.N., Poborcii, V.V., Holodkevici, S.V. and Shagin, S.I. (1983) *Pis'm'ma v JETF (Russ.)*, **38**, 532.

Buzdugan, A.I., Vataman, I.I., Dolghier, V.T., Indricean, K.M. and Popescu, A. (1989) *Proceedings of the International Conference on Non-Crystalline Semiconductors '89*, Uzhgorod, Ukraine, Vol. I, p. 175.

Chang, M.S. and Chen, J.K.T. (1978) *Appl. Phys. Lett.*, **33**, 892.

Chaudhari, P., Herd, S.R., Ast, D., Brodsky, M.H. and Von Gutfeld, R.J. (1972) *J. Non-Cryst. Solids*, **8/10**, 900.

de Neufville, J.P. (1975) In *Optical Properties of Solids—New Developments* (Ed, Seraphin, B.O.) North-Holland, Amsterdam, p. 437.

de Neufville, J.P., Moss, S.C. and Ovshinsky, S.R. (1973/1974) *J. Non-Cryst. Solids*, **13**, 191.

Dresner, J. and Stringfellow, G.B. (1968) *J. Phys. Chem. Solids*, **29**, 303.

Elliott, S.R. (1986) *J. Non-Cryst. Solids*, **81**, 71.

Elliott, S.R. (1991a) *J. Non-Cryst. Solids*, **130**, 85.

Elliott, S.R. (1991b) In *Chalcogenide Glasses Materials Science and Technology*, Vol. 9 (Ed, Zarzycki, J.) VCH, Weinheim, pp. 375–454.

Elliott, S.R. and Kolobov, A.V. (1991) *J. Non-Cryst. Solids*, **128**, 216.

Emelianova, E.V., Hertogen, P., Arkhipov, V.I. and Adriaenssens, G.J. (1999) *Homage Book Andrei Andriesh*, INOE&INFM Publ. House, Bucharest, p. 153; Emelianova, E.V., Hertogen, P., Arkhipov, V.I. and Adriaenssens, G.J. (1999) *Rom. Rep. Phys.*, **51**(3–4), 265.

Feinleib, J. (1970) *USA Patent No. 3636526.*

Feinleib, J., de Neufville, J.P., Moss, S.C. and Ovshinsky, S.R. (1971) *Appl. Phys. Lett.*, **18**(6), 254.

Firth, A.P., Ewen, P.J.S., Owen, A.E. and Huntley, C.M. (1986) *Adv. Resist. Technol. Process., Int. Soc. Optoelectron. Eng.*, **539**, 160.

Fritzsche, H. (1995) *Phys. Rev.*, **B52**, 15854.

Fritzsche, H., Said, V., Ugur, H. and Gaczi, P.J. (1981) *J. Phys. (Fr.)*, **42**, C4-699.

Frumar, M., Firth, A.P. and Owen, A.E. (1984) In *Proceedings of the International Conference on Amorphous Semiconductors '84*, Vol. 1 (Ed, Farchi-Vateva, A.) Buroff, Sofia, p. 216.

Frumar, M., Firth, A.P. and Owen, A.E. (1995) *J. Non-Cryst. Solids*, **192/193**, 447.

Gaszó, J., Hajtó, J. and Zénai, G. (1976) *Preprint*, Central Research Institute, Budapest, KFKI-76-4.

Goldsmidt, D. and Rudman, P.S. (1976) *J. Non-Cryst. Solids*, **22**, 229.

Hajto, J. and Ewen, P.J.S. (1979) *Phys. Status Solidi (a)*, **54**, 385.

Hajtó, J., Jánossy, I. and Forgács, G. (1982a) *Preprint KFKI-1982-29*, Budapest.

Hajto, J., Janossy, I. and Forgacs, G. (1982b) *J. Phys.*, **C15**, 6293.

Hamanaka, H., Minomura, S. and Tsuji, K. (1991) *J. Non-Cryst. Solids*, **137/138**, 977.

Hamanaka, H., Tanaka, K. and Iizima, S. (1977a) *Solid State Commun.*, **23**, 63.

Hamanaka, H., Tanaka, K. and Iizima, S. (1977b) *Solid State Commun.*, **23**, 63.

Hamanaka, H., Tanaka, K., Matsuda, A. and Iijima, S. (1976) *Solid State Commun.*, **19**, 499.

Haro, E., Xu, Z.S., Morhange, J.-F., Balkanski, M., Espinosa, G.P. and Phillips, J.C. (1985) *Phys. Rev.*, **B32**(15), 969.

Hisakuni, H. and Tanaka, K. (1994) *Appl. Phys. Lett.*, **65**(23), 2925.

Hisakuni, H. and Tanaka, K. (1995) *Science*, **270**, 974.

Janai, M. (1981) *J. Phys. (Paris)*, **42**, 1105; Janai, M. (1982) *Phys. Rev. Lett.*, **47**, 726.

Janai, M. and Rudman, P.S. (1974) In *Proceedings of the Fifth International Conference on Liquid and Amorphous Semiconductors* (Eds, Stuke, J. and Brenig, W.) Taylor & Francis, London, p. 425.

Jdanov, V.G., Kolomiets, B.T., Lyubin, V.M. and Malinovski, V.K. (1979) *Phys. Status Solidi (a)*, **52**, 621.

Jdanov, V.G. and Malinovski, V.K. (1980) *Proceedings of the International Conference on Amorphous Semiconductors '80*, Physical Phenomena in Non-Crystalline Semiconductors, p. 226.

Jdanov, V.G., Malinovski, V.K., Nikolova, L.P. and Todorov, T.A. (1979) *Opt. Commun.*, **30**(3), 329.

Jecu, D. (1986) *PhD Thesis*, University of Bucharest.

Jecu, D., Zamfira, S. and Popescu, M. (1989) *Proceedings of the International Conference on Non-Crystalline Semiconductors '89*, Uzhgorod Ukraine, Vol. II, p. 195.

Katayama, Y., Yao, M., Ajiro, Y., Inui, M. and Endo, H. (1989) *J. Phys. Soc. Jpn*, **58**, 1811.

Kawaguchi, T., Maruno, S. and Elliott, S.R. (1997) *J. Non-Cryst. Solids*, **211**, 187.

Kawamoto, Y., Agata, M. and Tsuchihashi, S. (1974) *J. Ceram. Assoc. Jpn*, **82**, 502.

Kawashima, K., Hosono, H. and Abe, Y. (1987) *J. Non-Cryst. Solids*, **95/96**, 741.

Keneman, S.A. (1971) *Appl. Phys. Lett.*, **19**, 205.

Keneman, S.A., Bordogna, J. and Zemel, J.N. (1978) *J. Appl. Phys.*, **49**, 4663.

Khrishnawami, S., Jain, H. and Miller, A.C. (2001) *J. Optoelectron. Adv. Mater.*, **3**(3), 695.

Kimura, K., Murayama, K. and Ninomiya, T. (1985) *J. Non-Cryst. Solids*, **77/78**, 1203.

Kimura, K., Nakata, H., Murayama, K. and Ninomya, T. (1981) *Solid State Commun.*, **40**, 551.

Kluge, G. (1987) *Phys. Status Solidi (a)*, **101**, 105.

Kokado, M., Shimizu, I. and Inoue, E. (1976) *J. Non-Cryst. Solids*, **20**, 131.

Kolobov, A.V. (1995) *J. Non-Cryst. Solids*, **189**, 297.

Kolobov, A.V., Lyubin, V., Yasuda, T., Klebanov, M. and Tanaka, K. (1997) *Phys. Rev. B, Condens. Matter*, **55**(14), 8788.

Kolobov, A.V. and Tanaka, Ka. (1999) *J. Optoelectron. Adv. Mater.*, **1**(2), 3.

Kolomiets, B.T., Lantratova, S.S., Lyubin, V.M., Puh, V.P. and Taghirdjanov, M.A. (1976) *Fiz. Tverd. Tela (Russ.)*, **18**, 1189.

Kolomiets, B.T., Lyubin, V.M. and Shilo, V.P. (1973) *Pis'm'ma v JETF (Russ.)*, **17**, 577.

Kolomiets, B.T., Lyubin, V.M. and Shilo, V.P. (1978) *Fiz. Him. Stekla (Russ.)*, **4**, 351.

Kolwicz, M.D. and Chang, M.S. (1980) *J. Electrochem. Soc.*, **127**, 135.

Koseki, H. and Odajima, A. (1982) *Jpn. J. Appl. Phys.*, **21**, 424.

Kostishin, M.T., Mihailovskaia, E.U. and Romanenko, P.F. (1966) *Fiz. Tverd. Tela (Russ.)*, **8**, 451.

Kotkata, M.F., Shamah, A.M., El-Den, M.B. and El-Mously, M.K. (1983) *Acta Phys. Hung.*, **54**, 49.

Krecmer, P., Moulin, A.M., Stephenson, R.J., Rayment, T., Weland, M.E. and Elliott, S.R. (1997) *Science*, **277**, 1799.

Kuzukawa, Y., Ganjoo, A. and Shimakawa, K. (1998) *J. Non-Cryst. Solids*, **227–230**, 715.

Kuzukawa, Y., Ganjoo, A. and Shimakawa, K. (1999) *Phil. Mag.*, **B79**(2), 249.

Lee, J.M. (1990) *PhD Thesis*, North-Carolina State University.

Lee, J.M., Pfeiffer, G., Paessler, M.A., Sayers, D.E. and Fontaine, A. (1989) *J. Non-Cryst. Solids*, **114**, 52.

Lyubin, V.M. (1985) In *Physics of Disordered Materials* (Eds, Adler, D., Fritzsche, H. and Ovshinsky, S.R.) Institute for Amorphous Studies, Plenum Press, New York, p. 673.

Lyubin, V.M. and Klebanov, M. (2001) *J. Optoelectron. Adv. Mater.*, **3**(2), 265.

Lyubin, V.M. and Tikhomirov, V.M. (1989) *J. Non-Cryst. Solids*, **114**, 133.

Lyubin, V.M. and Tikhomirov, V.K. (1990a) *Fiz. Tverd. Tela (Soviet Phys.: Solid State)*, **32**, 1069.

Lyubin, V.M. and Tikhomirov, V.K. (1990b) *Pis'm'ma v JETF*, **52**, 722.

Lyubin, V.M. and Tikhomirov, V.K. (1991) *J. Non-Cryst. Solids*, **135**, 37.

Lyubin, V., Yasuda, T., Kolobov, A.V., Tanaka, K., Klebanov, M. and Boehm, L. (1998) *Physica B, Condens. Matter*, **245**, 201.

Marquez, E., Gonzalez-Leal, J.M., Prieto-Alcon, R., Jimenez-Garay, R. and Vlcek, M. (1999) *J. Phys. D: Appl. Phys.*, **32**, 3128.

Matsuda, A. and Kikuchi, M. (1973) *Solid State Commun.*, **13**(3), 285.

Matsuda, O., Takeuchi, H., Wang, Y., Inoue, K. and Murase, K. (1997) Proceedings of the Seventh International Conference on the Structure of Non-Crystalline Materials, 15–19 September 1997, Sardegna, Italy.

Matsuda, A. and Yoshimoto, N. (1975) *J. Appl. Phys.*, **46**, 2334.

Meherun-Nessa, K., Shimakawa, A. and Ganjoo (2002) *J. Optoelectron. Adv. Mater.*, **4**(1), 93.

Mikhailov, M.D., Karpova, E.A., Cimpl, Z. and Kosek, F. (1990) *Phys. Status Solidi (a)*, **117**, 467.

Mott, N.F. and Stoneham, A.M. (1977) *J. Phys. C, Solid State Phys.*, **10**, 3391.

Murayama, K. (1987) In *Physics of Disordered Materials* (Eds, Kastner, M.A., Thomas, Q.A. and Ovshinsky, S.R.) Plenum Press, New York, p. 185.

Mytilineou, E., Taylor, P.C. and Davis, E.A. (1980) *Solid State Commun.*, **35**, 497.

Ohta, T. (2001) *J. Optoelectron. Adv. Mater.*, **3**(3), 609.

Oldale, J.M., Elliott, S.R. and Elliott, S.R. (1997) *J. Non-Cryst. Solids*, **211**, 187.

Onari, S., Asai, K. and Arai, T. (1985) *J. Non-Cryst. Solids*, **76**, 243.

Ovshinsky, S.R. (1968) *Phys. Rev. Lett.*, **21**, 1450.

Owen, A.E., Firth, A.P. and Ewen, P.J.S. (1985) *Phil. Mag.*, **B52**, 347.

Paribok-Alexandrovitch, I.A. (1969) *Fiz. Tverd. Tela*, **11**(7), 2019.

Pearson, A.D. and Bagley, B.G. (1971) *Mater. Res. Bull.*, **6**, 1041.

Pfeiffer, G., Paesler, M.A. and Aggarwal, S.C. (1991) *J. Non-Cryst. Solids*, **130**, 111.

Pontushka, W.M. and Taylor, P.C. (1981) *Solid State Commun.*, **38**, 573.

Popescu, M., Sava, F., Lorinczi, A., Skordeva, E., Koch, P.-J. and Bradaczek, H. (1998) *J. Non-Cryst. Solids*, **227–230**, 719.

Porter, E.J. and Sheldrick, G.M. (1972) *J. Chem. Soc. Dalton*, 1347.

Prieto-Alcón, R., Márquez, E. and Gonzales-Leal, J.M. (2000) *J. Optoelectron. Adv. Mater.*, **2**(2), 139.

Rajagopalan, S., Harshavardhan, K.S., Malhotra, L.K. and Chopra, K.L. (1982) *J. Non-Cryst. Solids*, **50**, 29.

Rennie, J.H.S., Elliott, S.R. and Jeynes, C. (1986) *Appl. Phys. Lett.*, **48**, 1430.

Roy, A., Kolobov, A.V. and Tanaka, Ka. (1998) *J. Appl. Phys.*, **83**(9), 4951.

Serbulenko, M.G., Tiscenko, Iu.N. and Nenashev, V.G. (1974) *Seventh Conference of Coherent and Non-linear Optics*, Tashkent, Abstract Booklet, p. 127.

Schwabb, G.-M. (1957) *Semiconductor Surface Physics*, University of Pennsylvania Press, Philadelphia, p. 283.

Schwartz, R.N. and Blair, G.R. (1989) *J. Appl. Phys.*, **65**, 710.

Shimakawa, K. and Elliott, S.R. (1988) *Phys. Rev.*, **B38**, 12479.

Shimakawa, K., Hattori, K. and Elliott, S.R. (1987) *Phys. Rev.*, **B36**, 7741.

Shimakawa, K., Inami, S., Kato, T. and Elliott, S.R. (1992) *Phys. Rev.*, **B46**, 10062.

Shimakawa, K., Kolobov, A.V. and Elliott, S.R. (1995) *Adv. Phys.*, **44**, 475.

Skordeva, E. (1995) *J. Non-Cryst. Solids*, **192/193**, 665.

Street, R.A. (1977) *Solid State Commun.*, **24**, 363.

Stuchlik, M., Krecmer, P. and Elliott, S.R. (2001) *J. Optoelectron. Adv. Mater.*, **3**(2), 361.

Tanaka, K. (1980a) *J. Non-Cryst. Solids*, **35/36**, 1023.

Tanaka, K. (1980b) *Solid State Commun.*, **34**, 201.

Tanaka, K. (1984) *Phys. Rev. B*, **30**, 4549.

Tanaka, K. (1986) *Jpn. J. Appl. Phys.*, **25**, 779.

Tanaka, K. (1987a) *Phys. Rev.*, **B36**, 9746.

Tanaka, K. (1987b) *J. Non-Cryst. Solids*, **90**, 363.

Tanaka, Ke. (1990) *Rev. Solid State Phys.*, **4**, 641.

Tanaka, Ke. (1996) In *Physics and Applications of Non-Crystalline Semiconductors in Optoelectronics (Proceedings of the NATO Advanced Research Workshop, Kishinau, Moldova)* NATO ASI Series, 3. High Technology, Vol. 36 (Eds, Andriesh, A. and Berolotti, M.) Kluwer, Dordrecht.

Tanaka, Ke., Gotoh, T. and Nakayama, H. (1999) *Appl. Phys. Lett.*, **75**(15), 2256.

Tanaka, K. and Itoh, M. (1994) *Optoelectronics*, **9**, 299.

Tanaka, Ke., Kasanuki, Y. and Odajima, A. (1984) *Thin Solid Films*, **117**, 251.

Tanaka, K. and Kikuchi, M. (1973) *Solid State Commun.*, **13**(6), 669.

Tanaka, Ke. and Nakayama, S. (2000) *J. Optoelectron. Adv. Mater.*, **2**(1), 5.

Tanaka, K. and Ohtsuka, Y. (1977) *J. Optique (Paris)*, **8**, 121.

Tanaka, Ke. and Ohtsuka, Y. (1978) *J. Appl. Phys.*, **49**, 6132.

Terao, M. (1972) *Jpn. J. Appl. Phys.*, **41**, 61.

Teteris, Ya.A. and Manika, I.P. (1990) *Fiz. Tverd. Tela (Sov. Phys. Solid State)*, **32**, 1441.

Tichú, L., Tríska, A., Tichá, H. and Frumar, M. (1986) *Phil. Mag.*, **B54**(3), 219.

Tichú, L., Tycha, H. and Bandler, K. (1987) *J. Non-Cryst. Solids*, **97/98**, 1227.

Tikhomirov, V.K. and Adriaenssens, G.J. (1997) *Phys. Rev.*, **B55**, R660.

Tikhomirov, V.K. and Elliott, S.R. (1994) *Phys. Rev.*, **B49**, 17476.

Tikhomirov, V.K. and Elliott, S.R. (1995a) *J. Phys.: Condens. Matter*, **7**, 1337.

Tikhomirov, V.K. and Elliott, S.R. (1995b) *Phys. Rev.*, **B51**(8), 5538.

Tikhomirov, V.K. and Elliott, S.R. (1995c) *J. Phys.: Condens. Matter*, **7**, 1737.

Tikhomirov, V., Hertogen, P., Adriaenssens, G., Krasteva, V., Sigel, G., Kirchhof, J., Kobelke, J. and Scheffler, M. (1997) *Proceedings of the International Conference on Amorphous and Microcrystalline Semiconductors, ICAMS 17, Budapest*; Tikhomirov, V., Hertogen, P., Adriaenssens, G., Krasteva, V., Sigel, G., Kirchhof, J., Kobelke, J. and Scheffler, M. (1998) *J. Non-Cryst. Solids*, **227**, 694.

Tikhomirov, V.K., Hertogen, P., Glorieux, C. and Adriaenssens, G.J. (1997) *Phys. Status Solidi (a)*, **162**, R1.

Utsugi, T. (1990) *Nature*, **347**, 747.

von Gutfeld, R.J. (1973) *Appl. Phys. Lett.*, **22**(5), 257.

Wágner, T. and Frumar, M. (1990) *J. Non-Cryst. Solids*, **116**, 269.

Wágner, T., Frumar, M. and Benes, L. (1990) *J. Non-Cryst. Solids*, **90**, 517.

Wagner, T., Kasap, S.O., Vlcek, M., Frumar, M. and Nesladek, M. (2001) *Appl. Surf. Sci.*, **175/176**, 117.

Weigert, F. (1920) *Ann. Phys. IV*, **63**, 681.

Weiser, K., Gambino, R.J. and Reinold, J.A. (1973) *Appl. Phys. Lett.*, **22**(1), 48.

Zenkin, S.A., Khirianov, K.V., Lobanov, A.B., Mihailov, M.D., Iusupov, Iu. and Iakovuk, O.A. (1989) *Proceedings of the International Conference on Non-Crystalline Semiconductors* '89, Uzhgorod, Ukraine, Vol. II, p. 186.

Zhou, W., Sayers, D.E. and Paesler, M.A. (1993) *Phys. Rev. B*, **47**, 686.

CHAPTER 6

RADIATION-INDUCED EFFECTS IN CHALCOGENIDE VITREOUS SEMICONDUCTORS

Oleg I. Shpotyuk

LVIV SCIENTIFIC RESEARCH INSTITUTE OF MATERIALS OF SRC "CARAT", 202, STRYJSKA STR., LVIV, UA-79031, UKRAINE

INSTITUTE OF PHYSICS OF PEDAGOGICAL UNIVERSITY, 13/15, AL. ARMII KRAJOWEJ, CZESTOCHOWA, 42201, POLAND

1. Introduction

Chalcogenide vitreous semiconductors (ChVSs) are chemical compounds of chalcogen S, Se or Te atoms with some elements of IV and V groups of the periodic table (typically As, Ge, Sb, Bi, etc.) obtained by melt-quenching (Borisova, 1982; Feltz, 1986; Minaev, 1991). One of the most attractive features of ChVSs is their sensitivity to the external influences. This unique ability has not been fully explained to date. But it is probably associated with a high steric flexibility proper to a glassy-like network with low average coordination; relatively large internal free volume; and specific lp-character of electronic states localized at a valence-band top (Elliott, 1986). The photoinduced effects of ChVSs are well known and are the basis for ChVSs-based optical memory systems (Berkes, Ing and Hillegas, 1971; DeNeufville, Moss and Ovshinsky, 1974; Gurevich, Ilyashenko, Kolomiets, Lyubin and Shilo, 1974; Elliott, 1985, 1986).

On the other hand, neither the radiation-induced effects (RIEs) were analyzed for a long time, nor were the changes of ChVSs' physical properties stimulated by high-energetic ($E > 1$ MeV) ionizing influences such as γ-quanta of ^{60}Co radioisotope, accelerated electrons, thermal neutrons and protons. Indeed, since the discovery of the semiconductor properties of ChVSs by Goryunova and Kolomiets nearly half a century ago (Goryunova and Kolomiets, 1955, 1956), they were expected to be usefully distinguished from their crystalline counterparts by a high radiation stability. It was supposed that these glassy materials, owing to positional (topological) and compositional (chemical) disorders frozen near a glass transition temperature T_g during melt-quenching, would not incur any additional structural defects by the irradiation treatment that would change their physical properties. Furthermore, the covalent-like built-in mechanism of ChVS structural network, keeping a full atomic saturation defined by $(8 - N)$ rule (Borisova, 1982; Feltz, 1986), speaks in favor of the above conclusion. Hence, high radiation stability, as well as non-doping ability were expected to be the most essential features of ChVSs.

This is why the findings of the first report on radiation tests in ChVS-based ovonic threshold switches by Ovshinsky et al. at the end of the 1960s (Ovshinsky, Eans, Nelson

ISBN 0-12-752187-9
ISSN 0080-8784

and Fritzsche, 1968), which declared the remarkable radiation hardness of ChVSs, were generally accepted to address a wide variety of ChVSs, despite some specific experimental limitations and disadvantages that include:

- the compositional restriction of the investigated samples by exceptional selection of Te-containing ChVSs with a high saturation of covalent-like chemical bonding and, consequently, a small defect formation ability;
- the technological restriction of the investigated samples by cathode-sputtered films with too small a thickness of about 10^{-6} m (only the bulk ChVS samples can accumulate a large overall RIE due to deep penetration ability of ionizing radiation);
- the limitation factors of radiation treatment (in spite of huge energies $E >> 1$ MeV of neutron flux, X-rays or γ-quanta, accompanied sometimes by a non-controlled radiation heating, the doses Φ were chosen without consideration of the minimum level of ChVSs' radiation sensitivity).

It should be noted that the above article (Ovshinsky et al., 1968) followed an earlier article by Edmond, Male and Chester (1968). The latter concerned the influence of reactor irradiation, created by γ-ray flux of 5×10^{13} MeV cm^{-2} s^{-1}, as well as fast and thermal neutron fluxes of 3×10^{13} cm^{-2} s^{-1}, on electrical properties of liquid semiconductors in the mixed As–S–Se–Te–Ge system. No changes were detected even at the fast neutrons doses up to 1.8×10^{20} cm^{-2}. But it remained unclear whether this irradiation did not produce significant damage, or that high-temperature thermal heating (at more than 470 K) was sufficient to anneal any damage.

All these circumstances strongly restricted the observation possibilities for RIEs in ChVSs. Nevertheless, the general conclusion on their unique radiation stability was repeated often and with enviable constancy, even in the mid-1970s (Zaharov and Gerasimenko, 1976).

2. Historical Overview of the Problem

The first announcements of Domoryad (Institute of Nuclear Physics, Tashkent, Uzbekistan) on the changes of ChVSs' mechanical properties, caused by ^{60}Co γ-irradiation, appeared in the early 1960s (Starodubcev, Domoryad and Khiznichenko, 1961; Domoryad, Kaipnazarov and Khiznichenko, 1963). Contrary to the above-discussed radiation tests of Ovshinsky in ChVS-based ovonic devices (Ovshinsky et al., 1968), the bulk samples of vitreous v-Se, v-As_2S_3, v-As_2Se_3 and some of their simplest quasi-binary compositions were chosen as investigated objects. As a result, it was established that the experimentally detectable RIEs in these glasses were observed at the absorbed doses of 10^5–10^6 Gy, and revealed changes of microhardness, Young's module, internal friction and geometrical dimensions (Starodubcev et al., 1961; Domoryad et al., 1963; Domoryad and Kaipnazarov, 1964; Domoryad, 1969). These changes were stable at room temperature over a long period after irradiation (4–7 months), but they were fully or partly restored after annealing to, respectively, low temperatures 20–30 K below the glass transition point T_g. These RIEs were reversible in multiple cycles of γ-irradiation and thermal annealing with slight damping component. In the 1970s, similar radiation-induced changes were observed in photoluminescence (Kolomiets, Mamontova, Domoryad and Babaev, 1971), photoconduction (Kolomiets, Domoryad, Andriesh, Iovu and Shutov, 1975) and dissolution (Domoryad, Kolomiets, Lyubin and Shilo, 1975) of the above ChVSs. However, no precise experimental results of the microstructural origin of these phenomena were observed.

Further study regarding RIEs was conducted in the early 1980s by some territorial scientific research centers in the former Soviet Union. The greatest success was achieved by the scientific group of Sarsembinov (Kazakh State University, Alma-Ata, Kazakhstan), which studied the influence of accelerated electrons on optical and electrical properties of As- and Ge-based ChVSs. It was shown that electron irradiation of these samples with 2 MeV energy, $10^{17}-10^{18}$ cm^{-2} fluences and 10^{13} cm^{-1} s^{-1} flux led to reduction of their optical transmission coefficient in the whole spectral region and long-wave shift of fundamental optical absorption edge, accompanied by character slope decrease (Sarsembinov, Abdulgafarov, Tumanov and Rogachev, 1980; Sarsembinov and Abdulgafarov, 1980a,b; Guralnik, Lantratova, Lyubin and Sarsembinov, 1982). These changes were reversible in multiple cycles of irradiation and annealing. The other ChVSs' properties such as microhardness, glass transition temperature T_g, dissolution rate, photoluminescence and photoconductivity were also sensitive to electron irradiation. Changes in electrophysical properties were attributed to electron-induced diffusion of metals, deposited at the surface of irradiated samples for electrical contacts. It must be emphasized that this scientific group was the first to put forward one of the most practically important ideas on the electron-induced modification of ChVSs (Sarsembinov and Abdulgafarov, 1981) and to study the physical nature of the observed RIEs, using IR spectroscopy (Sarsembinov and Abdulgafarov, 1980a,b), ESR (Sarsembinov Abdulga farov) and positron annihilation (Sarsembinov and Abdulgafarov, 1980a,b). Surface damages created with high-energetic accelerated electrons did not allow them to investigate the microstructural origin of these effects directly at short- and medium-range ordering levels.

Some attempts to study RIEs at the extra-high doses of γ- and reactor neutron irradiation were made by Konorova et al. at A.F. Ioffe Physical-Technical Institute (St.-Petersburg, formerly—Leningrad, Russia) (Konorova, Kim, Zhdanovich and Litovski, 1985, 1987; Konorova, Zhdanovich, Didik and Prudnikov, 1989). However, despite a great number of experimental measurements, their scientific significance remained relatively poor and speculative because of some essential complications in radiation-treatment conditions, such as uncontrolled thermal misbalance during irradiation resulting in specific structural transformations (crystallization, segregation and phase separation). The investigated samples (v-As_2S_3, v-AsSe and ternary v-AsGeSe or v-$AsGe_{0.2}Se$, additionally doped with Cu and Pb) were chosen too arbitrary, without any respect to their structural–chemical pre-history and technological quality. Sometimes, one could doubt the accuracy of the presented results, as optical transmission spectra of non-irradiated samples in the vicinity of the fundamental absorption edge contained the specific stretched bands, proper usually to light-scattering processes caused by technological macro-inhomogeneities, voids, cracks and impurities (Konorova et al., 1985). The only experimentally proven conclusion of this group was the confirmation of chemical interaction between intrinsic structural fragments of ChVSs and absorbed impurities, stimulated by prolonged irradiation.

A similar conclusion on radiation-induced impurity processes has been put forward recently by scientists from the National Centre for Radiation Research and Technology (Cairo, Egypt). It was shown, in part, that additional weak absorption bands associated with oxygen-based impurity complexes appeared in the powder of v-$Ge_{20}As_{30}Se_{50-x}Te_x$ ($x = 0–40$), irradiated by ^{60}Co γ-quanta with $\Phi = 0.25$ MGy dose (Maged, Wahab and El Kholy, 1998). In contrast to the previous research, the small γ-irradiation dose (no more than 0.34 MGy) did not allow to observe the stronger changes. However, a number of results concerning temperature dependence of steady-state conductivity in v-$As_4Se_2Te_4$ γ-irradiated near T_g (El-Fouly, El-Behay and Fayek, 1982), or γ-induced thermoluminescence in v-$Si_xTe_{60-x}As_{30}Ge_{10}$ (El-Fouly et al., 1982) appear to be quite interesting. The authors of these publications maintain that microstructural origin of the observed RIEs is connected with specific defect centers (broken or dangling bonds, vacancies, non-bridging atoms, chain ends, etc.) created by atomic displacements at a high temperature by secondary electrons from γ-quanta (El-Fouly et al., 1982; Kotkata, El-Fouly, Fayek and El-Hakim, 1986).

Other important research in this field includes:

- effect of γ-induced electrical conductivity in v-As–S(Se)–Te studied by Minami, Yoshida and Tanaka (1972);
- X-ray diffraction study of γ-induced structural transformations in v-As_2S_3 and v-As_2Se_3 by Poltavtsev and Pozdnyakova (1973);
- first observation of electron-induced long-wave shift of fundamental optical absorption edge in v-As_2S_3 and v-As_2Se_3 by Moskalonov (1976);
- effects of thermally stimulated conductivity in γ-irradiated v-$AsS_{3.5}Te_{2.0}$ investigated by Minami, Honjo and Tanaka (1977);
- neutron-induced effects in v-GeS_x and v-As_2S_3 observed by Macko and Mackova (1977), Macko and Doupovec (1978), Durnij, Macko and Mackova (1979) and Lukasik and Macko (1981);
- ESR study of paramagnetic counterparts of radiation-induced defects in ChVS by Taylor, Strom and Bishop (1978), Kumagai, Shirafuji and Inuishi (1984), Chepeleva (1987) and Zhilinskaya, Lazukin, Valeev and Oblasov (1990, 1992);
- γ-induced structural relaxation in v-Se studied by Calemczuk and Bonjour (1981a,b);
- RIEs in ChVS-based optical fibers observed by Andriesh, Bykovskij, Borodakij, Kozhin, Mironos, Smirnov and Ponomar (1984) and Vinokurov, Garkavenko, Litinskaya, Mironos and Rodin (1988);
- electron-induced crystallization in the ternary Ge–Sb–Se glasses investigated by Kalinich, Turjanitsa, Dobosh, Himinets and Zholudev (1986).

In the early 1980s, the complex and comprehensive experimental investigations of RIEs in As_2S_3-based ChVSs, caused by ^{60}Co γ-irradiation, were initiated at the Institute of Materials of Scientific-Research Company 'Carat' (Lviv, Ukraine). Apart from a great number of experimental measurements of RIEs (their compositional, dose, temperature and spectral dependences) (Shpotyuk, 1985, 1987a,b, 1990, 2000; Shpotyuk and

Savitsky, 1989, 1990; Shpotyuk, Savitsky and Kovalsky, 1989; Shpotyuk, Kovalsky, Vakiv and Mrooz, 1994; Shpotyuk, Matkovskii, Kovalsky and Vakiv, 1995; Shpotyuk, Skordeva, Golovchak, Pamukchieva, Kovalskij, Vateva, Arsova and Vakiv, 1999a,b; Savitsky and Shpotyuk, 1990; Matkovsky, 1992; Shpotyuk and Matkovskii, 1994a,b; Skordeva, Arsova, Pamukchieva, Vateva, Golovchak and Kovalskiy, 2000), the physical nature of the observed radiation-structural transformations was treated using IR Fourier spectroscopy (Shpotyuk, 1993a,b, 1994a,b; Balitska and Shpotyuk, 1998), EPR (Shvec, Shpotyuk, Matkovskij, Kavka and Savitskij, 1986; Budinas, Mackus, Savytsky and Shpotyuk, 1987) and mass-spectrometry data (Shpotyuk and Vakiv, 1991). An example was shown of v-As_2S_3—the typical glass-forming model compound with high radiation sensitivity and well-studied structural parameters. The observed RIEs were explained by two interconnected processes. The first one was the process of coordination topological defects (CTDs) associated with covalent chemical bond switching (Shpotyuk, 1993a,b, 1996, 2000; Shpotyuk and Balitska, 1997), and the other one, an effect of radiation-induced chemical interaction between intrinsic ChVS structural fragments and absorbed impurities (Shpotyuk, 1987a,b; Shpotyuk and Vakiv, 1991). Having developed the model of radiation-induced CTDs (Shpotyuk, 1993a,b, 1996, 2000; Shpotyuk and Balitska, 1997), the theoretical principles of topological simulation for destruction-polymerization transformations in the complex ChVS-based systems were presented for the first time (Shpotyuk, Shvarts, Kornerlyuk, Shunin, Pirogov, Shpotyuk, Vakiv, Kornelyuk and Kovalsky, 1991a,b). Among the important practical results of these investigations, the previously stated idea of radiation modification took on a new sense (Shpotyuk et al., 1991a,b; Matkovsky, 1992), as well as the possibilities for ChVSs use in industrial dosimetry (Shpotyuk, Vakiv, Kornelyuk and Kovalsky, 1991a,b; Shpotyuk, 1995). In the late 1990s, this scientific group was close to a resolution of the actual problem of compositional description of RIEs in physically different multicomponent ChVS systems (Shpotyuk et al., 1999a,b; Shpotyuk, 2000; Skordeva et al., 2000; Balitska, Filipecki, Shpotyuk, Swiatek and Vakiv, 2001).

3. Methodology of RIEs Observation

We used ChVS samples of various chemical compositions prepared from high-purity elemental constituents by direct synthesis in evacuated quartz ampoules using the standard rocking furnace technique followed by air quenching (Borisova, 1982; Feltz, 1986; Minaev, 1991). After synthesis all ingots were air-annealed at a temperature of 420–430 K for 3–5 h and cut into plates about 1–2 mm in thickness. The sample surfaces were polished with 1 μm alumina. Samples for acousto-optical measurements were cut into rectangular $10 \times 10 \times 15\ mm^3$ parallelepipeds. X-ray diffraction measurements confirmed that phase separation and crystallization did not occur.

The microhardness H was measured with PMT-3 device. The optical absorption spectra were obtained with 'Specord M-40' spectrophotometer in the wavelength region from 200 to 900 nm. The IR absorption measurements were carried out using 'Specord 75 IR' spectrophotometer (2.5–25 μm wavelengths). The acousto-optical properties of the prepared parallelepipeds (the longitudinal acoustic velocity V, the frequency-normalized acoustic loss coefficient α/f^2 and the acousto-optical figure of merit M_2) were measured with acoustic interference and Bragg diffraction methods (Gusev and Kludzin). The ESR spectra of ChVS powder were obtained at 77 K in a nitrogen ambient with a standard X-band bridge 'Varian E-9' spectrometer.

The radiation-structural transformations were studied using 'differential' Fourier-spectroscopy technique in a long-wave IR region (400–100 cm^{-1}). The observed radiation-structural transformations were associated with reflectivity changes, ΔR, in the main vibrational bands of the investigated ChVS samples. The positive values of $\Delta R > 0$ correspond to complexes appearing after irradiation, and negative ones of $\Delta R < 0$, correspond to complexes disappearing after irradiation. The advantage of this experimental technique lies in the fact that only a small part of the vibrational spectrum induced by γ-irradiation is investigated, but not the whole spectrum. Multiple accumulation of this additional reflectivity signal, when fast Fourier transformation is used, allows us to achieve the sensitivity at the breaking bonds level of ~1%.

We consider the v-As_2S_3 model from the point of detection and identification of γ-induced destruction-polymerization transformations. This ChVS composition is chosen because of good resolution of vibrational bands for structural complexes with covalent chemical bonds of different types. In particular, the pyramidal AsS_3 units (335–285 cm^{-1}) with heteropolar As–S bonds, as well as molecular products with 'wrong' homopolar As–As (379, 340, 231, 210, 168, 140 cm^{-1}) and S–S chemical bonds (243 and 188 cm^{-1}) are demonstrated (Scott, McCullough and Kruse, 1964; Solin and Papatheodorou, 1977; Strom and Martin, 1979; Mori, Matsuishi and Arai, 1984).

The RIEs can be produced in the bulk ChVS samples by high-energetic ($E > 1$ MeV) ionizing irradiation of different kinds, but γ-quanta irradiation by ^{60}Co radioisotope has a number of significant advantages over other methods. These advantages are as follows (Pikaev, 1985):

- the average energy of ^{60}Co γ-quanta (1.25 MeV) is greater than the dual rest energies of electrons (1.02 MeV), which determines the high-energetic character of the observed RIEs;
- the γ-irradiation is characterized by high penetration ability and, consequently, a high uniformity of the produced structural changes throughout the sample thickness;
- the γ-irradiation does not cause the direct atomic displacements resulting in surface macro-damages, craters or cracks, proper to high-energetic corpuscular radiation (accelerated electrons, protons, neutrons, etc.);
- the nuclear transmutations, induced by thermal neutrons, do not take place during γ-irradiation.

Hence, attention has to be focused in the discussion of the RIEs stimulated by ^{60}Co γ-irradiation. This radiation treatment is performed in the normal conditions of stationary radiation field, created in a closed cylindrical cavity by concentrically established ^{60}Co ($E = 1.25$ MeV) radioisotope capsules. The accumulated doses of $\Phi = 0.1–10.0$ MGy were chosen with due account of the previous results of Domoryad's investigations (Starodubcev et al., 1961; Domoryad et al., 1963, 1975; Domoryad and Kaipnazarov, 1964; Domoryad, 1969; Kolomiets et al., 1971). The absorbed dose power P was chosen from a few up to 25 Gy s^{-1}. This P value determined the maximum temperature of accompanying thermal heating in irradiating chamber. This temperature did not exceed 310–320 K during prolonged γ-irradiation (more than 10 days), provided the dose power $P < 5-10$ Gy s^{-1}. However, it reached approximately even 380–390 K at the dose power of ~25 Gy s^{-1}.

4. Remarkable Features of RIEs

The investigated RIEs in ChVSs reveal themselves through related changes of their physical–chemical properties under high-energetic irradiation. The other known phenomena, such as γ-induced electrical conductivity (Minami et al., 1972) or electron-induced anisotropy (dichroism) (Shpotyuk and Balitska, 1998), will not be considered here owing to their specific physical nature (they are the subjects of other scientific reviews and will be analyzed in detail elsewhere). Let us consider the most distinguished features of RIEs.

4.1. Sharply Defined Changes of Physical Properties

These changes, associated with the discussed RIEs, are especially well pronounced in the bulk samples of v-As_2S_3.

4.1.1. Microhardness

Microhardness is one of the most γ-sensitive parameters for ChVS, the first changes being experimentally measured at the beginning of the 1960s by Domoryad et al. (Starodubcev et al., 1961; Domoryad, 1969). In the case of v-As_2S_3, the maximal microhardness increase or, in other words, the γ-induced hardening effect reaches 20–25% depending on technical parameters of radiation treatment (the value of absorbed dose Φ and dose power P, in the first hand) (Domoryad, 1969; Guralnik et al., 1982; Shpotyuk, 1985; Matkovsky, 1992; Shpotyuk and Matkovskii, 1994a,b; Shpotyuk et al., 1994, 1995). The dose threshold for these microhardness changes lies near the critical dose of 0.5 MGy.

4.1.2. Fundamental Optical Absorption

The considerable changes in optical properties of v-As_2S_3 also appear after γ-irradiation at the absorbed doses of $\Phi > 0.5$ MGy (Sarsembinov, 1980; Sarsembinov and Abdulgafarov, 1980a,b; Guralnik et al., 1982; Shpotyuk, 1985, 1987a,b, 1990; Shpotyuk and Savitsky, 1989, 1990; Matkovsky, 1992; Shpotyuk and Matkovskii, 1994a,b; Shpotyuk et al., 1994, 1995).

The γ-induced long-wave shift of the optical transmission coefficient curve $\tau(h\nu)$, or the so-called radiation-induced darkening effect, was observed in the fundamental optical absorption edge region of As_2S_3-based ChVSs (Fig. 1). This shift was nearly parallel for the most ChVS compositions. However, in the case of some glasses with increased

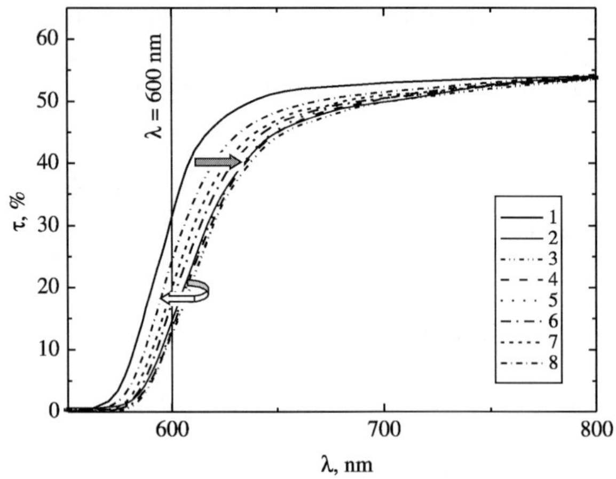

FIG. 1. Optical transmission spectra of the v-As_2S_3 ($d = 1$ mm) before (curve 1) and after (curve 2) γ-irradiation ($\Phi = 10.0$ MGy, $P = 25$ Gy s^{-1}), as well as with further thermal annealing at 330 (curve 3), 370 (curve 4), 380 (curve 5), 395 (curve 6), 420 (curve 7) and 440 K (curve 8).

spatial dimensionality, characterized by a high average coordination number $z > 2.7$ (the number of covalent chemical bonds per one atom of glass formula unit), the slope of the $\tau(h\nu)$ curves additionally decreased after radiation treatment (Shpotyuk et al., 1999a,b; Shpotyuk, 2000; Skordeva et al., 2000). Apart from this shift, some changes in the optical transmittance (in the long-wave spectral range just behind optical absorption edge) are often observed in γ-irradiated ChVS samples (Sarsembinov, 1980; Sarsembinov and Abdulgafarov, 1980a,b; Guralnik et al., 1982; Konorova et al., 1985, 1987; Shpotyuk et al., 1999a,b; Shpotyuk, 2000; Skordeva et al., 2000). The latter effect can be positive (transmittance increase) or negative (transmittance decrease) depending on glass composition and its thermal pre-history.

By plotting $\Delta\tau(h\nu)$ dependence for non-irradiated and γ-irradiated v-As_2S_3 ($d = 1$ mm), we obtained an asymmetric bell-shaped curve with more or less pronounced maximum $\Delta\tau_{max}$, observed at fixed photons energy $h\nu_{max}$, sharp high-energetic edge and more extended low-energetic 'tail' (Fig. 2). It is obvious that the shorter low-energetic tail corresponds to parallel-like $\tau(h\nu)$ shift in the investigated optical transmission spectra. As to the spectral position of $\Delta\tau_{max}$ (the $h\nu_{max}$ values), it is tightly connected with ChVS band gap energy E_g, being directly proportional to the latter.

The optical absorption spectra of the investigated v-As_2S_3 before and after γ-irradiation are presented in Figure 3. The values of absorption coefficient α were calculated, using the well-known formula for high-absorbed substances with approximately constant index of reflection (no sufficient changes in reflectivity of the irradiated samples were observed) (Uhanov, 1977). It is obvious that the $\alpha(h\nu)$ curve is shifted towards lower energies (in long-wave spectral region) after γ-treatment. Two linear sections with different slopes σ and η may be defined from the ln α dependence on the photon energy $h\nu$ (Fig. 3). The region of exponential broadening of the fundamental optical absorption edge (or the so-called Urbach absorption 'tail') at $\alpha = 10^1 - 10^2$ cm^{-1} ($\sigma = 17.6$ eV^{-1}) is related to the fluctuations of internal electric fields (Mott and Davis,

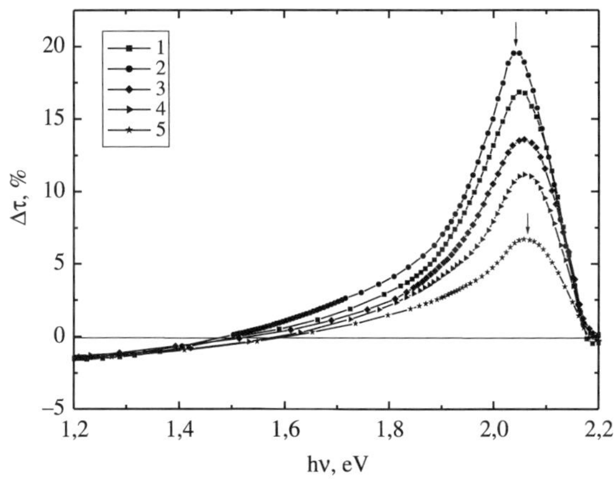

FIG. 2. Spectral dependences of optical transmission differences $\Delta\tau$ in the γ-irradiated ($\Phi = 10.0$ MGy, $P = 25$ Gy s^{-1}) v-As_2S_3 samples (curve 1) and with further thermal annealing at 330 (curve 2), 370 (curve 3), 380 (curve 4), 395 (curve 5), 420 (curve 6) and 440 K (curve 7).

1979), in particular, the fields of charged defects to be considered below (Sarsembinov, 1982; Babacheva, Baranovsky, Lyubin, Tagirdzhanov and Fedorov, 1984). The value of σ decreases ($\Delta\sigma/\sigma = -15\%$) in the γ-irradiated glasses ($\Phi = 10.0$ MGy) owing to the changes in the density of defect states. The second slope η ($\eta \approx 3$–$4\ eV^{-1}$) in the range of $h\nu < 2.0$ eV ($\alpha < 2$–$3\ cm^{-1}$) is less sensitive to γ-irradiation. This part of the optical absorption spectra, associated with different types of macroscopic bulk and surface inhomogeneities (Uhanov, 1977; Borisova, 1982; Feltz, 1986; Minaev, 1991), shows a nearly parallel long-wave shift as a result of radiation treatment.

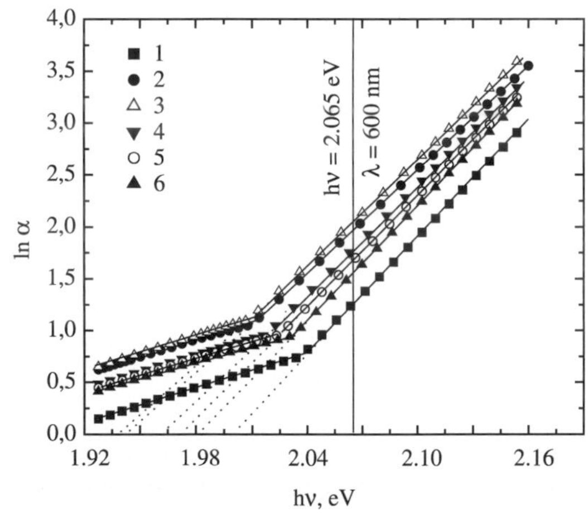

FIG. 3. Optical absorption spectra of the v-As_2S_3 before (curve 1) and after (curve 2) γ-irradiation ($\Phi = 10.0$ MGy, $P = 25$ Gy s^{-1}), as well as with further thermal annealing at 330 (curve 3), 395 (curve 4), 420 (curve 5) and 440 K (curve 6) (Shpotyuk and Matkovskii, 1994a,b).

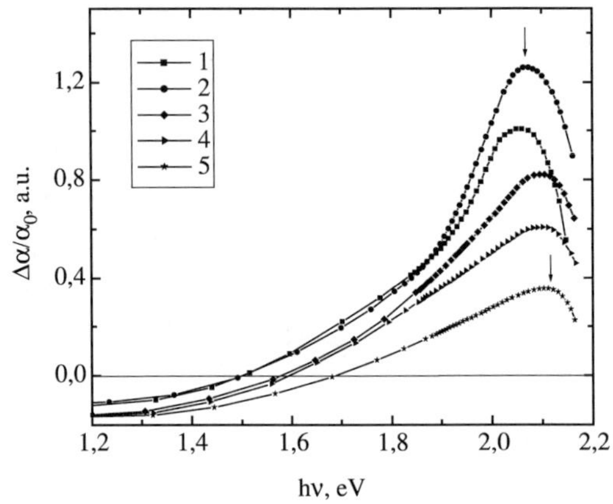

FIG. 4. Spectral dependences of relative optical absorption coefficient increase $\Delta\alpha/\alpha_o$ in the γ-irradiated ($\Phi = 10.0$ MGy, $P = 25$ Gy s^{-1}) v-As_2S_3 samples (curve 1) and with further thermal annealing at 330 (curve 2), 370 (curve 3), 380 (curve 4), 395 (curve 5), 420 (curve 6) and 440 K (curve 7).

The asymmetric bell-shaped $\Delta\alpha/\alpha_o(h\nu)$ curve, calculated for non-irradiated and γ-irradiated v-As_2S_3 samples, can be drawn analogously to the $\Delta\tau(h\nu)$ dependence (Fig. 4). It can be characterized by a well-defined $(\Delta\alpha/\alpha_o)_{max}$ value, an extended low-energetic 'tail' and a sharp high-energetic edge, interrupted at the photon energies $h\nu$ corresponding to non-detectable optical transmittance in γ-irradiated samples.

4.1.3. Acousto-optical Properties

The acoustic velocity V and the frequency-normalized acoustic loss coefficient α/f^2 in non-irradiated v-As_2S_3 are, respectively, as high as 2.58×10^3 m s^{-1} and 0.84 dB cm^{-1} MHz^{-2} in accordance with the well-known experimental data of other investigators (Pinnow, 1970; Sheloput and Glushkov, 1973).

Irradiation with $\Phi = 1.0$ MGy dose ($P = 25$ Gy s^{-1}) changes these values (Savitsky and Shpotyuk, 1990). The longitudinal acoustic velocity V increases up to 2.69×10^3 m s^{-1}, while the α/f^2 coefficient falls down to 0.56 dB cm^{-1} MHz^{-2}. The γ-induced changes of the acousto-optical figure of merit M_2 are smaller. The maximum decrease in this value is not larger than 8% in the case of γ-irradiation with the above dose. It was shown previously that this effect is mainly caused by mutual changes in refractive index and acoustic velocity (Savitsky and Shpotyuk, 1990).

4.1.4. Electron Spin Resonance

We have no ESR signal in non-irradiated v-As_2S_3 samples in accordance with known experimental data (Taylor et al., 1978; Gaczi, 1982; Liholit, Lyubin, Masterova and Fedorov, 1984; Chepeleva, 1987; Zhilinskaya and Lazukin, 1990; Zhilinskaya, 1992; Bishop, Strom and Taylor, 1975). It appears only after γ-irradiation and measurements performed at quite low temperatures. This γ-induced ESR spectrum, observed

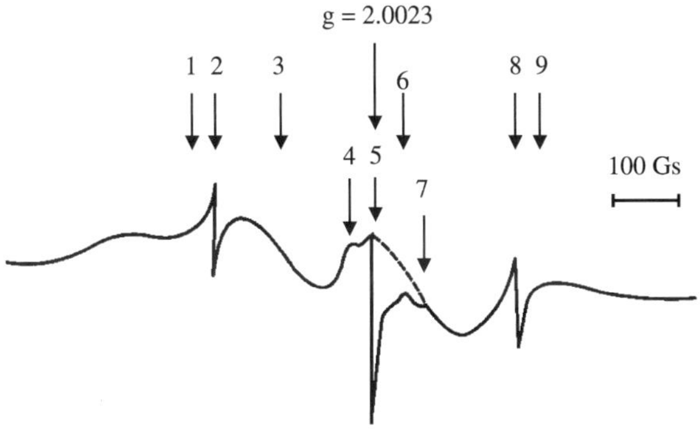

FIG. 5. ESR spectrum of the v-As_2S_3 γ-irradiated with $\Phi = 0.5$ MGy dose (Shpotyuk and Matkovskii, 1994a,b).

experimentally in nitrogen-cooled v-As_2S_3 ($T = 77$ K), is a multicomponent one shown in Figure 5 (Shvec et al., 1986; Budinas et al., 1987).

Slight bends in the central part of the ESR signal, defined as 4 and 7 components, are typical of freshly quenched ChVS samples. It is established that the g-factor of this single signal is equal to 1.970 and its width is $\Delta B = 140$ Gs. The obtained ESR components are bleached at $T > 150$ K and are not observed in the second and all following γ-treatment cycles.

The ESR signal containing components 2 and 8 ($g_2 = 2.183$, $g_8 = 1.847$, × $\Delta B_2 = \Delta B_8 < 1$ Gs) is bleached at $T > 160$ K. Both the above signals denoted by components 4–7 and 2–8 do not appear in γ-irradiated samples at the repeated cooling.

The component 5 in Figure 5 ($g_5 = 2.002$, $\Delta B_5 = 7.5$ Gs) is visible also at the room temperature, but its identification under these conditions is difficult because of a high noise level.

A detailed study of temperature and composition dependences (within As_2S_3–Sb_2S_3 system) for ESR-responses 1, 3, 6 and 9 shows that they correspond to one type of paramagnetic centers simultaneously formed by γ-irradiation (Fig. 5). The total width of this signal is near 1000 Gs, $g_3 = 2.065$, $\Delta B_3 = 90$ Gs, $g_6 = 1.950$ and $\Delta B_6 = 110$ Gs (an exact identification of components 1 and 9 appearing as slight and broad bends at the wings of the whole ESR signal is impossible under these conditions). This signal is bleached at $T > 210$ K.

We identify the observed low-temperature γ-induced paramagnetic centers in v-As_2S_3, taking into account the previous results on photoinduced ESR study (Bishop et al., 1975; Gaczi, 1982; Liholit et al., 1984).

Because of its spectroscopic characteristics, splitting parameters, composition and temperature dependences, the four-component signal, including ESR components 1, 3, 6 and 9 (Fig. 1), is related to paramagnetic=As–S$^\cdot$ and S_2–As$^\cdot$ defects (the unpaired spin is marked by a dot). Both defects appear as a result of γ-treatment when the heteropolar As–S bond is broken. The unpaired electron is localized on a p-like orbital. However, if photoinduced ESR of these defects is associated with a formation of new paramagnetic centers on existing diamagnetic ones, the γ-irradiation stimulates the additional defect formation processes due to chemical bonds destruction. It leads to a high localization degree of radiation-induced paramagnetic centers in comparison with photoinduced ones. Accordingly, the structural features of γ-induced ESR signal are revealed more distinctly.

The single ESR response with $g = 1.970$ and $\Delta B = 140$ Gs observed in freshly quenched v-As_2S_3 after radiation treatment was previously identified in light-irradiated thin films as the signal of an unpaired electron localized on As atom near a disturbed As–As bond (Bishop et al., 1975). This indicates that a certain small concentration of 'wrong' homopolar chemical bonds is maintained in ChVSs too.

The doublet signal with resonance splitting of $A = 502$ Gs, containing the two asymmetric mutually inverted lines 2 and 8 (Fig. 5) corresponds to ESR signal of hydrogen atoms ($g_{av} = 2.05$, $\Delta B_2 = \Delta B_8 < 1$ Gs). This same signal was previously obtained in vitreous silica (Amosov, Vasserman, Gladkih, Pryanishnikov and Udin, 1970). The source of hydrogen atoms may be impure S–H and As–OH complexes, as well as H_2O molecules adsorbed during ChVS preparation or γ-treatment.

The sharply defined ESR signal 5 is connected with a hole-like paramagnetic center having an unpaired electron localized on impurity ions of Fe or O (Pontuschka and Taylor, 1981). The latter version is more probable, as the Fe concentration in v-As_2S_3 does not exceed 10^{-4}%, while oxygen atoms in the form of As_4O_6, SO_2 and =As–O–As= complexes are usually present in all investigated samples obtained by direct synthesis (Borisova, 1982; Feltz, 1986; Shpotyuk, 1987a,b; Minaev, 1991).

4.1.5. *Impurity IR Absorption*

The observed IR absorption bands in v-As_2S_3 (4000–400 cm^{-1}) are due to vibration modes of some impurity complexes; for example, molecular As_4O_6 (1340, 1265, 1050 and 785 cm^{-1}), SO_2 (1150 and 1000 cm^{-1}), H_2S (2470 cm^{-1}), H_2O (3650–3500, 1580 cm^{-1}), structural groups of =As–OH (3470–3420 cm^{-1}), as well as homopolar –S–S– (940 and 490 cm^{-1}) and pyramidal $AsS_{3/2}$ units (750–600 cm^{-1}) (Zorina, Dembovsky, Velichkova and Vinogradova, 1965; Kirilenko, Dembovsky and Poliakov, 1975; Ma, Danielson and Moyniham, 1980; Savage, 1982; Tadashi and Yukio, 1982; Minaev, 1991). These complexes take part in γ-induced mass-transfer chemical interactions with intrinsic structural fragments, resulting in an increase in intensity for all impurity bands (Fig. 6) (Shpotyuk, 1987a,b; Shpotyuk and Vakiv, 1991).

4.1.6. *Intrinsic IR Absorption*

The signal of additional reflectivity $\Delta R(\nu)$ in v-As_2S_3 induced by third γ-irradiation cycle ($\Phi = 10^7$ Gy, $P = 25$ Gy s^{-1}) after two first irradiation–annealing cycles is shown in Figure 7 (Shpotyuk, 1993a,b, 1994; Balitska and Shpotyuk, 1998). It can be seen distinctly that irradiation leads to an increase in 379, 230, 168 and 140 cm^{-1} vibrational bands and a decrease in 335–285, 243 and 188 cm^{-1} ones. On the general background of the 335–285 cm^{-1} spectral band, one can point out weak features at 324 and 316 cm^{-1} as well as sharp peaks at 308, 301 and 288 cm^{-1}, properly correlating with results of the factor group analysis for As_2S_3 pyramidal units in crystalline As_2S_3 (Mori et al., 1984).

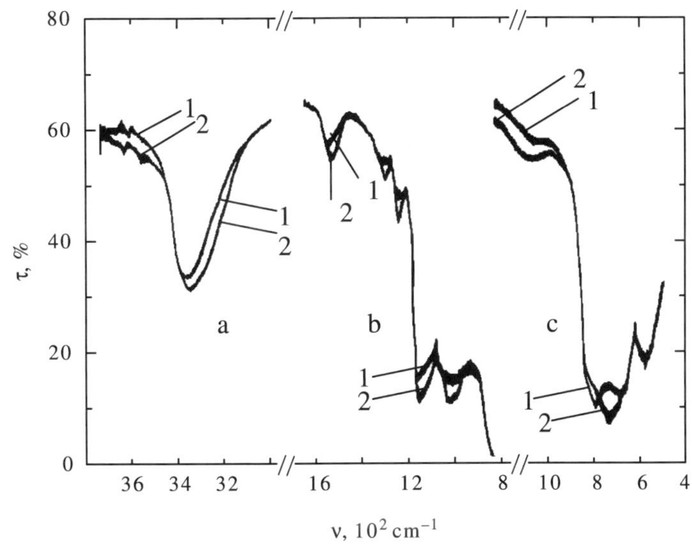

FIG. 6. IR transmission spectra of the v-As_2S_3 before (curve 1) and after (curve 2) γ-irradiation (Φ = 10.0 MGy, P = 25 Gy s^{-1}) with sample thickness of 12 (a, b) and 1.5 mm (c) (Shpotyuk and Matkovskii, 1994a,b).

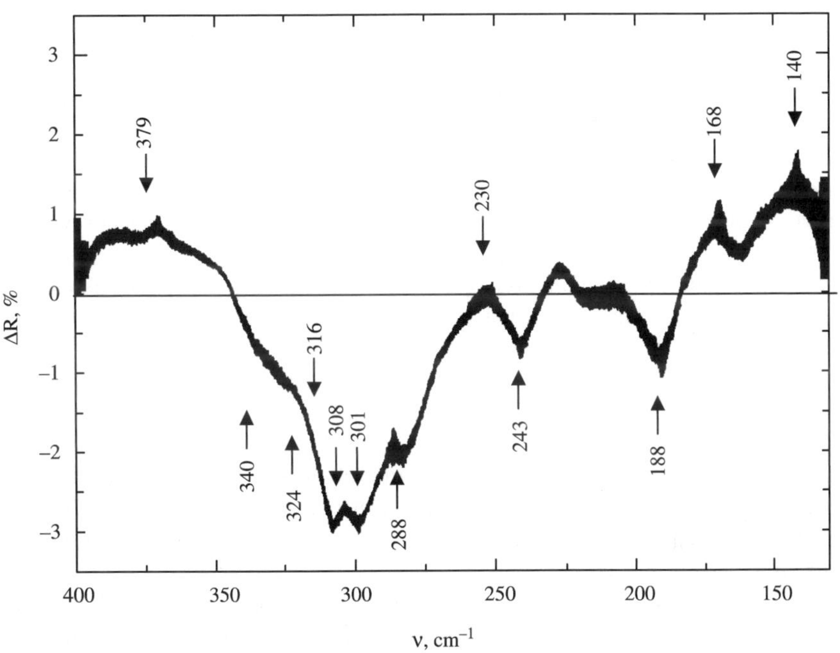

FIG. 7. Signal of additional reflectivity in the v-As_2S_3 induced by the third γ-irradiation cycle (Φ = 10.0 MGy, P = 25 Gy s^{-1}) (Shpotyuk and Matkovskii, 1994a,b).

Taking into account the structural relevance of the main vibrational bands in v-As_2S_3 (Scott et al., 1964; Solin and Papatheodorou, 1977; Strom and Martin, 1979; Mori et al., 1984), we can conclude that the observed RIEs are accompanied by transformations of structural fragments containing homopolar S–S and heteropolar As–S covalent chemical bonds into ones containing heteropolar As–S and homopolar As–As bonds, respectively:

$$(\text{As–S}) \rightarrow (\text{As–As}), \quad (1)$$

$$(\text{S–S}) \rightarrow (\text{As–S}). \quad (2)$$

Resultant data (Fig. 7) show that the statistical weight of reaction (1) is larger in comparison with that of reaction (2). Subsequent annealing ($T > 400$ K) causes the opposite changes in bond distribution. So the observed structural transformations are reversible in multiple cycles of γ-irradiation and thermal annealing.

4.2. Dose Dependence

It was stated above that ChVSs' dose sensitivity to γ-induced changes lies near 0.5 MGy.

The γ-irradiation of v-As_2S_3 by absorbed doses of $\Phi = 0.5–10.0$ MGy is followed by increase in its microhardness as is shown in Figure 8a (Shpotyuk, 1985; Matkovsky, 1992). Typically, the first initial part of this dose dependence at $P = 25$ Gy s^{-1} is a sharp one up to the doses of 1–2 MGy, when relative saturation of these changes begins (Fig. 8a, curve 2). The next decreasing region in $\Delta H/H_0(\Phi)$ dependence is attributed to the partial restoration of the observed changes owing to samples annealing at the prolonged γ-irradiation. The maximum temperature of this spontaneous heating in the irradiating chamber of ^{60}Co source increases with the dose power P, and as a result, the observed effect of microhardness reduction is enhanced. This effect can be fully excluded owing to radiation treatment at stabilized temperature or small dose power of $P < 5$ Gy s^{-1} (Fig. 8a, curve 1).

The dose dependences of the relative γ-induced changes in optical absorption of v-As_2S_3 at different regimes of radiation treatment are shown in Figure 8b (Shpotyuk, 1985, 1990; Shpotyuk and Savitsky, 1989; Matkovsky, 1992). All calculations were performed for the wavelength of $\lambda = 600$ nm corresponding to the middle linear-increased part of optical transmission spectra in the fundamental absorption edge region (this wavelength is distinguished in Figs. 1 and 3). These dependences are similar to the ones obtained for γ-induced microhardness increase (Fig. 8a). At the low dose power P, the $\Delta\alpha/\alpha_0$ values rise linearly with absorbed doses Φ (Fig. 8b, curve 1), whereas the sharp visible saturation effect appears at the higher dose power $P = 25$ Gy s^{-1} (Fig. 8b, curve 2).

The γ-induced changes of acousto-optical properties depend on the value of absorbed radiation dose Φ in the same way (Savitsky and Shpotyuk, 1990). Thus, in the case of $\Phi = 1.0$ MGy, the relative increase in acoustic velocity V in v-As_2S_3 is almost twice as large as that for $\Phi = 5.0$ MGy ($P = 25$ Gy s^{-1}).

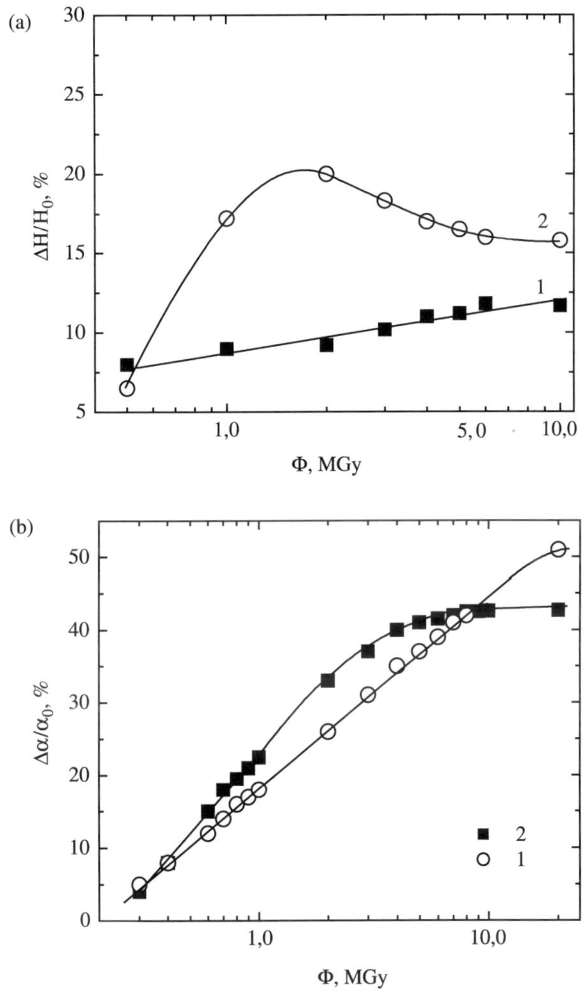

FIG. 8. Dose dependences of the relative γ-induced microhardness (a) and optical absorption at $\lambda = 600$ nm (b) changes in the v-As_2S_3 at the dose power P of 5 Gy s^{-1} (curve 1) and 25 Gy s^{-1} (curve 2) (Shpotyuk and Matkovskii, 1994a,b).

4.3. THICKNESS DEPENDENCE

It is quite understandable that γ-irradiation, owing to its high-penetration ability (Pikaev, 1985), leads to more significant changes in those ChVSs' properties that are determined by sample thickness. This conclusion is demonstrated by the radiation–optical properties of v-As_2S_3.

Taking the γ-induced energetic shift of fundamental optical absorption edge ΔE (estimated at the level of $\tau = 15\%$) in v-As_2S_3 as a controlled parameter, it can be shown that its magnitude rises with sample thickness d according to the next formula

(Shpotyuk and Savitsky, 1990):

$$\Delta E = C_0(1 - \exp(-\kappa d)), \qquad (3)$$

where C_0 and κ are some material-related constants dependent on irradiation parameters.

The sample thickness d in these measurements varied from 1 to 14 mm. This is why the γ-induced optical changes are practically undetectable in very thin ChVS samples and films.

4.4. Thermal Threshold of Restoration

The investigated γ-induced changes of physical properties in v-As_2S_3 can be restored by subsequent thermal annealing at the temperatures close to glass transition point T_g.

With temperature increasing, the microhardness of γ-irradiated samples restores smoothly, beginning from 340 to 350 K up to 400–450 K, when the first thermally induced macroscopic damages appear at the sample surface (Shpotyuk, 1987a,b; Shpotyuk and Savitsky, 1989; Matkovsky, 1992) (Fig. 9a). The time dependence of this thermal annealing is difficult to establish experimentally because of its specific character (each annealing cycle includes the isothermal exposure followed by air quenching of the investigated sample to the room temperature). Whatever the case, more than 70–80% of γ-induced changes can be restored at the first 30–60 min of annealing. The kinetic investigations show that $\Delta t = 2$ h duration of thermal annealing is quite sufficient for microhardness stabilization in v-As_2S_3.

In contrast to this result, the process of thermal restoration of fundamental optical absorption edge of γ-irradiated v-As_2S_3 has a well-pronounced threshold character (Shpotyuk, 1987a,b; Shpotyuk and Savitsky, 1989; Matkovsky, 1992) (Fig. 9b).

At the temperatures of up to 385 K, the spectral position of this edge shifts slightly towards long wavelengths (Fig. 1). It can be supposed that only relatively slight thermally activated stabilization transformations take place at these temperatures, without any significant decrease in concentration of γ-induced defects. The bleaching of γ-irradiated v-As_2S_3 begins in the temperature range from 390 to 410 K, and later rises with annealing temperature T up to the glass transition point T_g, showing a nearly linear $\ln(\Delta\alpha/\alpha_o) = f(10^3/T)$ dependence. So we can describe this process by some effective activation energy E_a, this parameter being close to 0.50 eV for $390 < T < 410$ K and 0.27 eV for $T > 410$ K. By tending close to T_g, the thermal damages appear in the investigated bulk samples. The temperature of 390–400 K, by analogy with the thermal bleaching threshold of photodarkened thin ChVS films (Averianov, Kolobov, Kolomiets and Lyubin, 1979), can be considered as the thermal bleaching threshold of γ-irradiated v-As_2S_3 (Shpotyuk and Savitsky, 1989).

The similar thermally induced changes are proper to the slope of fundamental optical absorption edge (Fig. 3), and to the acousto-optical properties of the investigated samples. It should be noted that microhardness and optical properties of v-As_2S_3 can only be partly restored by thermal annealing (compare curves 1 and 8 in Fig. 1), the irreversible component of this process increasing sufficiently with absorbed dose Φ of γ-irradiation.

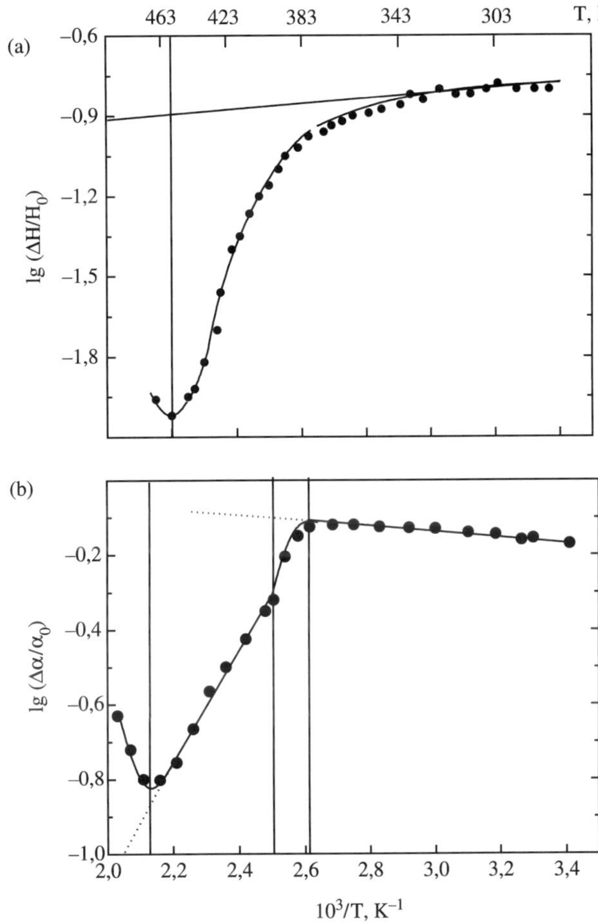

FIG. 9. Relative microhardness (a) and optical absorption at $\lambda = 600$ nm (b) changes in the γ-irradiated ($\Phi = 10.0$ MGy, $P = 25$ Gy s^{-1}) v-As_2S_3 with dependence on post-irradiation thermal annealing temperature.

4.5. Reversibility

The observed changes in physical properties of v-As_2S_3 are reversible in multiple cycles of γ-irradiation and post-irradiation thermal annealing, revealing different sensitivity to slow irreversible structural transformations (damping component) (Matkovsky, 1992; Shpotyuk and Matkovskii, 1994a,b). In the case of microhardness, this damping component does not exceed 2–3%, beginning with the second irradiation–annealing cycle (Fig. 10). However, it can jump up to 25% in the first irradiation–annealing cycle.

Analogous changes were observed in optical absorption of v-As_2S_3 under repeated cycles of γ-irradiation and thermal annealing. It should be noted that these changes were still noticeable in cycles 5 and 6, while the acousto-optical properties did not change

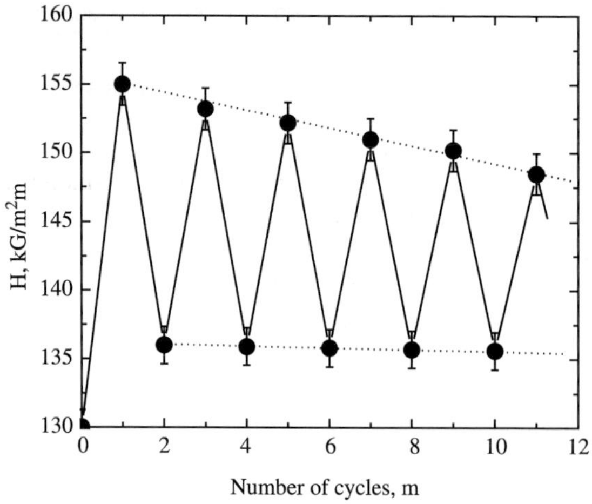

FIG. 10. Reversible changes of microhardness of the v-As_2S_3 in the cycles of γ-irradiation ($m = 1, 3, 5, 7, 9, 11$; $\Phi = 5.0$ MGy, $P = 25$ Gy s^{-1}) and thermal annealing ($m = 2, 4, 6, 8, 10, 12$; $T = 423$ K, $\Delta t = 12$ h) (Shpotyuk and Matkovskii, 1994a,b).

sufficiently at this stage (Savitsky and Shpotyuk, 1990). Our measurements indicate the acousto-optical properties of v-As_2S_3 are more sensitive to irreversible γ-induced structural transformations.

4.6. Compositional Dependence

Despite a great amount of experimental research, the compositional dependence of RIEs remains the most controversial area in this field. These effects, which are well pronounced as a rule in v-As_2S_3 and some other simple As_2S_3-based ChVS compositions, are strongly determined by structural–chemical peculiarities of a glassy-like network, including photoinduced optical changes in thin ChVS films (Elliott, 1986; Berkes et al., 1971; DeNeufville et al., 1974; Gurevich et al., 1974; Elliott, 1985). When these peculiarities have been determined, a comparable analysis of RIEs can be derived for different ChVS systems. At the same time, we have to admit that a number of faults in the selection of the investigated sample compositions were often permitted in the previous experimental research (Edmond et al., 1968; Ovshinsky et al., 1968; Konorova et al., 1985, 1987, 1989; Maged and Wahab, 1998).

Considering compositional dependences for the following sequence of different ChVS species:

- quasi-binary stoichiometric sulphide systems;
- non-stoichiometric sulfide systems with wide deviation of average coordination number Z (calculated as a number of covalent chemical bonds per one atom of glass formula unit);
- v-As_2Se_3 and quasi-binary As_2Se_3-based ChVSs.

RIEs are relatively well researched in *stoichiometric ChVS systems*, formed by two (or more) glass-forming units of the simplest binary sulfide-based compounds. Such systems have been studied since the first experiments of Domoryad et al. (Domoryad and Kaipnazarov, 1964). The main conclusion with respect to these objects is that the quantitative features of RIEs change smoothly with their chemical composition. It can be easily proved using the spectral dependences of γ-induced ($\Phi = 1.66$ MGy, $P \simeq 1$ Gy s^{-1}) optical transmission decrease $\Delta\tau(h\nu)$ in the fundamental absorption edge region for the bulk $(As_2S_3)_x(Sb_2S_3)_{1-x}$ glasses (Starodubcev et al., 1961) (Fig. 11). It is obvious that $\Delta\tau_{max}$ values (or, in other words, the top of bell-shaped $\Delta\tau(h\nu)$ spectral dependence) decay slowly as measured 1 day after γ-irradiation, with Sb_2S_3 content falling from 6.8% for v-As_2S_3 to 1.1% for v-$(As_2S_3)_{0.7}(Sb_2S_3)_{0.3}$. The low-energetic 'tail' of this curve approaches 0% for v-As_2S_3 and then changes its sign tending to −2% in the glasses with maximal Sb_2S_3 concentration ($x = 0.7$). The latter feature has an irreversible character. It is caused by a mixed radiation–thermal influence, resulting in some atomic displacements towards more homogeneous state without crystallization (disappearing of technological imperfections frozen at melt-quenching, in part).

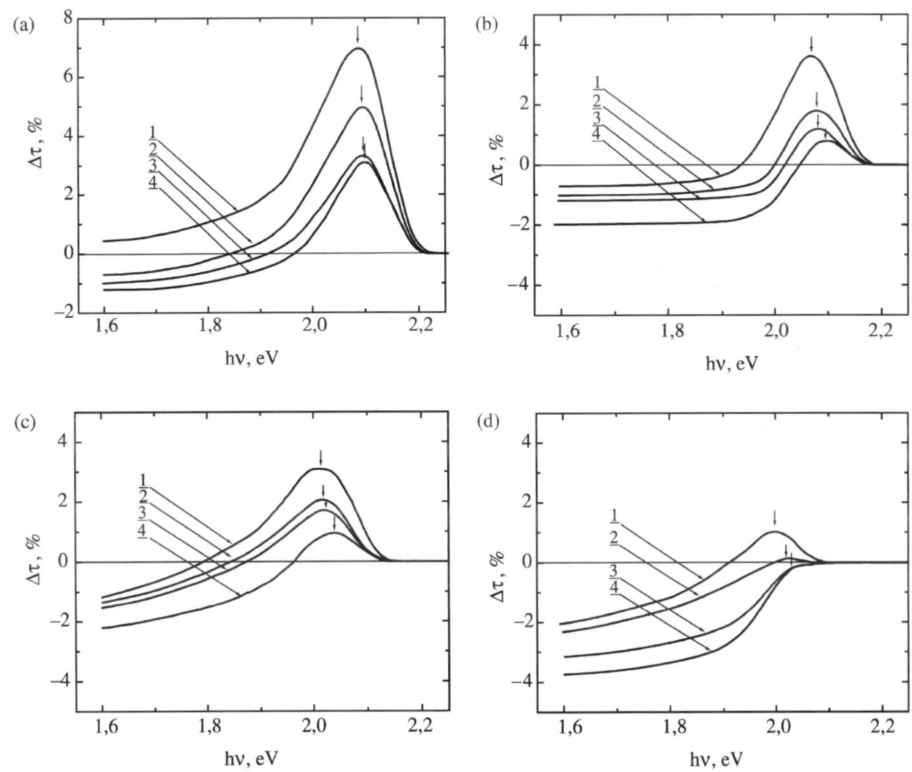

FIG. 11. Spectral dependences of optical transmission differences $\Delta\tau$ in the v-$(As_2S_3)_x(Sb_2S_3)_{1-x}$ glasses ($d = 0.7$ mm) measured 1 (curve 1), 3 (curve 2), 5 (curve 3) and 40 (curve 4) days after γ-irradiation ($\Phi =$ 1.66 MGy, $P \sim 1$ Gy s^{-1}) : $a - x = 1.0$; $b - x = 0.9$; $c - x = 0.8$; $d - x = 0.7$.

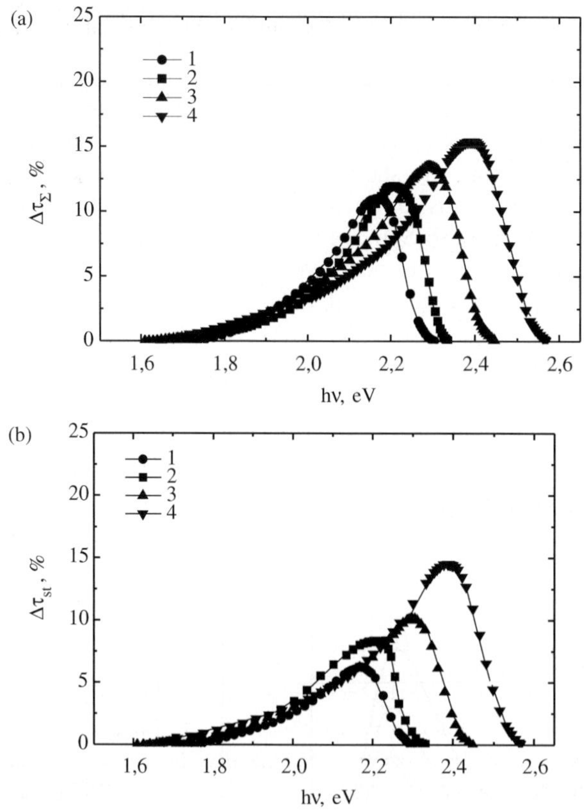

FIG. 12. Spectral dependences of optical transmission differences $\Delta\tau$ in the v-$(As_2S_3)_x(GeS_2)_{1-x}$ glasses ($d = 1$ mm) measured 1 day (a) and 2 months (b) after γ-irradiation ($\Phi = 2.2$ MGy, $P \sim 1$ Gy s^{-1}) : curve $1 - x = 0.2$; curve $2 - x = 0.4$; curve $3 - x = 0.6$; curve $4 - x = 0.8$.

The similar γ-induced changes are observed in another quasi-binary ChVS system formed by structurally different glass-forming units—layer-like $AsS_{3/2}$ pyramids and cross-linked $GeS_{4/2}$ tetrahedra (Fig. 12a, Table I) (Shpotyuk, 2000). The radiation–optical effects increase smoothly with $GeS_{4/2}$ concentration in these ChVSs, the most

TABLE I

QUANTITATIVE CHARACTERISTICS OF γ-INDUCED DARKENING EFFECTS ($\Phi = 2.2$ MGY, $P \sim 1$ GY/S) IN QUASIBINARY $(As_2S_3)_x(GeS_2)_{1-x}$CHVSS

Glass composition		The total RIE		The static RIE		The dynamic RIE
x	Z	$(h\nu_{max})_{\Sigma}$, eV	$(\Delta\tau_{max})_{\Sigma}$, a.u.	$(h\nu_{max})_{st}$, eV	$(\Delta\tau_{max})_{st}$, a.u.	$(\Delta\tau_{max})_{dyn}$, a.u.
0.8	2.43	2.16	0.110	2.18	0.065	0.045
0.6	2.48	2.21	0.120	2.22	0.080	0.040
0.4	2.52	2.30	0.135	2.31	0.100	0.035
0.2	2.59	2.39	0.155	2.40	0.145	0.010

considerable changes occurring only in the fundamental optical absorption edge region similar to v-As_2S_3.

The ChVS samples of *non-stoichiometric sulfide systems* also typically demonstrate a well-pronounced γ-induced long-wave shift of their fundamental optical absorption edge. Let us consider the quantitative features of this shift at the example of $(As_2S_3)_x(Ge_2S_3)_{1-x}$ ChVSs characterized by a wide range of average coordination number Z from $Z = 2.4$ (v-As_2S_3, $x = 1.0$) to $Z = 2.8$ (v-Ge_2S_3, $x = 0$) (Shpotyuk et al., 1999a,b; Shpotyuk, 2000). The ChVS compositions with $Z < 2.67$ can be conditionally accepted as those having 2D layer-like structure, while the ChVS compositions with $Z > 2.67$ are considered as 3D cross-linked glasses.

It is shown that the value and character of the above long-wave shift of the fundamental optical absorption edge of these glasses depend strongly on their structural dimensionality and γ-irradiation parameters. Thus, for non-stoichiometric 2D-like ChVS samples ($Z < 2.67$), the parallel shift of optical transmission $\tau(h\nu)$ edge is observed. However, in the case of 3D-like glasses ($Z > 2.67$), this edge shifts with an additional decrease in a slope.

The above peculiarity is clearly expressed in recalculated $\Delta\tau(h\nu)$ dependences (Fig. 13). The more extended low-energetic tail is observed for 3D-like ChVS samples. Similar behavior is revealed also in γ-induced relative changes of optical absorption $(\Delta\alpha/\alpha_o)_{max}$.

In general, the $\Delta\tau_{max}$ or $(\Delta\alpha/\alpha_o)_{max}$ values achieve a local maximum with glass composition near the 'magic' point of $Z = 2.67$. However, at the prolonged γ-irradiation accompanied with more essential uncontrolled thermal annealing of the investigated glasses, this effect can be changed by the opposite one. This feature is illustrated by the typical concentration dependences of $(\Delta\alpha/\alpha_o)_{max}$ parameter of these glasses for 1.0 and 4.4 MGy doses (Fig. 14). At the small dose (1.0 MGy) accompanied with non-essential thermal heating ($T < 310$ K), the sharply expressed maximum is visible in the above concentration dependence (Fig. 14a, curve 1). At the higher doses (4.4 MGy), the temperature in the irradiating chamber rises and, as a result, a slight minimum is revealed in the above concentration dependence (Fig. 14b, curve 1).

The ChVS samples with $Z = 2.67$ are the most sensitive to the influence of both γ-irradiation and accompanying thermal annealing.

There has been no exact explanation for the above concentration feature in ChVS up to now. The origin of this 'magic' point at $Z = 2.67$ is sometimes connected with topological phase transition from 2D to 3D glassy like network (Tanaka, 1989), as in the case of floppy-rigid on-set at $Z = 2.4$ (Phillips, 1985; Thorpe, 1985). Another microstructural explanation is related to the specific redistribution of covalent chemical bonds (Tichy and Ticha, 1999) or possible phase segregation (Boolchand, Feng, Selvanathan and Bresser, 1999). Whatever the case, the described RIEs in non-stoichiometric ChVS systems show the evident sharp anomalies in the vicinity of this point ($Z = 2.67$), similar to the analogous concentration behavior of other physical–chemical parameters (Mahadevan and Giridhar, 1992; Srinivasan, Madhusoodanan, Gopal and Philip, 1992; Arsova, Skordeva and Vateva, 1994; Skordeva and Arsova, 1995). Taking into account the fact that atomic compactness drops to the minimum in this range of average coordination numbers Z (Feltz, 1986; Skordeva and Arsova, 1995), we

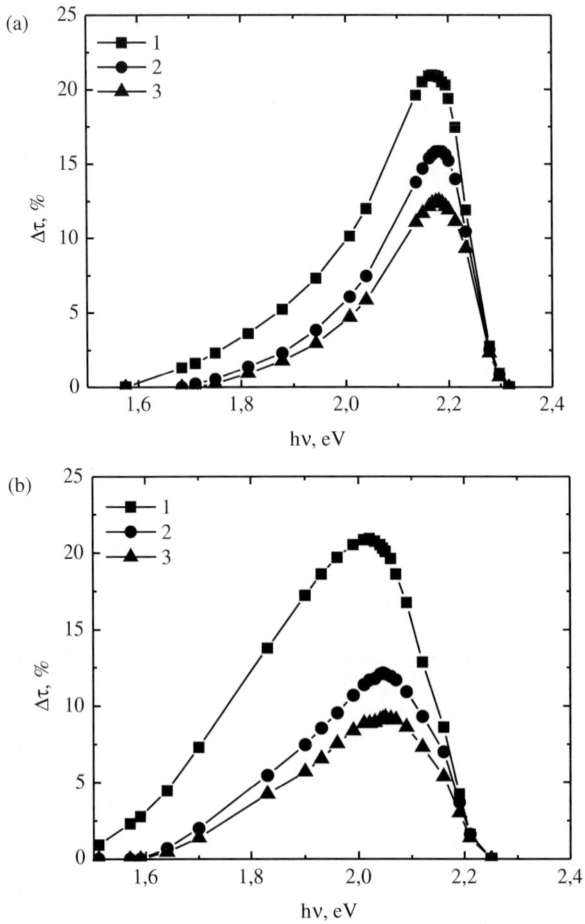

FIG. 13. Spectral dependences of optical transmission differences $\Delta\tau$ in the v-$(As_2S_3)_x(Ge_2S_3)_{1-x}$ glasses ($d = 2$ mm) with typical 2D ($x = 0.6$; (a)) and 3D ($x = 0.1$; (b)) structure measured at 1 day (1), 1 month (2) and 2 months (3) after γ-irradiation ($\Phi = 4.4$ MGy, $P \sim 1$ Gy s^{-1}).

suppose the origin of this anomaly is linked with a high stability of created radiation defects, owing to effective blocking of backward transformations in conditions of a sparse atomic network.

As for v-As_2Se_3 and quasi-binary As_2Se_3-based ChVSs, their optical properties are more sensitive to thermal conditions of γ-irradiation and absorbed dose Φ (Shpotyuk et al., 1989; Shpotyuk, 1990). The γ-induced darkening effect is observed only at relatively low doses of $\Phi < 1.5–2.0$ MGy, the maximal magnitude of this effect being nearly four times smaller as in v-As_2S_3. The greatest changes in v-As_2Se_3 optical properties are observed at $\Phi \approx 0.5–0.7$ MGy. At more prolonged γ-irradiation ($\Phi >$ 3–5 MGy), this γ-induced darkening effect fully decays and subsequently transfers into the γ-induced bleaching one. The above transition takes place due to uncontrolled thermal annealing of the irradiated samples. The critical absorbed dose of γ-irradiation

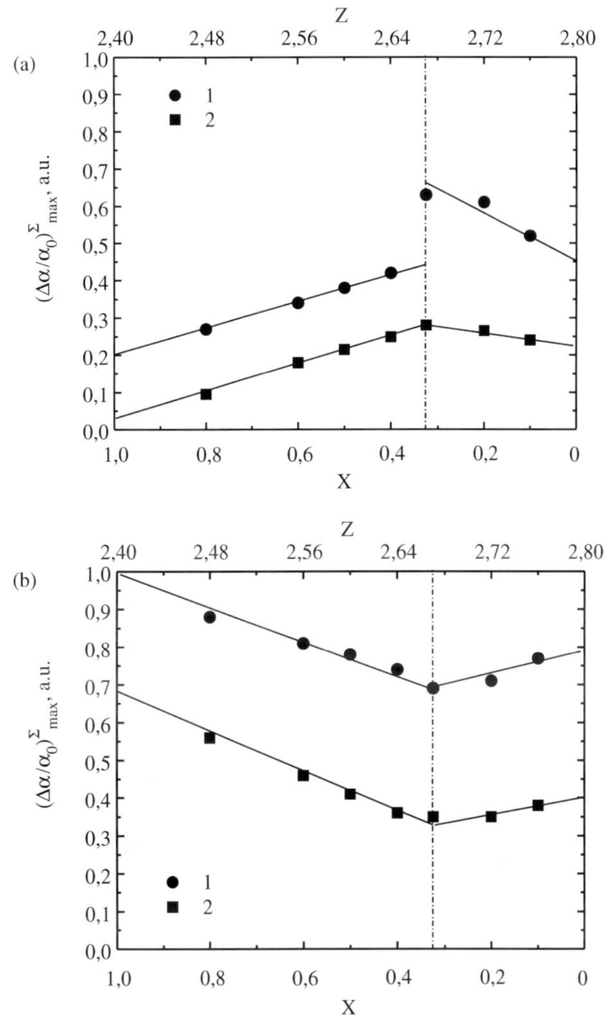

FIG. 14. Compositional dependences of the maximum value of relative absorption coefficient increase $(\Delta\alpha/\alpha)_{max}$ in the v-$(As_2S_3)_x(Ge_2S_3)_{1-x}$ glasses for total (1) and static (2) RIEs at absorbed dose Φ of 1.0 (a) and 4.4 MGy (b).

for this transition can be replaced towards higher Φ values by keeping the relatively low temperature in γ-source chamber or by accumulating the total absorbed dose in small separate cycles with prolonged pauses between them. It should be noted that the described short-wave shift of the fundamental optical absorption edge in v-As_2Se_3 at high γ-irradiation doses is in good agreement with the same behavior of spectral position of radiative recombination maximum observed in this glass after γ-irradiation previously (Kolomiets et al., 1971).

The long-wave shift of the fundamental optical absorption edge of v-As_2Se_3 after γ-irradiation with $\Phi = 1.0$ MGy dose is accompanied by relative slope decrease of

$\Delta\sigma/\sigma = 15\%$. This value falls down to only 4.6% after γ-irradiation with $\Phi = 5.0$ MGy dose, which causes the opposite γ-induced bleaching effect. The subsequent thermal annealing at $T = 423$ K ($\Delta t = 2-3$ h) enhances the σ value additionally by 2.5%. After these treatments, however, the spectral position of the fundamental optical absorption edge of v-As_2Se_3 becomes non-sensitive to the next cycles of γ-irradiation and thermal annealing.

These anomalies reveal themselves yet more distinguishably in ChVS samples of quasi-binary $(As_2Se_3)_x(Sb_2Se_3)_{1-x}$ system characterized by smaller T_g values (Shpotyuk et al., 1989).

4.7. Post-irradiation Instability

It has been pointed out in the first scientific works of Domoryad et al. (Domoryad, 1969) that the observed changes in ChVSs' mechanical properties caused by ^{60}Co γ-quanta are unchangeable for long periods after radiation treatment of up to 5–7 months, provided the irradiated samples are kept at the normal temperature conditions ($T \approx 290-310$ K). In other words, any post-irradiation effects have been accounted as negligible ones in γ-irradiated ChVSs at the room temperature independently on glass composition. Thermal annealing of γ-irradiated samples at the temperatures of 20–30 K below glass transition point T_g has been accepted as the only way to restore their initial physical properties (Domoryad, 1969). However, the accuracy of this statement has not been experimentally verified since the end of the 1960s.

This is the first announcement on the self-restoration effect for γ-induced changes in optical properties of the ternary As–Ge–S ChVSs. By the end of the 1990s (Shpotyuk and Skordeva, 1999), this result has been accepted as a real surprise. It has been shown, in part, that the experimentally studied RIEs observed just after γ-irradiation are unstable in time at room temperature, gradually restoring to some residual value (associated with the static RIE component) during a certain period of up to 2–3 months (Shpotyuk et al., 1999a,b; Shpotyuk, 2000; Skordeva et al., 2000). Hence, the total RIE in as-irradiated ChVS samples consists of two components—the static one, remaining constant for a long time after γ-irradiation, and the dynamic one, gradually decaying with time after γ-irradiation. In this connection, it should be noted that γ-induced changes of ChVSs' physical properties previously discussed in Sections 4.1–4.6 belong to the typical static RIEs.

The 'dynamic component' definition is not appropriate with respect to the observed post-irradiation instability. Sometimes it concerns the RIEs measured directly in the stationary radiation field, such as γ-induced electrical conductivity (Minami et al., 1972) or structural relaxation (Calemczuk and Bonjour, 1981a,b). But we shall use this definition in the above context, taking into account only its close relation to the decaying behavior of the investigated RIEs.

The post-irradiation instability effects are sharply determined in γ-induced changes of ChVSs' optical properties shown in Figures 11a–d (curves 1–3), 12a,b (curves 1–4) and 13a,b (Shpotyuk et al., 1999a,b; Shpotyuk, 2000; Skordeva et al., 2000). It should be emphasized that $\Delta\tau(h\nu)$ dependences obtained 6 months after irradiation are very similar to those denoted by curve 3 in Figure 11, curves 1–4 in Figure 12b and curve 3 in Figure 13. The following conclusions can be drawn from a detailed inspection of these figures:

- The effect of post-irradiation instability reveals itself most sharply in the fundamental optical absorption edge region of the investigated ChVSs, but sometimes, as in the case of quasi-binary $As_2S_3-Sb_2S_3$ glasses (Fig. 11), it extends into the low-energy spectral region of optical transmittance.
- The observed self-restoration effect (or the dynamic RIE component) is described by time-dependent decrease of $\Delta\tau_{max}$ value with simultaneous high-energetic shift of its spectral position $h\nu_{max}$.
- This effect embraces only part of the total RIE.
- The tendency towards saturation of this effect with time is present.
- The $\Delta\tau(h\nu)$ curves take more symmetric shape after restoration (the greatest changes take place in the low-energy spectral region).

Therefore, in terms of the spectral dependence of γ-induced optical transmission decrease in the fundamental optical absorption edge region $\Delta\tau(h\nu)$, the observed total RIEs in ChVSs $\Delta\tau_{\Sigma}(h\nu)$ can be decomposed in two components owing to the next expression:

$$\sum \Delta\tau_{\Sigma}(h\nu) = \Delta\tau_{dyn}(h\nu) + \Delta\tau_{st}(h\nu), \tag{4}$$

where the subscript denotes total (Σ), static (st) and dynamic (dyn) RIEs, respectively.

The quantitative characteristics of γ-induced darkening ($\Phi = 2.2$ MGy, $P \sim 1$ Gy s^{-1}) in quasi-binary $As_2S_3-GeS_2$ ChVSs (Fig. 12) with dependence on their chemical composition are presented in Table I, as determined by x parameter and average coordination number Z. It is obvious that amplitude of total $(\Delta\tau_{max})_{\Sigma}$ and static $(\Delta\tau_{max})_{st}$ RIEs enhances with GeS_2 content, while amplitude of dynamic RIE $(\Delta\tau_{max})_{dyn}$ decreases. These features correspond entirely to well-known compositional dependence of free volume in this ChVS system (Feltz, 1986). The larger the free volume fraction (e.g., GeS_2-enriched ChVS compositions (Miyauchi, Qiu, Shojiya, Kawamotoa and Kitamura, 2001; Takebe, Maeda and Morinaga, 2001)), the more sharply defined the total and static RIEs. The opposite statement for dynamic RIEs in these glasses is obvious too. The more compact a glassy-like network is (which is proper to As_2S_3-enriched ChVSs), the greater is the post-irradiation relaxation and, as a consequence, the amplitude of dynamic RIEs. The similar compositional dependence was observed recently in the changes of v-$As_2S_3-GeS_2$ optical properties caused by hydrostatic pressure (Onary, Inokuma, Kataura and Arai, 1987).

To quantitatively describe the kinetics of the observed dynamic RIEs, all possible mathematical variants of post-irradiation relaxation processes in ChVS must be taken into account. By accepting the $(\Delta\alpha/\alpha_o)_{max}$ value as the controlled relaxation parameter ψ, the next general differential equation can be written for the rate of its decaying after γ-irradiation $\chi = \Delta\psi/\psi_{\infty}$ (ψ_{∞} corresponds to $(\Delta\alpha/\alpha_o)_{max}$ value in the stationary state after full finishing of post-irradiation decaying or, in other words, to static RIE component) (Balitska et al., 2001):

$$\frac{d\chi}{dt} = -\lambda\chi^{\alpha}t^{\beta}, \tag{5}$$

where $\chi(t)$ is a controlled relaxation parameter, λ, α and β are material-specific constants.

The solution of the above differential Equation (5) is a relaxation function, which fulfills the following conditions of the dynamic RIE observation:

$$\begin{cases} t \to 0 \Rightarrow \chi \to \chi_0 = \text{const} \\ t \to \infty \Rightarrow \chi \to 0 \end{cases} \tag{6}$$

It was shown previously that there were five different relaxation functions, listed in Table II, which satisfy these conditions (6) (Balitska et al., 2001).

In the case of $\alpha = 1$ and $\beta = 0$, we observe a well-known monomolecular relaxation process, expressed by simple exponential dependence on time t (function 3 in Table II). If the dynamic component is caused by recombination of specific defect pairs such as electrons and holes, vacancies and interstitials, etc., the underlying kinetics is determined by bimolecular function (function 4 in Table II obtained at $\alpha = 2$ and $\beta = 0$). The full solution of the above differential Eq. (5) at $\beta = 0$ gives the relaxation function 2, exhibited 'stretched' behavior owing to standard αth order kinetics of degradation. This function is often used for description of post-irradiation thermal effects in some oxide glasses (Griscom, Gingerich and Friebele, 1993). In the case of $\beta \neq 0$ and $\alpha = 1$, the relaxation process is described by stretched exponential function 5, which is most suitable for quantitative description of structural, mechanical and electrical degradation processes in glasses or other solids with the so-called dispersive nature of relaxation (Mazurin, 1977). This function was first introduced by De Bast and Gilard (1963), as

TABLE II

THE MAIN DIFFERENTIAL EQUATIONS AND THEIR SOLUTIONS (THE RELAXATION FUNCTIONS) FOR DYNAMIC RIEs IN ChVSs

Differential equation of post-irradiation decaying	Relaxation function and conditions for its determination
(1) $\frac{d\chi}{dt} = -\lambda\chi^{\alpha}t^{\beta}$ $(0 \leq \beta \leq 1)$	$\chi = \frac{\chi_0}{\left(1+\left(\frac{t}{\tau}\right)^{\kappa}\right)^{r}}$, $r = \frac{1}{\alpha-1}$; $\kappa = 1+\beta$; $\tau = \left(\frac{c}{\lambda}\frac{1+\beta}{\alpha-1}\right)^{1/1+\beta}$; $\chi_0 = c^{1/1-\alpha}$ (c–constant of integration, $\alpha \neq 1$, $\beta \neq -1$, $\lambda \neq 0$).
(2) $\frac{d\chi}{dt} = -\lambda\chi^{\alpha}$	$\chi = \frac{\chi_0}{\left(1+\frac{t}{\tau}\right)^{\kappa}}$, $\chi_0 = c^{1/\alpha-1}$; $\tau = \frac{c}{\lambda(\alpha-1)}$; $\kappa = \frac{1}{\alpha-1}$ (c–constant of integration, $\alpha \neq 1$, $\lambda \neq 0$).
(3) $\frac{d\chi}{dt} = -\lambda\chi$	$\chi = \chi_0 e^{t/\tau}$, $\chi_0 = e^{c}$; $\tau = \frac{1}{\lambda}$ (c–constant of integration, $\lambda \neq 0$).
(4) $\frac{d\chi}{dt} = -\lambda\chi^{2}$	$\chi = \frac{\chi_0}{1+\frac{t}{\tau}}$, $\chi_0 = e^{c}$; $\tau = \frac{1}{\lambda}$; (c–constant of integration, $\lambda \neq 0$).
(5) $\frac{d\chi}{dt} = -\lambda\chi t^{\beta}$	$\chi = \chi_0 \exp\left[-\left(\frac{t}{\tau}\right)^{\kappa}\right]$, $\tau = \left(\frac{1+\beta}{\lambda}\right)^{1/1+\beta}$; $\kappa = 1+\beta$; $\chi_0 = e^{c}$ (c–constant of integration, $\beta \neq -1$, $\lambda \neq 0$).

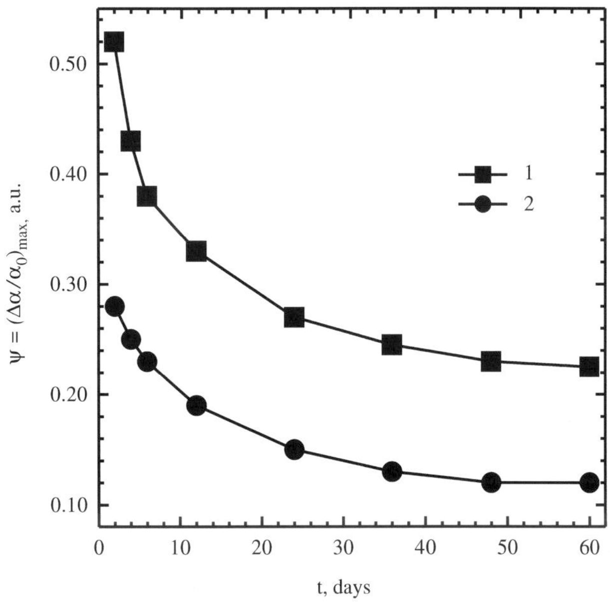

FIG. 15. Typical time-dependent curves of decaying kinetics for the controlled $\psi(t)$ relaxation parameter in γ-irradiated $(As_2S_3)_{0.1}(Ge_2S_3)_{0.9}$ (curve 1) and $(As_2S_3)_{0.8}(Ge_2S_3)_{0.2}$ (curve 2) glasses (Balitska et al., 2001).

well as by Williams and Watts (1970). The exact general solution of Eq. (5) with arbitrary α and β values differed from 0 or 1 can be presented by function 1 (Table II). It contains four fitting parameters (χ_o, r, κ and τ) and, in this connection, is rarely used for mathematical description of the real degradation processes. Other kinds of decaying processes are hypothetical ones, because the correspondent functions do not fulfill the above conditions (6).

For adequate mathematical modeling of dynamic RIEs kinetics, the χ_o, r, κ, and τ were calculated in such a way as to minimize the mean-square deviation error of experimentally measured $\chi(t)$ points from the ones defined by the above relaxation functions listed in Table II. This procedure was fulfilled for ChVSs of both stoichiometric $(As_2S_3)_y(GeS_2)_{1-y}$, and non-stoichiometric $(As_2S_3)_x(Ge_2S_3)_{1-x}$ systems (Fig. 15) (Balitska et al., 2001).

It was established that decaying kinetics of dynamic RIEs can be satisfactorily developed with the greatest accuracy on the basis of bimolecular relaxation function 4 (Table II). In this case, the low values of error are achieved at the minimum number of fitting parameters (χ_o and τ), and their concentration dependences are smooth ones for stoichiometric glasses and more complicated extremum-like (at $Z = 2.7$) for non-stoichiometric ChVSs.

5. Microstructural Nature of RIEs

The microstructural transformations responsible for the observed RIEs in ChVSs can be classified into two large groups according to their behavior in multiple cycles of γ-irradiation and thermal annealing (at the temperature of 20–30 K below glass transition temperature T_g):

- The reversible ones, which can be repeated, multiply in two opposite directions due to the type of external influence ('positive' and 'negative' structural changes).
- The irreversible ones, show a slow damping component with a total number of cycles.

5.1. On the Origin of Reversible Radiation-Structural Transformations

5.1.1. Main Principles of Topological–Mathematical Simulation of Radiation-Induced Destruction-Polymerization Transformations in ChVSs

The experimental results obtained using IR Fourier spectroscopy of additional reflectivity $\Delta R(\nu)$ (Section 4.1.6, Fig. 7) testify that the microstructural origin of reversible RIEs in v-As_2S_3 is connected with the so-called destruction-polymerization transformations due to bond-switching reactions (1) and (2). One covalent chemical bond is broken, but another one is formed in place of the former in its nearest vicinity under γ-irradiation. As a result, two atoms of a glassy-like network obtained a local atomic coordination, which did not comply with the well-known $(8 - N)$ rule (Feltz, 1986; Minaev, 1991). These atoms, over- and under-coordinated ones, create a diamagnetic CTD-pair, because, in addition to the 'wrong' coordination, they have the opposite electrical charges (positive charge excess in the case of extra-coordination and negative one in the under-coordinated state).

Electronic configurations of the above CTDs were proposed, taking into account Anderson's postulate on negative U-centers in ChVSs (Anderson, 1975). It was assumed that all states in the forbidden band gap corresponded to double-paired carriers with opposite spins, their energies forming a quasi-continuous spectrum. According to this postulate, Mott, Davis and Street (1975) put forward the model of CTDs in the form of D-centers or unsaturated 'dangling' bonds. Later, the model of C-centers or valence alternation pairs (VAPs) was developed by Kastner (1976) and, finally, Kastner's model of intimate valence alternation pairs (IVAPs), considering the Coulomb interaction between opposite charged CTDs, was proposed (Kastner, 1978). Soon after, Street used the CTD concept in order to explain the reversible photostructural effects in thin layers of ChVSs, connecting their origin with exciton self-trapping (Street, 1977, 1978).

It must be noted that effective correlation energy for electrons of various CTD configurations was adopted to be a negative one according to ESR-signal absence in ChVSs. However, the consistent theoretical calculations of this parameter were not carried out in the mid-1970s. Moreover, in the beginning of the 1980s, it was shown that this condition did not satisfy some kinds of native point-like CTDs in amorphous Se (Vanderbilt and Joannopoulos, 1983). But these restrictions did not deal with induced effects in ChVSs. Consequently, the CTD concept is accepted to be quite meaningful for RIEs identification.

With this in mind, one should accept the following rules and designations, in other words, the main principles of topological–mathematical simulation of radiation-induced destruction-polymerization transformations in ChVSs:

1. The whole variety of all statistically possible radiation-induced transformations in ChVS of the given chemical composition must be taken into account. They are conveniently described by the following bond-switching scheme or reaction:

$$\text{broken bond} \rightarrow \text{created bond}. \tag{7}$$

   Only one bond switching at a time is adopted as initiated act of the experimentally observed γ-induced structural changes (for example, schemes (1) and (2) in Section 4.1.6).
2. Since the high-energetic γ-irradiation introduces an additional disorder in a glassy-like network associated with some deviations in the existing thermally established distribution of covalent chemical bonds, it can be concluded that the weaker 'wrong' bonds appear instead of the stronger ones as a result of radiation-structural disturbances. It means that only such radiation-induced bond-switching processes should be considered, which are accompanied with a negative difference ΔE in dissociation energies for created E_c and destructed E_d covalent chemical bonds:

$$\Delta E = E_c - E_d < 0. \tag{8}$$

   This condition (8) determines, in turn, the low-energetic shift of fundamental optical absorption edge in γ-irradiated ChVSs, as the decrease in the character dissociation energies of main glass-forming units leads to a narrowing in band-gap width of the correspondent glasses (Kastner, 1973). The greater the ΔE is, the more essential is the energetic barrier between initial and final metastable states in the structural–configurational diagram of the investigated glassy-like system and, consequently, the more stable is the created CTD pair.
3. The final choice of physically real destruction-polymerization transformations in ChVSs can be made on the basis of experimentally established bond-switching scheme (7). By applying a method of IR Fourier spectroscopy of additional γ-induced reflectivity, the left-side component of the above bond-switching reaction (destructed bond) can be determined owing to vibrational bands of negative intensities $\Delta R(\nu) < 0$ and, vice versa, the right-side component of the above bond-switching reaction (created bond) is accepted to be attributed with bands of positive intensities $\Delta R(\nu) > 0$. As a result, the bond-switching schemes (1) and (2) are identified as responsible for the reversible RIEs in v-As_2S_3 (Section 4.1.6).
4. The CTD formation, being a sufficiently atomic-dynamic process, is accompanied by structural changes at short- and medium-range ordering levels in strong dependence on ChVS compactness. If we have a close-packed glass network with a high-atomic density, only bond-switching processes with a large $|\Delta E|$ occur. However, this rule is evidently not fulfilled in ChVSs of low-atomic compactness owing to a high content of intrinsic native microvoids, which prevent the backward annihilation of the created CTDs.

Let us subsequently apply these principles in order to develop the CTD concept for the microstructural explanation of reversible RIEs in v-As_2S_3.

5.1.2. CTD Model for Reversible RIEs in v-As_2S_3

The existence of several types of paramagnetic defects in γ-irradiated v-As_2S_3 glasses at low temperatures and their destruction when increasing the temperature up to room temperature (Section 4.1.4, Fig. 5) make one think that the process of radiation-induced structural transformation takes place in two stages. As a result of bond breakage, the paramagnetic defect centers are formed in the first stage. They are stable at a low temperature of $T < 100$ K. Then with temperature increase in the second stage ($T = 300$ K), they annihilate between each other or transform into diamagnetic CTDs by forming new covalent chemical bonds.

First, we shall analyze all statistically possible topological variants of CTDs formation in v-As_2S_3, taking into account the main types of its initial structural units: heteropolar As–S covalent chemical bonds in the framework of pyramidal AsS_3 or bridge As–S–As complexes, as well as homopolar As–As or S–S covalent chemical bonds within different fully or partially polymerized molecular fragments (Borisova, 1982; Feltz, 1986; Minaev, 1991).

Since the final ChVS state depends on both broken bonds, and their nearest neighbors (it means that two initial structural units take place in one elementary act of CTDs formation), there are 16 topological schemes of the considered destruction-polymerization transformations for four initial units mentioned above. In other words, the overall number of all statistically possible variants of CTDs formation is equal to the permutations of four taken two at a time (for four initial structural units proper to v-As_2S_3). These topological schemes showing homopolar (schemes 1–8) and heteropolar (schemes 9–16) bond-switching processes are presented in Figure 16 (Shpotyuk, 1993a,b, 1994; Balitska and Shpotyuk, 1998). We maintain that the absence of such statistical consideration for ChVSs of defined chemical composition (v-As_2S_3, for example) leads to incorrect conclusions on the possible mechanisms of induced structural transformations, especially in the cases of multiple external influences such as photoexposure (or high-energetic irradiation) and thermal annealing.

Each scheme in Figure 16 corresponds to one CTD pair. The upper index in the defect signature (superscript) means the charge electrical state of the atom, and the lower one (subscript)—the coordination number. The CTDs appear in a glassy-like structural network in pairs (under- and over-coordinated, negative and positive ones), providing the full conservation of sample's electroneutrality. The whole variety of CTDs in v-As_2S_3 is marked as S_1^-, S_3^+, As_2^- and As_4^+.

Schemes 1–4 in Figure 16 are connected with homopolar-to-heteropolar bond-switching, and schemes 9–12 with heteropolar-to-homopolar bond-switching. Schemes 5–8 and 13–16 in Figure 16 do not change the chemical bond type (one bond is broken, but the same appears instead of it again). Such destruction-polymerization transformations are the most difficult for experimental identification, because only intermediate-range ordering changes can be associated with them. However, in the previous cases

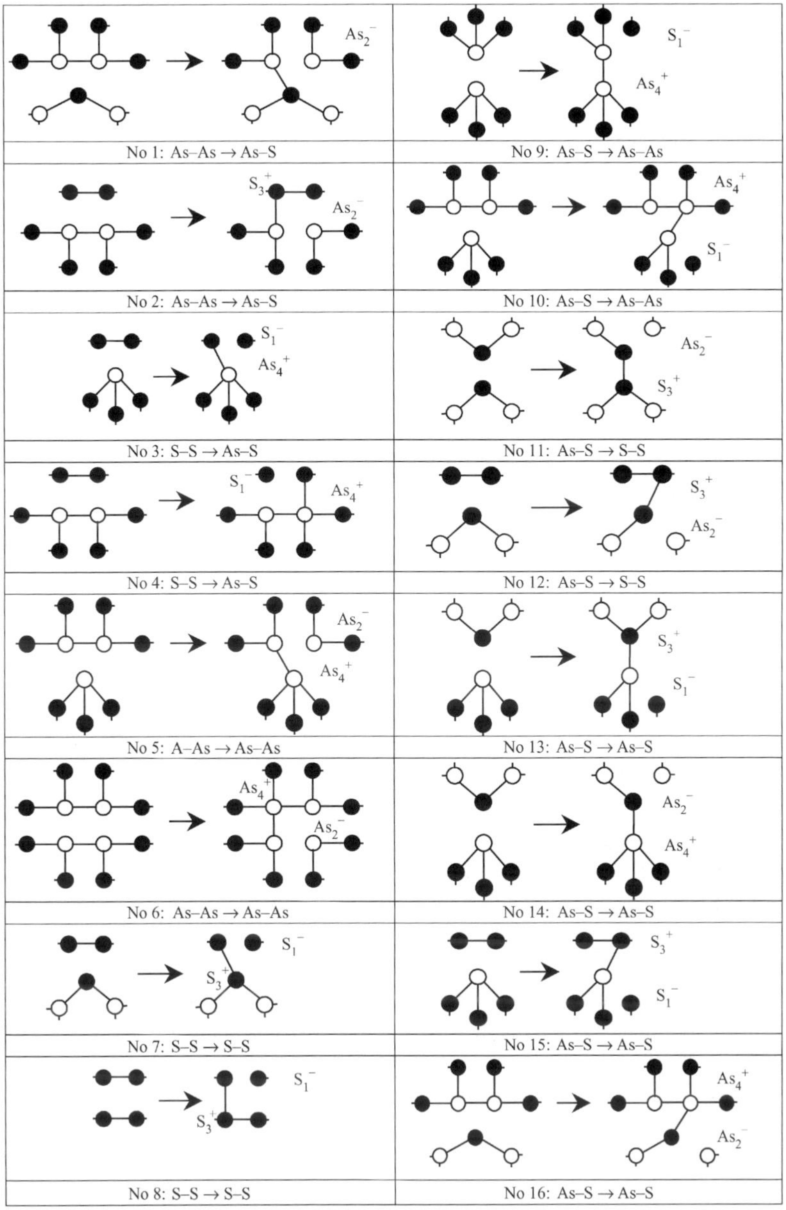

FIG. 16. Statistically possible topological schemes of CTDs formation in v-As_2S_3 associated with homopolar (1–8) and heteropolar (9–16) chemical-bonds switching (Balitska and Shpotyuk, 1998).

(topological schemes 1–4 and 9–12 in Fig. 16), we deal with microstructural transformations at the level of short-range ordering and, consequently, some changes in the correspondent vibrational bands are expected to be experimentally detectable in IR Fourier spectra of radiation-induced reflectivity.

The above CTD pairs can be characterized not only by electrical charge and coordination number, but also by 'wrong' homopolar covalent chemical bonds in the vicinity of anomalously coordinated atoms. The number of such 'wrong' bonds determines the so-called ordering of CTD pair. Thus, for example, (As_2^-, S_3^+) CTD pair presented by the first topological scheme in Figure 16 is of 0-ordering as there are no homopolar chemical bonds in its nearest vicinity. At the same time, (As_2^-, S_3^+) CTD pair presented by the second topological scheme in Figure 16 is of 1-ordering as S_3^+ defect has one S–S 'wrong' homopolar bond nearby.

Taking into consideration all statistically possible variants of CTDs in v-As_2S_3, shown in Figure 16, the following conclusions can be drawn:

1. If the bond type is not changed (topological schemes 5–8 and 13–16 in Fig. 16), the appearing CTD pair is homoatomic or, in other words, both CTDs have the same chemical nature (S_1^- and S_3^+, As_2^- and As_4^+). If the bond type changes during irradiation (schemes 1–4 and 9–12 in Fig. 16), the CTD pair is heteroatomic (S_1^- and As_4^+, as well as As_2^- and S_3^+).
2. There are 4 CTD pairs of 0-ordering (schemes 1, 3, 13 and 14), 8 CTD pairs of 1-ordering (schemes 2, 4, 5, 7, 9, 11, 15 and 16) and 4 CTD pairs of 2-ordering (schemes 6, 8, 10 and 12) in v-As_2S_3 among all 16 statistically possible destruction-polymerization transformations (Fig. 16).

Having analyzed these topological variants of destruction-polymerization transformations in v-As_2S_3, we are able to choose those among them, which correspond exactly to the experimentally observed radiation-induced changes in IR Fourier spectra of additional reflectivity (Fig. 7). Thus, we conclude that only four types of CTDs formation processes shown in Figure 16 are in full agreement with the obtained experimental data: the topological schemes 9 and 10 are described by bond-switching reaction (1), while the topological schemes 3 and 4 are described by bond-switching reaction (2). However, the initial concentration of bridge $=S_2As–AsS_2=$ structural complexes with 'wrong' homopolar As–As covalent chemical bonds in bulk v-As_2S_3 is so small (Borisova, 1982; Feltz, 1986; Minaev, 1991) that the topological schemes 4 and 10 may be excluded from further consideration. So only topological schemes 3 and 9 correspond to the physically real CTDs formation in v-As_2S_3, the statistical weight of the former reaction being smaller because of low S–S covalent bonds concentration.

In both cases (topological schemes 3 and 9 in Fig. 16), the (As_4^+, S_1^-) CTDs appear in a glassy-like network as a result of radiation-induced bond switching. The energetic activation barrier ΔE for these transformations, estimated as bond energy differences after and before γ-irradiation owing to Eq. (8), is negative. Taking into account the well-known balance of bond energies in the binary As–S system ($E_{As-As} = 2.07$ eV, $E_{As-S} = 2.48$ eV and $E_{S-S} = 2.69$ eV (Rao and Mohan, 1981)), it can be estimated that ΔE values reach -0.21 and -0.41 eV for topological schemes 3 and 9 (Fig. 16), respectively. Another essential difference between schemes 3 and 9 in Figure 16 is that the (As_4^+, S_1^-) CTD pair has the 0-ordering in the first case (there are no homopolar chemical bonds in the vicinity of this CTD-pair) and the 1-ordering in the second one (one As–As homopolar chemical bond appears near this CTD pair).

As was pointed out in Section 4.1.6, the subsequent thermal annealing of the investigated v-As_2S_3 samples caused completely opposite changes in the bonds arrangement. Thus, the above destruction-polymerization transformations are really reversible.

5.1.3. Medium-Range Ordering Structural Transformations in γ-Irradiated v-As_2S_3

Accepting the CTD concept, put forward in the second half of the 1970s (Mott et al., 1975; Kastner, Adler and Fritzsche, 1976; Street, 1977, 1978; Kastner, 1978) only as a starting point in our work, it is important to consider the concept of destruction-polymerization transformations in covalent-bonded topologically disordered solids, as developed by Zakis, 1984. As a result, we describe the radiation-induced CTDs formation in a glassy-like network at the levels of both short- and medium-range ordering (Shpotyuk, 1993a,b, 1994; Balitska and Shpotyuk, 1998).

According to the first concept, the CTDs appear in a glassy-like network in the form of positively and negatively charged atoms with an abnormal number of nearest neighbors (under- and over-coordinated ones) due to the bond-switching processes; or in other words, the local deviations in covalent bonds arrangement given by $(8 - N)$ rule (Borisova, 1982; Feltz, 1986; Minaev, 1991).

The second concept (the concept of destruction-polymerization transformations) describes a spatial extension of such structural transformations in the ChVS atomic network. It is assumed that equilibrium states for many atoms (atomic transfer) are changed in any elementary bond-switching act, being a cooperative process of configuration-deformation disturbances at the level of short- and medium-range ordering (Zakis, 1984). The process of bond switching effectively occurs in some convenient atomic configurations, which prevent the backward annihilation reaction for the created CTD-pair. Analyzing the possible radiation-induced destruction-polymerization transformations in ChVSs, we must pay attention to the structural fragments with the least atomic compactness, which have the largest open volumes frozen technologically at melt-quenching. These fragments are most suitable for the above CTDs formation.

The process of medium-range ordering structural transformations in the vicinity of as-appeared (S_1^-, As_4^+) CTD pair in v-As_2S_3 (topological scheme 9 in Fig. 16) is shown schematically in Figure 17. It is clear that the formation of As–As covalent chemical bond instead of broken As–S one nearby the positively charged As_4^+ defect leads to the local densification of atomic package, while in the vicinity of the negatively charged S_1^- CTD the atomic network is distorted with open volume formation (is crosshatched in Fig. 17). In other words, the lack of one covalent chemical bond at the negatively charged CTD and its shift along existing bond towards neighboring directly bonded atom leads to the appearance of open-volume microvoid. These microvoids, associated with negatively charged CTDs, can be effective traps for positrons, giving a reasonable explanation for positron annihilation lifetime measurements in ChVSs (Shpotyuk and Filipecki, 2001).

The backward structural transformations, accompanied by CTDs disappearing or self-annihilating, can be quite sufficient, provided that the described medium-range ordering

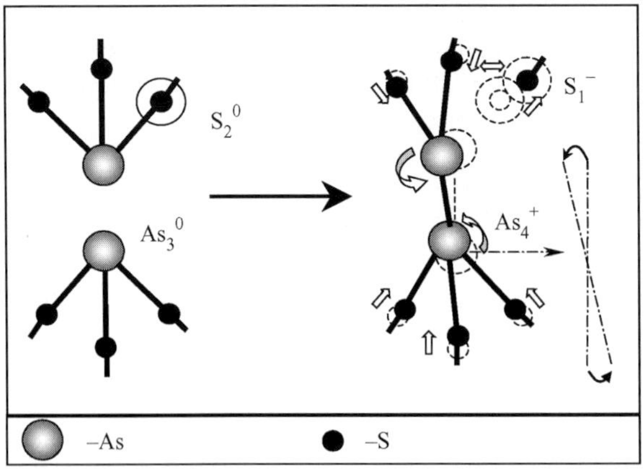

FIG. 17. Topological scheme illustrating the medium-range ordering structural transformations in the vicinity of the negatively charged S_1^- CTD in v-As_2S_3.

changes are too slight to preserve the created metastability. This situation is proper to one-type bond-switching processes (when bond type does not change during switching), or to CTD formation in ChVSs with a more rigid covalent-linked bond network. The previously observed dynamic RIEs in bulk ternary ChVSs (Shpotyuk et al., 1999a,b; Shpotyuk, 2000; Skordeva et al., 2000) probably originate from such structural transformations.

The medium-range ordering transformations associated with open-volume microvoids formation offer the necessary conditions for CTDs stabilization in a glassy-like network, preventing a possibility of their disappearing. This condition can be easily fulfilled, provided over- and under-coordinated atoms of radiation-induced CTD pair in its final metastable state (Fig. 17) are separated by a number of structural fragments that are formed due to additional bond-switching acts (state 3 in Fig. 18). As a rule, these CTD-conserved bond-switching processes do not change the type of the covalent bond, which are characterized by a negligible potential barrier (transition from state 2 to state 3 in Fig. 18). But they keep the essential configuration-deformation disturbances in the vicinity of as-created CTDs at the medium-range ordering level, leading to the formation of induced CTD-based microvoids.

5.2. On the Origin of Irreversible Radiation-Structural Transformations

The irreversible radiation-structural transformations in ChVSs, owing to their microstructural nature, are of two types (Shpotyuk, 1987a,b; Shpotyuk and Vakiv, 1991; Matkovsky, 1992):

- the intrinsic ones, associated with destruction-polymerization transformations having a positive difference ΔE (Eq. (8)) in dissociation energies for created E_c and broken E_d covalent chemical bonds, as well as
- the impurity ones, caused by chemical interaction between intrinsic structural complexes of a glassy-like network and absorbed impurities.

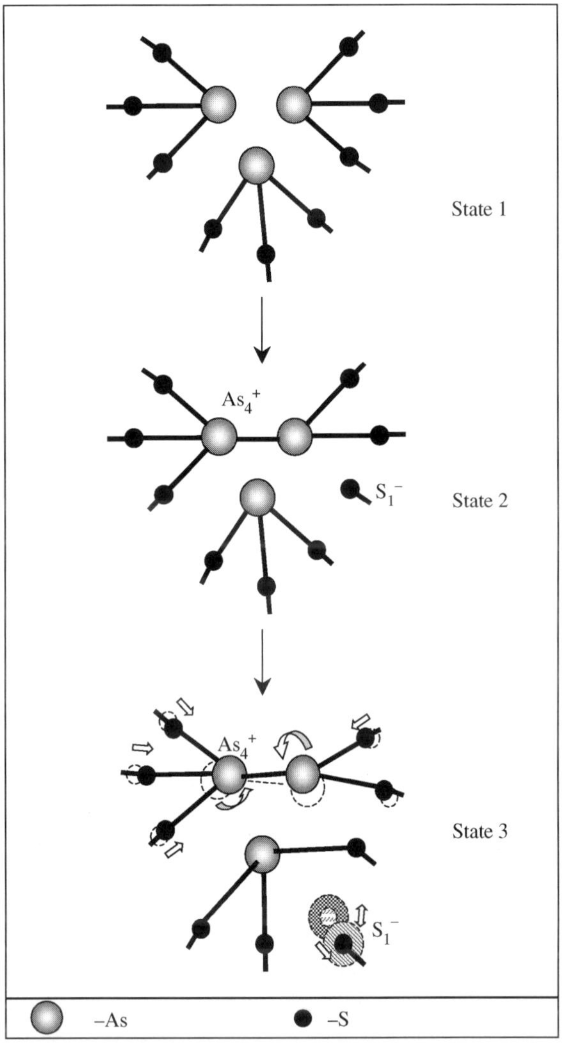

FIG. 18. Topological schemes of CTD-conserved bond-switching in v-As_2S_3, accompanied by a simultaneous separation of anomalously coordinated atoms with an additional open volume appearance.

5.2.1. *Intrinsic Destruction-Polymerization Transformations Induced by γ-Irradiation*

These transformations with positive ΔE values do not change the average coordination number Z, but enhance the mean energetic linking of a glassy-like network. They reveal themselves through two types of microstructural changes.

The first type is described by previously discussed CTDs formation (or bond-switching) processes with $\Delta E > 0$. In the case of v-As_2S_3 (Fig. 16), only topological reactions 1, 2, 11 and 12 correspond to positive balance in dissociation energies of switching covalent bonds. The first two reactions are accompanied by switching of homopolar As–As bond into heteropolar As–S one, while the second two reactions—

by switching of heteropolar As–S bond into homopolar S–S one. The energetic balance ΔE is 0.41 eV for the first bond-switching reaction and 0.21 eV for the second one. Therefore, the final structural state of v-As_2S_3 after bond-switching owing to topological schemes 1, 2, 11 and 12 (Fig. 16) is characterized by the higher total energy in comparison with non-irradiated state. This is why these destruction-polymerization transformations are irreversible in the subsequent irradiation—annealing cycles. It is quite understandable that they are more sufficient in as-prepared thin ChVS films obtained by thermal deposition because of a high concentration of 'wrong' homopolar covalent chemical bonds. However, in bulk ChVS samples of stoichiometric chemical composition, these bonds exist only as slight remainders on the background of energetically more favorable heteropolar covalent chemical bonds, giving a source for the above irreversible radiation-structural transformations.

The second type of γ-induced irreversible intrinsic destruction-polymerization transformations is proposed under suggestion on simultaneous switching of two covalent chemical bonds (bond pair) in ChVSs. In this case, the observed changes in the vibration band intensities, appearing in IR Fourier absorption (reflection) (Shpotyuk, 1993a,b, 1994; Balitska and Shpotyuk, 1998) or Raman scattering spectra (Frumar, Polak, Cernosek, Vlcel and Frumarova, 1997), are explained only by covalent chemical bonds redistribution without CTDs formation at the final stage. The typical topological scheme for such transformations, involving a simultaneous switching of homopolar S–S and As–As covalent bonds into two heteropolar As–S ones is shown by Figure 19. The correspondent energetic balance ΔE for such twofold bond switching in v-As_2S_3 is near 0.2 eV.

This scheme is often used for microstructural explanation of irreversible thermo- (Solin and Papatheodorou, 1977) or photoinduced (Strom and Martin, 1979) polymerization in as-deposited As_2S_3 thin films. The reason is that these freshly deposited ChVS films have a high concentration of molecular products with 'wrong' homopolar As–As and S–S covalent chemical bonds (Feltz, 1986).

Not refuting this variant of CTD-free bond switching, we believe, nevertheless, it has a very small probability for the real destruction-polymerization transformations in bulk ChVS samples. The fact is that the simultaneous two-fold bond-switching processes are proper only for some rare atomic configurations, which satisfy close co-existing of two heteropolar and homopolar covalent chemical bonds. In other words, all four atoms forming the initial pair of homopolar covalent chemical bonds (As–As and S–S bonds with lengths of 2.49 and 2.20 Å, respectively (Feltz, 1986; Elliott, 1986)) must be located in such sites of a glassy-like network, where the other bond pair can be created simultaneously (two As and two S atoms in the framework of two neighboring $AsS_{3/2}$

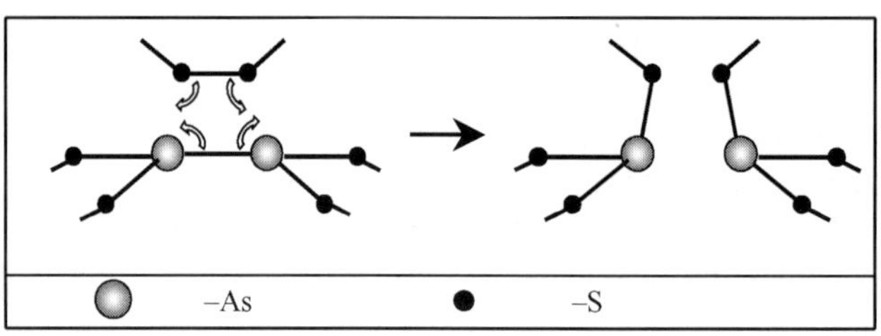

FIG. 19. Topological scheme of irreversible CTD-free two-fold bond-switching in v-As_2S_3.

pyramidal structural units in Fig. 19) without significant atomic displacements. It is more probable that this condition is satisfied mainly for one covalent bond itself (as in the case of CTDs formation in Fig. 16), but not for two bonds.

5.2.2. *Chemical Interaction with Absorbed Impurities Induced by γ-Irradiation*

The essential irreversible changes in physical properties caused by radiation-induced impurity processes of oxidation, hydrogenization, hydratation, carbonization and hydrocarbonization are observed in ChVSs, as a rule, after prolonged γ-irradiation. Let us consider these processes in the example of v-As_2S_3 (Shpotyuk, 1987a,b; Shpotyuk and Vakiv, 1991).

The radiation-induced oxidation of v-As_2S_3 is sharply expressed at the absorbed doses of γ-irradiation more than 5 MGy (Shpotyuk, 1987a,b). This conclusion has been proved by experimental results of IR spectroscopy in 4000–400 cm^{-1} region shown in Figure 6. It is obvious that intensities of all impurity bands associated with oxygen-containing complexes such as molecular As_4O_6 (1340, 1265, 1050 and 785 cm^{-1}) and SO_2 (1150 and 1000 cm^{-1}) increase after γ-irradiation with $\Phi = 10.0$ MGy dose ($P = 25$ Gy s^{-1}). The oxidation strength is so intensive at the doses close to 10.0 MGy that white layer of molecular As_4O_6 covering the sample's surface is observed.

According to the results of electron microprobe analysis ('CAMEBAX' micro-analyser), the As : S ratio at the surface of non-irradiated v-As_2S_3 glasses is near 1.56. After γ-irradiation, this parameter ($\Phi > 5$ MGy) increases in the near-surface layer by ~4%, but it does not change essentially in the depth of the bulk sample. By grinding of the created oxygen-enriched layer from the surface of the investigated glass, we can restore the intensities of 1340, 1265, 1050 and 785 cm^{-1} vibrational bands (Fig. 6), but the intensities of 1125 and 640 cm^{-1} bands corresponding to =As–O–As= structural groups are renewed only partially.

The radiation-induced oxidation of v-As_2S_3 is evidently due to chemical interaction of air-absorbed oxygen with intrinsic structural units, destroyed by a high-energetic γ-irradiation. It is established that this interaction can be sufficiently blocked, provided one of the following conditions is fulfilled:

- the investigated v-As_2S_3 samples are placed into evacuated (10^{-1}–10^{-2} Pa) quartz ampoules established then in ^{60}Co γ-irradiating cavity;
- the total γ-irradiation dose Φ accumulated in the normal conditions of stationary radiation field (without samples evacuation) does not exceed 1 MGy;
- the dose power P of γ-irradiation in the normal conditions (without evacuation) is less than 5–10 Gy s^{-1}.

The first condition limits an additional source of oxygen access needed for radiation-induced oxidation, while the next two give the necessary technical parameters correspondent to molecular As_4O_6 formation.

At the microstructural level of v-As_2S_3, the process of radiation-induced oxidation can be divided into three subsequent elementary stages. First, the metallic arsenic renews from paramagnetic $=As^{\cdot}$ states by joining some radiolysis products of absorbed moisture (atomic hydrogen, hydrostatic electrons, etc.). Owing to a high pressure of gas phase, this metallic As is extracted from the sample interior onto its surface during the second stage. Then the direct chemical interaction with atmospheric oxygen produces the observed surface As_4O_6 layer. So the oxidation process is enhanced by accompanied thermal annealing of the irradiated v-As_2S_3 samples at high absorbed doses and dose powers, or, alternatively, by additional thermal annealing at the temperatures near T_g performed just after radiation treatment.

The air-absorbed oxygen can be built in a glassy-like network through a simple substitution reaction in bridge =As–S–As= structural fragments. This reaction is quite possible in the sample interior and on its surface too. But the surface-laid =As–O–As= complexes, possessing an additional ability of chemical interaction, are easily transformed in molecular As_4O_6.

Similar considerations are proper to the process of SO_2 formation in γ-irradiated v-As_2S_3.

The radiation-induced hydrogenization is a process of chemical interaction of atomic hydrogen, created owing to radiolysis of absorbed moisture and air, with intrinsic destructed complexes. The hydrogen atoms easily join the dangling bonds, created in ChVSs by γ-irradiation, leading to their saturation. The molecular H_2S is the main product of hydrogenization. Its appearance in γ-irradiated v-As_2S_3 is clearly evident from the increase in vibration band intensity at 2470 cm^{-1} in IR spectra shown in Figure 6. The same conclusion results from low-temperature ESR measurements: the doublet signal 2–8 with resonance splitting $A = 502$ Gs in Figure 5 is attributed to the ESR signal of hydrogen atoms (Shvec et al., 1986). The results of laser mass-spectrometry (LAMMA-1000 'Leybold–Herraeus' spectrometer) of γ-irradiated v-As_2S_3 samples in Figure 20 (Shpotyuk and Vakiv, 1991), showing an increase in the intensities of positively charged

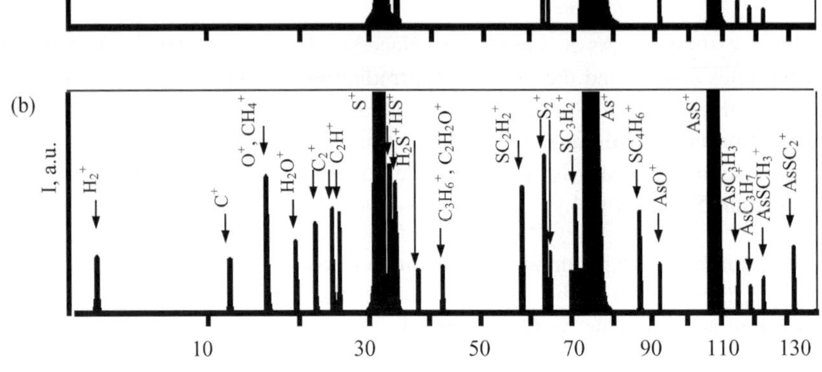

FIG. 20. Fragment of mass-spectrum obtained from the surface of non-irradiated (a) and γ-irradiated ($\Phi =$ 10.0 MGy, $P = 25$ Gy s^{-1}) in the third cycle v-As_2S_3 (b).

HS^+ (33 amu) and H_2S^+ (34 amu) responses, are utilized as an additional confirmation for the process of radiation-induced hydrogenization.

The increase in the content of impurity hydroxide O_nH_m groups linked with intrinsic structural units in γ-irradiated v-As_2S_3 samples is the matter of the irreversible *radiation-induced hydratation process*. It was first pointed out in Konorova et al. (1985) that ^{60}Co γ-irradiation sufficiently affected the spectrum of stretch and bend vibrations of these groups owing to their stronger interaction with radiation-modified glassy-like network. Similar experimental results have been obtained (Shpotyuk, 1987a,b), but we have put forward another interpretation, using a more detailed analysis of vibrational OH-group spectrum in ChVSs (Tadashi and Yukio, 1982). We believe particularly that the 3450 cm^{-1} band in v-As_2S_3 is attributed to =As–OH structural fragments, while the 3420 cm^{-1} band to –S–OH ones. Therefore, the observed increase in the intensity of 3600–3300 cm^{-1} vibrational band with simultaneous long-wave shift of its maximum (Fig. 6) testifies not only in favor of radiation-induced absorption of molecular water, but also on its radiolysis with a subsequent joining of the created products to the intrinsic structural units of a glassy-like network. So, the =As–OH complexes are the dominant ones among impurity products of radiation-induced hydratation in γ-irradiated v-As_2S_3.

If radiation–thermal treatment of ChVSs is performed multiply at high doses of γ-irradiation ($\Phi \sim$ 10 MGy) and thermal annealing temperature near T_g, the new kinds of irreversible impurity processes are revealed. The first one is the *radiation-induced carbonization* (chemical interaction of γ-destructed intrinsic structural units with absorbed carbon atoms), and the second one—*the radiation-induced hydrocarbonization* (chemical interaction of γ-destructed intrinsic structural units with absorbed hydrocarbon C_nH_m groups). Their main products in γ-irradiated v-As_2S_3 can be identified with mass-spectrometry technique in the form of positively charged C^+ (12 amu), CH_4^+ (16 amu), C_2^+ (24 amu), C_2H^+ (25 amu), $C_3H_6^+$ (42 amu), $SC_2H_2^+$ (58 amu), $SC_3H_2^+$ (70 amu), $SC_4H_6^+$ (86 amu), $AsC_3H_3^+$ (114 amu), $AsC_3H_7^+$ (118 amu), $AsSCH_3^+$ (122 amu) and $AsSC_2^+$ (131 amu).

The ChVSs containing organic components were synthesized first by Hermann et al. (Herman, 1978; Herman, Lemnitzer and Mankeim, 1978). It was shown that organic polymeric chains were linked to inorganic ones in the synthesized mixed glasses through bridge arsenic–carbon and sulfur–carbon units, with double-fold and triple-fold chemical bonds being accepted as quite possible.

The similar linking fragments (≡C–As=C≡, =As–C≡, S=C=, –S–C≡, etc.) can be formed during multiple radiation–thermal treatment of ChVSs. So, these glasses can be conditionally presented by v-As_2S_3–C_nH_m chemical formula.

It is established that the above irreversible γ-induced impurity transformations decay with concentration of chemical elements having a high level of bond saturation (chemical analogs of As–Sb, Bi) in the framework of quasi-binary $As_2S(Se)_3$–$Sb(Bi)_2S(Se)_3$ cross-sections (Shpotyuk and Vakiv, 1991).

6. Some Practical Applications of RIEs

The above-described RIEs can be successfully used in high-energetic dosimetry of ionizing irradiation and in some technological developments devoted to the modification of ChVSs' physical properties.

6.1. ChVS-Based Optical Dosimetric Systems

Registration of high levels of ionizing irradiation is one of the most important problems in the field of solid dosimetric systems of industrial application. The coloring oxide glasses are conventionally used to resolve this problem (Frank and Shtoltz, 1973; Pikaev, 1975). They are sufficiently simple to use and manufacture, resistant to external influences, but do not allow determination of the absorbed doses more than 1 MGy. Additional inconveniences of these materials are connected with the necessity of high-temperature annealing for restoration of their initial optical properties (800–1000 K).

The above disadvantages can be easily eliminated in ChVS-based optical dosimetric systems (Shpotyuk et al., 1991a,b; Shpotyuk, 1995), especially those containing the plane-parallel v-As_2S_3 plate as a radiation-sensitive element. If all measurements are carried out at the wavelength correspondent to the middle point of the upward part of optical transmission edge (in the fundamental optical absorption edge region) and optical density D is used as controlled dose-sensitive parameter, then the following linear expression is valid for the absorbed γ-irradiation doses in the 0.5–10.0 MGy range:

$$\Delta D/D_o = S \log \Phi + A, \tag{9}$$

where $\Delta D/D_o$ is the relative increase in optical density, caused by γ-irradiation, D_o is the optical density of non-irradiated sample and S and A are some material-related constants.

It is obvious that the sensitivity of ChVS-based dosimetric system to the absorbed dose Φ is determined by S constant. The thickness of the plane-parallel dose-sensitive element likely varies from 1 up to 2 mm. In this case, the relative increase of optical density $\Delta D/D_o$ is calculated at the wavelength of helium–neon laser ($\lambda = 633$ nm), the sensitivity S in the freshly prepared bulk v-As_2S_3 being close to 0.30 (Shpotyuk, 1995). This value is lowered only by 0.05 in the second and all following irradiation–annealing cycles. Therefore, for a high accuracy of dose registration, it is advisable to do the first 'idle' cycle of γ-irradiation and thermal annealing of as-prepared ChVS samples.

The ChVS-based dosimeters have a number of advantages in comparison with the ones using the oxide glasses as radiation-sensitive elements. Optical properties of ChVSs are stable over a long period after γ-irradiation for no less than 10 years, provided the temperature is smaller than thermal bleaching threshold (Shpotyuk, 1987a,b; Shpotyuk and Savitsky, 1989). Additionally, the dosimetric characteristics of ChVSs are not dependent considerably on the dose power P, when the temperature in the irradiating chamber is not higher than 330–340 K. The latter condition is satisfied, when the dose powers are less than 15 Gy s^{-1} or the total γ-irradiation dose is accumulated in the ChVS sample by separate steps with 3–5 kGy, keeping the average temperature at the level of 310–320 K.

6.2. Radiation Modification of ChVSs Physical Properties

The first experiments on the use of high-energetic ionizing irradiation to modify the ChVSs physical properties were carried out by the scientific group of Sarsembinov

(Kazakh State University, Alma-Ata, Kazakhstan) in the early 1980s (Sarsembinov and Abdulgafarov, 1981). However, this practically important conclusion dealt exceptionally with irradiation effects caused by accelerated electrons ($E > 2$ MeV).

The experimental results, presented above and characterized by the sharply defined γ-induced changes in v-As_2S_3 (Section 4.1), testify that radiation treatment of bulk ChVS samples, performed in the normal conditions of stationary ^{60}Co γ-irradiation field ($E = 1.25$ MeV) with accumulated doses of $\Phi = 0.5-10.0$ MGy, is an effective alternative way to improve their main exploitation parameters. Possessing a number of sufficient advantages over corpuscular irradiation, first of all, a high uniformity of the produced structural changes throughout the sample thickness (Pikaev, 1985), the ^{60}Co γ-irradiation changes the ChVSs microhardness (Section 4.1.1), spectral position and slope of the fundamental optical absorption edge (Section 4.1.2), acoustic velocity and acousto-optical figure of merit (Section 4.1.3), etc.

7. Final Remarks

In spite of their complicated nature, the RIEs possess remarkable features and have been in the sphere of special interest of many scientific groups. To thoroughly understand their essence, more precise microstructural research using the technical possibilities of new in situ and non-traditional experimental techniques, such as positron annihilation, EXAFS, etc., must be carried out. We hope these investigations will be successful in the near future.

References

Amosov, A.V., Vasserman, I.M., Gladkih, A.A., Pryanishnikov, V.P. and Udin, D.M. (1970) Paramagnetic centers in vitreous silica, *Zh. Prikl. Spectroscopii*, **13**(1), 142–148.

Anderson, P.W. (1975) Model for the electronic structure of amorphous semiconductors, *Phys. Rev. Lett.*, **34**, 953–955.

Andriesh, A.M., Bykovskij, U.A., Borodakij, U.V., Kozhin, A.F., Mironos, A.V., Smirnov, V.L. and Ponomar, V.V. (1984) Stability of IR fibers based on ChVS under a large dose neutron irradiation, *Pisma v ZhTF*, **10**(9), 547–549.

Arsova, D., Skordeva, E. and Vateva, E. (1994) Topological threshold in $Ge_xAs_{40-x}Se_{60}$ glasses and thin films, *Solid State Commun.*, **90**(5), 299–302.

Averianov, V.L., Kolobov, A.V., Kolomiets, B.T. and Lyubin, V.M. (1979) Thermally-optical transitions at photostructural transformations in chalcogenide vitreous semiconductors, *Pisma v ZhETF*, **30**(9), 621–624.

Babacheva, M., Baranovsky, S.D., Lyubin, V.M., Tagirdzhanov, M.A. and Fedorov, V.A. (1984) Influence of photostructural transformations in As_2S_3 films on Urbach absorption edge, *Fiz. Tverd. Tela*, **26**(7), 2194–2196.

Balitska, V., Filipecki, J., Shpotyuk, O., Swiatek, J. and Vakiv, M. (2001) Dynamic radiation-induced effects in chalcogenide vitreous compounds, *J. Non-Cryst. Solids*, **287**, 329–332.

Balitska, V.O. and Shpotyuk, O.I. (1998) Radiation-induced structural transformations in vitreous chalcogenide semiconductors, *J. Non-Cryst. Solids*, **227–230**, 723–727.

Berkes, J.S., Ing, S.W. Jr. and Hillegas, W.J. (1971) Photodecomposition of amorphous As_2Se_3, *J. Appl. Phys.*, **42**(12), 4908–4916.

Bishop, S.G., Strom, U. and Taylor, P.C. (1975) Optically induced localized paramagnetic states in chalcogenide glasses, *Phys. Rev. Lett.*, **34**(21), 1346–1350.

Boolchand, P., Feng, X., Selvanathan, D. and Bresser, W.J. (1999) Rigidity transition in chalcogenide glasses. In *Rigidity Theory and Applications* (Eds, Thorpe, MF and Duxbury, PM), Kluwer Academic/Plenum Publishers, New York, pp. 1–17.

Borisova, Z.U. (1982) *Chalcogenide Vitreous Glasses*, Leningrad University, Leningrad, p. 344.

Budinas, T., Mackus, P., Savytsky, I.V. and Shpotyuk, O.I. (1987) Spectroscopic investigations of induced processes in arsenic sulphide chalcogenide glasses, *J. Non-Cryst. Solids*, **90**(1–3), 521–523.

Calemczuk, R. and Bonjour, E. (1981a) Gamma-ray induced relaxation in selenium glass, *J. Non-Cryst. Solids*, **43**(3), 427–432.

Calemczuk, R. and Bonjour, E. (1981b) Structural relaxation of glassy selenium under gamma-irradiation, *J. Phys. Lett.*, **42**(22), L501–L502.

Chepeleva, I.V. (1987) Paramagnetic radiation defects in chalcogenide glasses and glassy Se, *J. Non-Cryst. Solids*, **97–98**(Pt 2), 1179–1182.

De Bast, J. and Gilard, P. (1963) Variation of the viscosity of glass and relaxation of stresses during stabilization, *Phys. Chem. Glasses*, **4**(4), 117–128.

DeNeufville, J.P., Moss, S.C. and Ovshinsky, S.R. (1974) Photostructural transformations in amorphous As_2Se_3 and As_2S_3 films, *J. Non-Cryst. Solids*, **13**, 191–223.

Domoryad, I.A. (1969) Hardness of γ-irradiated As_2S_3 glass. *Radiacyonno-Stimulirovannye Processy v Tverdych Telach.*, FAN, Tashkent, pp. 57–59.

Domoryad, I.A. and Kaipnazarov, D. (1964) Effect of γ-irradiation on elastic properties of chalcogenide glasses of As_2Se_3–As_2Te_3 system, *Izv. AN Uzb. SSR. Ser. Fiz.-Mat. Nauk*(3), 67–70.

Domoryad, I.A., Kaipnazarov, D. and Khiznichenko, L.P. (1963) Effect of gamma-irradiation on elastic properties of vitreous arsenic trisulphide, *Izv. AN Uzb. SSR. Ser. Fiz.-Mat. Nauk*(5), 87–89.

Domoryad, I.A., Kolomiets, B.T., Lyubin, V.M. and Shilo, V.P. (1975) Effect of γ-irradiation on vitreous As_2Se_3, *Izv. AN SSSR. Neorganicheskie Materialy*, **11**(4), 743–744.

Durnij, R., Macko, P. and Mackova, V. (1979) Paramagnetic centers in neutron irradiated glassy semiconductors, *Solid State Commun.*, **31**(3), 173–174.

Edmond, J.T., Male, J.C. and Chester, P.F. (1968) On the suitability of liquid semiconductors for resistance thermometry in nuclear reactor, *J. Sci. Instr. (J. Phys. E)*, **1**, 373–377.

El-Fouly, M.H., El-Behay, A.Z. and Fayek, S.A. (1982) Effect of γ-radiation on the electrical properties of bulk amorphous $As_4Se_2Te_4$, *Rad. Phys. Chem.*, **19**(1), 57–61.

Elliott, S.R. (1985) Photostructural changes in amorphous chalcogenide materials, *Glass...Curr. Issues Proc. NATO Adv. Study Inst., Generife, Apr. 2–13, 1984*, Dordrecht e.a., pp. 399–407.

Elliott, S.R. (1986) A unified model for reversible photostructural effects in chalcogenide glasses, *J. Non-Cryst. Solids*, **81**, 71–98.

Feltz, A. (1986) *Amorphous and Vitreous Inorganic Solids*, Mir, Moscow, p. 556.

Frank, M. and Shtoltz, B. (1973) *Solid State Dosimetry of Ionising Radiation*, Atomizdat, Moscow, pp. 247–888.

Frumar, M., Polak, Z., Cernosek, Z., Vlcel, M. and Frumarova, B. (1997) Photoinducedeffects in amorphous chalcogenides. In *Physics and Applications of Non-Crystalline Semiconductors in Optoelectronics* NATO ASI Series 3. High Technology, Vol. 36 (Eds, Andreish, A. and Bertolotti, M.), Kluwer Academic Publishers, Dordrecht, pp. 123–139.

Gaczi, P.J. (1982) As and S centred paramagnetic species in sulphide glasses, *Phil. Mag. B*, **45**(2), 241–259.

Goryunova, N.A. and Kolomiets, B.T. (1955) Invention diploma No. 98 USSR. Glasses Semiconductors. Appl. No. OT-7460 from 26/11/1969. Prior. Jan. 1955.

Goryunova, N.A. and Kolomiets, B.T. (1956) New vitreous semiconductors, *Izv. AN USSR. Ser. Fiz.*, **20**(12), 1496–1500.

Griscom, D.L., Gingerich, M.E. and Friebele, E.J. (1993) Radiation-induced defects in glasses: origin of power-law dependences of concentration on dose, *Phys. Rev. Lett.*, **71**(7), 1019–1022.

Guralnik, R.M., Lantratova, S.S., Lyubin, V.M. and Sarsembinov, Sh.Sh. (1982) Electron-stimulated reversible changes of optical properties and microhardness of vitreous As_2S_3, *Fiz. Tverd. Tela*, **24**(5), 1334–1337.

Gurevich, S.B., Ilyashenko, N.N., Kolomiets, B.T., Lyubin, V.M. and Shilo, V.P. (1974) Photostimulated changes of optical properties and structural of amorphous Se–As films, *Phys. Status Solidi*, **26**(2), K127–K130.

Gusev, O.B. and Kludzin, V.V. (1987) *Acoustoptical Measurements*, Leningrad. University, Leningrad, p. 150.

Herman, D. (1978) Chalkogenidglaser mit organischen Komponenten. I. Uber Azobenzolgruppen enthaltende Glaser der Reiche $R_xSe_{60-x}As_{20}Br_{20}$, *Z. Anorg. Allg. Chem.*, **438**(1), 75–82.

Herman, D., Lemnitzer, M. and Mankeim, W. (1978) Chalkogenidglaser mit organischen Komponenten. II. Uber Glaser der zusammensetzung Selen-Arsen-Jod-Biphenylen, *Z. Anorg. Allg. Chem.*, **438**(1), 83–89.

Kalinich, S.I., Turjanitsa, I.I., Dobosh, M.V., Himinets, V.V. and Zholudev, S.M. (1986) Effect of the influence of high-energetic electron irradiation on the crystallization of Ge–Sb–Se glasses, *Ukr. Fiz. Zh.*, **31**(11), 1724–1726.

Kastner, M. (1973) Compositional trends in the optical properties of amorphous lone-pair semiconductors, *Phys. Rev. B*, **7**(12), 5237–5252.

Kastner, M. (1978) Defect chemistry and states in the gap of lone-pair semiconductors, *J. Non-Cryst. Solids*, **31**, 223–240.

Kastner, M.A., Adler, D. and Fritzsche, H. (1976) Valence-alternation model for localized gap states in lone-pair semiconductors, *Phys. Rev. Lett.*, **37**, 1504–1507.

Kirilenko, V.V., Dembovsky, S.A. and Poliakov, Yu.A. (1975) Optical properties of glasses in As_2S_3–Tl_2S and As_2Se_3–Tl_2Se systems, *Izv. AN SSR. Neorgan. Mater.*, **11**(11), 1923–1928.

Kolomiets, B.T., Domoryad, I.A., Andriesh, A.M., Iovu, M.S. and Shutov, S.D. (1975) Effect of gamma-irradiation on photoconductivity of vitreous arsenic sulphide. *Fizika i Khimia Slozhnych Poluprovodnikov.*, Shtinca, Kishinev, pp. 149–152.

Kolomiets, B.T., Mamontova, T.N., Domoryad, I.A. and Babaev, A.A. (1971) Photoluminescence in γ-irradiated vitreous and monocrystalline As_2S_3 and As_2Se_3, *Phys. Status Solidi A*, **7**, K29–K32.

Konorova, L.F., Kim, T.I., Zhdanovich, N.S. and Litovski, M.A. (1985) Effect of γ-quanta on IR absorption in chalcogenide vitreous alloys, *Zh. Techn. Fiz.*, **55**(4), 788–791.

Konorova, L.F., Kim, T.I., Zhdanovich, N.S. and Litovski, M.A. (1987) Effect of γ-quanta on IR absorption in vitreous AsGeSe, contained the metals, *Fiz. Techn. Poluprov.*, **21**(7), 1300–1302.

Konorova, L.F., Zhdanovich, N.S., Didik, V.A. and Prudnikov, I.M. (1989) On structural changes in vitreous AsGaSe, irradiated by high doses of neutrons, *Fiz. Techn. Poluprov.*, **23**(4), 706–709.

Kotkata, M.F., El-Fouly, M.H., Fayek, S.A. and El-Hakim, S.A. (1986) The effect of Tl addition on the electrical and thermal transport properties of amorphous As_2Se_3, *Semicond. Sci. Technol.*, **1**, 313–319.

Kumagai, N., Shirafuji, J. and Inuishi, Y. (1984) Photo-induced ESR and its annealing behaviour in melt-quenched Ge–S glasses, *Jpn. J. Appl. Phys.*, **23**(10), 1287–1293.

Liholit, I.L., Lyubin, V.M., Masterov, V.F. and Fedorov, V.A. (1984) Investigation of photostructural transformations in As_2S_3 films by photoinduced paramagnetic resonance method, *Fiz. Tverd. Tela*, **26**(1), 172–178.

Lukasik, D. and Macko, P. (1981) Creation of thermal spikes in the GeS_{11} glassy system by fast neutron irradiation, *Phys. Status Solidi*, **64**(2), 677–682.

Ma, D.S., Danielson, P.S. and Moyniham, C.T. (1980) Bulk and impurity infrared absorption in $0.5As_2Se_3$–$0.5GeSe_2$, *J. Non-Cryst. Solids*, **37**(2), 181–190.

Macko, P. and Doupovec, J. (1978) Thermal and electrical properties of neutron irradiated GeS_{11} glass, *Phys. Status Solidi*, **50**, K51–K55.

Macko, P. and Mackova, V. (1977) Change of GeS_x parameters after neutron irradiation, *Chech. J. Phys. B*, **27**, 1139–1145.

Maged, A.F., Wahab, L.A. and El Kholy, I.A. (1998) Effect of Te addition on some physical properties of the Se_3S_2 system and a study of the response to γ-irradiation, *J. Mater. Sci.*, **33**, 3331–3335.

Mahadevan, S. and Giridhar, A. (1992) Floppy to rigid transition and chemical ordering in Ge–Sb(As)–Se glasses, *J. Non-Cryst. Solids*, **143**(14), 52–58.

Matkovsky, A.O., Sugak, D.Yu., Ubizsky, S.B., Shpotyuk, O.I., Cherny, E.A., Vakiv, N.M. and Mokrytsky, V.A. (1992) *Influence of Ionizing Irradiation on Materials for Electronics*, Svit, Lvov, pp. 212–555.

Mazurin, O.V. (1977) Relaxation phenomena in glass, *J. Non-Cryst. Solids*, **25**(1–s3), 130–169.

Minaev, V.S. (1991) *Vitreous Semiconductor Alloys*, Metallurgia, Moscow, p. 376.

Minami, T., Honjo, K. and Tanaka, M. (1977) Thermally stimulated conductivity in $AsS_{3.5}Te_{2.0}$ glass irradiated with gamma-rays, *J. Non-Cryst. Solids*, **23**, 431–434.

Minami, T., Yoshida, A. and Tanaka, M. (1972) Gamma-ray induced conductivity of vitreous semiconductors in the systems As–S–Te and As–Se–Te, *J. Non-Cryst. Solids*, **7**, 328–336.

Miyauchi, K., Qiu, J., Shojiya, M., Kawamoto, Y. and Kitamura, N. (2001) Structural study of GeS_2 glasses permanently densified under high pressures up to 9 GPa, *J. Non-Cryst. Solids*, **279**, 186–195.

Mori, T., Matsuishi, K. and Arai, T. (1984) Vibrational properties and network topology of amorphous As–S system, *J. Non-Cryst. Solids*, **65**, 269–283.

Moskalonov, A.V. (1976) Radiation effects in arsenic sulphides and selenides, *Izv. AN LatvSSR. Ser. Fiz. i Tehn. Nauk*(5), 64–69.

Mott, N.F. and Davis, E.A. (1979) *Electron Processes in Non-crystalline Materials*, Clarendon Press, Oxford, p. 663.

Mott, N.F., Davis, E.A. and Street, R.A. (1975) States in the gap and recombination in amorphous semiconductors, *Phil. Mag.*, **32**, 961–996.

Onary, S., Inokuma, T., Kataura, H. and Arai, T. (1987) Absorption edge of the amorphous $(GeS_2)_x(As_2S_3)_{1-x}$ system under hydrostatic pressure, *Phys. Rev. B*, **35**(9), 4373–4379.

Ovshinsky, S.R., Eans, E.J., Nelson, D.L. and Fritzsche, H. (1968) Radiation hardness of ovonic devices, *IEEE Trans. Nucl. Sci.*, **15**(6), 311–321.

Phillips, J.C. (1985) "Phase transitions" in disordered solids. In *Physics of Disordered Materials* (Eds, Adler, D., Fritzsche, H. and Ovshinsky, S.R.), Plenum Press, New York, pp. 159–173.

Pikaev, A.K. (1975) *Dosimetry in Radiation Chemistry*, Nauka, Moscow, p. 312.

Pikaev, A.K. (1985) *Modern Radiation Chemistry. Main Principles, Experimental Technique and Methods*, Nauka, Moscow, p. 375.

Pinnow, D.A. (1970) Guide lines for the selection of acoustoptic materials, *IEEE J. Quant. Electron.*, **6**(4), 223–238.

Poltavtsev, Yu.G. and Pozdnyakova, V.M. (1973) Structural changes of vitreous arsenic trisulphide and triselenide under the influence of penetration radiation, *Ukr. Fiz. Zh.*, **18**(1), 1148–1150.

Pontuschka, W.M. and Taylor, P.C. (1981) ESR in X-irradiated As_2O_3 glass, *Solid State Commun.*, **38**(7), 573–577.

Rao, K.J. and Mohan, R. (1981) Chemical bond approach to determining conductivity band gaps in amorphous chalcogenides and pnictides, *Solid State Commun.*, **39**(10), 1065–1068.

Sarsembinov, Sh.Sh. (1982) On the nature of exponential edge of optical absorption in chalcogenide vitreous semiconductors, *Vest. AN KazSSR*(1), 55–62.

Sarsembinov, Sh.Sh. and Abdulgafarov, E.E. (1980a) Electron-stimulated changes of optical properties of chalcogenide vitreous semiconductors, *Izv. AN Kaz. SSR. Ser. Fiz.-Mat.*(6), 60–64.

Sarsembinov, Sh.Sh. and Abdulgafarov, E.E. (1980b) Positron annihilation in vitreous As_2S_3, treated by electron bombardment, *Vest. AN Kaz. SSR*(2), 48–55.

Sarsembinov, Sh.Sh. and Abdulgafarov, E.E. (1981) Electron-stimulated ESR in vitreous Ge_2S_3, *Fiz. Tverd. Tela*, **23**(1), 322–324.

Sarsembinov, Sh.Sh. and Abdulgafarov, E.E. (1981) Electron-stimulated modification of electrical properties of chalcogenide vitreous semiconductors, *Vest. AN Kaz.SSR*(5), 51–57.

Sarsembinov, Sh.Sh., Abdulgafarov, E.E., Tumanov, M.A. and Rogachev, N.A. (1980) Radiation-induced defects influence on the electrical, photoelectrical and optical properties of chalcogenide glasses, *J. Non-Cryst. Solids*, **35–36**(2, Pt 2), 877–882.

Savage, J.A. (1982) Optical properties of chalcogenide glasses, *J. Non-Cryst. Solids*, **47**(1), 101–116.

Savitsky, I.V. and Shpotyuk, O.I. (1990) Gamma-stimulated changes of acoustooptical properties of vitreous arsenic trisulphide, *Akust. Zh.*, **36**(2), 344–347.

Scott, D.W., McCullough, J.P. and Kruse, F.H. (1964) Vibrational assigment and force constants of S_8 from a normal-coordinate treatment, *J. Mol. Spectrosc.*, **13**(3), 313–320.

Sheloput, D.V. and Glushkov, V.F. (1973) Acoustooptical characteristics of chalcogenide glasses, *Izv. AN SSSR. Neorgan Mater.*, **9**(7), 1149–1152.

Shpotyuk, O.I. (1985) Influence of gamma-irradiation on optical transmission spectra and microhardness of vitreous As_2S_3–Sb_2S_3, *Visn. Lviv Univ.*, **19**, 25–31.

Shpotyuk, O.I. (1987a) Temperature stability of gamma-irradiated effects in arsenic trisulphide, *Zh. Prikl. Spetroscopii*, **46**(1), 122–126.

Shpotyuk, O.I. (1987b) Spectroscopic investigations of penctrating radiation influence on the surface of vitreous arsenic chalcogenides in IR region, *Ukr. Phys. Zh.*, **32**(4), 509–512.

Shpotyuk, O.I. (1990) Gamma-stimulated changes of optical properties of chalcogenide vitreous semiconductors, *Ukr. Phyz. Zh.*, **35**(10), 1545–1549.

Shpotyuk, O.I. (1993a) On the mechanism of reversible radiation-structural transformations in chalcogenide vitreous semiconductors, *Zh. Prikl. Spectroskopii*, **59**(5–6), 550–553.

Shpotyuk, O.I. (1993b) Induced effects in chalcogenide vitreous semiconductors and the concept of destrution-polymerization transformations, *Latv. J. Phys. Tech. Sci.*(4), 32–43.

Shpotyuk, O.I. (1994a) Reversible radiation effects in vitreous As_2S_3. 2. Mechanism of structural transformations, *Phys. Status Solidi A*, **145**(1), 69–75.

Shpotyuk, O.I. (1994b) Mechanism of radiation-structural transformations in amorphous As_2S_3, *Rad. Effects Defects Solids*, **132**(4), 393–396.

Shpotyuk, O.I. (1995) Amorphous chalcogenide semiconductors for dosimetry of high-energy ionizing radiation, *Rad. Phys. Chem.*, **46**(4–6), 1279–1282.

Shpotyuk, O.I. (1996) Photo- and radiation-induced coordination defects in amorphous chalcogenides, *Proc. SPIE*, **2968**, 246–255.

Shpotyuk, O.I. (2000) Radiation-induced effects in chalcogenide glasses: topological mechanism and application, *Nucl. Instr. Methods Phys. Res. B*, **166–167**, 525–528.

Shpotyuk, O.I. and Balitska, V.O. (1997) Coordination defects in vitreous As_2S_3 induced by γ-irradiation, *Acta Phys. Pol. A.*, **92**(3), 577–583.

Shpotyuk, O.I. and Balitska, V.O. (1998) Electron-induced dichroism in vitreous As_2S_3: physical features and microstructural mechanism, *Phys. Status Solidi A*, **165**, 295–302.

Shpotyuk, O.I., Filipecki, J., Golovchak, R.Ya., Kovalskiy, A.P. and Hyla, M. (2001) Radiation defects in amorphous As–Ge–S studied by positron annihilation technique, *J. Optoelectron. Adv. Mater.*, **3**(2), 329–332.

Shpotyuk, O.I., Golovchak, R.Ya., Kavetsky, T.S. and Kovalskiy, A.P. (2000) Radiation-optical effects in glassy Ge–As(Sb)–S systems, *Nucl. Instr. Methods Phys. Res. B*, **166–167**, 517–520.

Shpotyuk, O.I., Kovalskiy, A.P., Skordeva, E., Vateva, E., Arsova, D., Golovchak, R.Ya. and Vakiv, M.M. (1999b) Effect of gamma-irradiation on the optical properties of $Ge_xAs_{40-x}S_{60}$ glasses, *Phys. B. Condens. Matter*, **271**, 242–247.

Shpotyuk, O.I., Kovalsky, A.P., Vakiv, M.M. and Mrooz, O.Ya. (1994) Reversible radiation effects in vitreous As_2S_3. 1. Changes of physical properties, *Phys. Status Solidi A*, **144**(2), 277–283.

Shpotyuk, O.I. and Matkovskii, A.O. (1994a) Radiation-stimulated processes in vitreous arsenic trisulphide, *J. Non-Cryst. Solids*, **176**, 45–50.

Shpotyuk, O.I. and Matkovskii, A.O. (1994b) Radiation-optical properties of vitreous As_2S_3, *Opto-Electron. Rev.*, **2**(4), 100–103.

Shpotyuk, O.I., Matkovskii, A.O., Kovalsky, A.P. and Vakiv, M.M. (1995) Radiation-induced changes of amorphous As_2S_3 physical properties, *Rad. Effects Defects Solids*, **133**(1), 1–4.

Shpotyuk, O.I. and Savitsky, I.V. (1989) Radiation-thermal effects in vitreous semiconductors of As_2S_3–Sb_2S_3 system, *Ukr. Phyz. Zh.*, **34**(6), 894–898.

Shpotyuk, O.I. and Savitsky, I.V. (1990) Some features of radiation-optical effects in vitreous arsenic chalcogenides, *Visn. Lviv Univ.*, **23**, 91–93.

Shpotyuk, O.I., Savitsky, I.V. and Kovalsky, A.P. (1989) Physical features of radiation-stimulated effects in vitreous arsenic selenides, *Phys. Electron. (Lvov)*, **39**, 59–64.

Shpotyuk, O.I., Shvarts, K.K., Kornerlyuk, V.N., Shunin, Yu.N. and Pirogov, F.V. (1991a) Destruction-polymerization transformations in chalcogenide vitreous semiconductors. Part 1. Induced processes and defects. Part 2. Mathematical modeling. AN LatvSSR. Inst. Phys. Preprint xxxx-163, 105.

Shpotyuk, O.I., Skordeva, E.R., Golovchak, R.Ya., Pamukchieva, V.D. and Kovalskij, A.P. (1999a) Radiation-induced changes of transmission of As_2S_3–Ge_2S_3 chalcogenide glasses, *Zh. Prikl. Spetroscopii*, **66**(5), 657–660.

Shpotyuk, O.I. and Vakiv, M.M. (1991) Irreversible radiation-structural transformations in vitreous arsenic chalcogenides (a review). *Radioelectronic Materials Science, Part III*, pp. 92–106.

Shpotyuk, O.I., Vakiv, M.M., Kornelyuk, N.M. and Kovalsky, A.P. (1991b) Solid sensors of ionizing radiation based on amorphous arsenic chalcogenides, *Metrologija*(6), 15–18.

Shvec, D.I., Shpotyuk, O.I., Matkovskij, A.O., Kavka, I.I. and Savitskij, I.V. (1986) ESR of radiation-stimulated paramagnetic centers in vitreous arsenic trisulphide, *Ukr. Phyz. Zh.*, **31**(8), 1226–1228.

Skordeva, E.R. and Arsova, D.D. (1995) A topological phase transition in ternary chalcogenide films, *J. Non-Cryst. Solids*, **192–193**, 665–668.

Skordeva, E., Arsova, D., Pamukchieva, V., Vateva, E., Golovchak, R., Kovalskiy, O. and Shpotyuk, O. (2000) γ-Induced changes in Ge–As–S glasses, *J. Optoelectron. Adv. Mater.*, **2**(3), 259–266.

Solin, S.A. and Papatheodorou, G.N. (1977) Irreversible thermostructural transformations in amorphous As_2S_3 films: a light scattering study, *Phys. Rev.*, **B15**(4), 2084–2090.

Srinivasan, A., Madhusoodanan, K.N., Gopal, E.S.R. and Philip, J. (1992) Observation of a threshold behavior in the optical band gap and thermal diffusivity of Ge–Sb–Se glasses, *Phys. Rev. B*, **45**(14), 8112–8115.

Starodubcev, S.V., Domoryad, I.A. and Khiznichenko, L.P. (1961) Changes of mechanical characteristics of amorphous selenium under gamma-ray influence, *Dokl. AN SSSR*, **139**(3), 594–595.

Street, R.A. (1977) Non-radiative recombination in chalcogenide glasses, *Solid State Commun.*, **24**, 363–365.

Street, R.A. (1978) Recombination in amorphous semiconductors, *Phys. Rev. B*, **17**, 3984–3995.

Strom, U. and Martin, T.P. (1979) Photo-induced changes in the infrared vibrational spectrum of evaporated As_2S_3, *Solid State Commun.*, **29**(7), 527–530.

Tadashi, M. and Yukio, T. (1982) Optical transmission loss of As–S glass fiber in 1.0–5.5 μm wavelength region, *Jpn. J. Appl. Phys.*, **21**(2), 75–76.

Takebe, H., Maeda, H. and Morinaga, K. (2001) Compositional variation in the structure of Ge–S glasses, *J. Non-Cryst. Solids*, **291**, 14–24.

Tanaka, K. (1989) Structural phase transition in chalcogenide glasses, *Phys. Rev. B*, **39**(2), 1270–1279.

Taylor, P.C., Strom, U. and Bishop, S.G. (1978) Paramagnetic states induced in amorphous As, Se, and As_2Se_3 by electron bombardment at 77 K, *Phys. Rev. B*, **18**(1), 511–515.

Thorpe, M.F. (1985) Rigidity percolation. In *Physics of Disordered Materials* (Eds, Adler, D., Fritzsche, H. and Ovshinsky, S.R.), Plenum Press, New York, pp. 55–61.

Tichy, L. and Ticha, H. (1999) Is the chemical threshold in certain chalcogenide glasses responsible for the threshold at the mean coordination number of approximately 2.7?, *Phil. Mag. B*, **79**(2), 373–380.

Uhanov, Yu.I. (1977) *Optical Properties of Semiconductors*, Nauka, Moscow, p. 54–101.

Vanderbilt, D. and Joannopoulos, J.D. (1983) Total energies in Se. III. Defects in the glass, *Phys. Rev. B*, **27**, 6311–6321.

Vinokurov, A.Ya., Garkavenko, A.N., Litinskaya, L.I., Mironos, A.V. and Rodin, A.M. (1988) Radiation influence on the optical properties of IR fibers, *Avtometria*(5), 28–31.

Williams, G. and Watts, D.C. (1970) Non-symmetrical dielectric relaxation behavior arising from a simple empirical decay function, *Trans. Faraday Soc.*, **66**, 80–85.

Zaharov, V.P. and Gerasimenko, V.S. (1976) *Structural Peculiarities of Semiconductors in Amorphous State*, Naukova dumka, Kiev, pp. 255–256.

Zakis, Yu.R. (1984) *Defects in Vitreous State of Substance*, Zinatne, Riga, p. 202.

Zhilinskaya, E.A., Lazukin, V.N., Valeev, N.Kh. and Oblasov, A.K. (1990) Ge_xS_{1-x} glasses. Defects, density and structure, *J. Non-Cryst. Solids*, **124**, 48–62.

Zhilinskaya, E.A., Valeev, N.Kh. and Oblasov, A.K. (1992) Ge_xS_{1-x} glasses. II. Synthesis conditions and defect formation, *J. Non-Cryst. Solids*, **146**, 285–293.

Zorina, E.L., Dembovsky, S.A., Velichkova, V.B. and Vinogradova, G.Z. (1965) Infrared absorption of vitreous As_2Se_3, As_2Se_5 and $AsSe_4$, *Izv. AN SSSR. Neorgan. Mater.*, **1**(11), 1889–1891.

Index

Contents of Volumes in This Series

Volume 1 **Physics of III–V Compounds**

C. Hilsum, Some Key Features of III–V Compounds

F. Bassani, Methods of Band Calculations Applicable to III–V Compounds

E. O. Kane, The *k-p* Method

V. L. Bonch–Bruevich, Effect of Heavy Doping on the Semiconductor Band Structure

D. Long, Energy Band Structures of Mixed Crystals of III–V Compounds

L. M. Roth and P. N. Argyres, Magnetic Quantum Effects

S. M. Puri and T. H. Geballe, Thermomagnetic Effects in the Quantum Region

W. M. Becker, Band Characteristics near Principal Minima from Magnetoresistance

E. H. Putley, Freeze-Out Effects, Hot Electron Effects, and Submillimeter Photoconductivity in InSb

H. Weiss, Magnetoresistance

B. Ancker-Johnson, Plasma in Semiconductors and Semimetals

Volume 2 **Physics of III–V Compounds**

M. G. Holland, Thermal Conductivity

S. I. Novkova, Thermal Expansion

U. Piesbergen, Heat Capacity and Debye Temperatures

G. Giesecke, Lattice Constants

J. R. Drabble, Elastic Properties

A. U. Mac Rae and G. W. Gobeli, Low Energy Electron Diffraction Studies

R. Lee Mieher, Nuclear Magnetic Resonance

B. Goldstein, Electron Paramagnetic Resonance

T. S. Moss, Photoconduction in III–V Compounds

E. Antoncik and J. Tauc, Quantum Efficiency of the Internal Photoelectric Effect in InSb

G. W. Gobeli and I. G. Allen, Photoelectric Threshold and Work Function

P. S. Pershan, Nonlinear Optics in III–V Compounds

Volume 3 **Optical Properties of III–V Compounds**

Volume 4 **Physics of III–V Compounds**

Volume 5 **Infrared Detectors**

Volume 6 **Injection Phenomena**

Volume 7 **Application and Devices**

Part A

Part B

Volume 8 **Transport and Optical Phenomena**

Volume 9 **Modulation Techniques**

Volume 15 Contacts, Junctions, Emitters

B. L. Sharma, Ohmic Contacts to III–V Compounds Semiconductors

A. Nussbaum, The Theory of Semiconducting Junctions

J. S. Escher, NEA Semiconductor Photoemitters

Volume 16 Defects, (HgCd)Se, (HgCd)Te

H. Kressel, The Effect of Crystal Defects on Optoelectronic Devices

C. R. Whitsett, J. G. Broerman, and C. J. Summers, Crystal Growth and Properties of $Hg_{1-x}Cd_xSe$ Alloys

M. H. Weiler, Magnetooptical Properties of $Hg_{1-x}Cd_xTe$ Alloys

P. W. Kruse and J. G. Ready, Nonlinear Optical Effects in $Hg_{1-x}Cd_xTe$

Volume 17 CW Processing of Silicon and Other Semiconductors

J. F. Gibbons, Beam Processing of Silicon

A. Lietoila, R. B. Gold, J. F. Gibbons, and L. A. Christel, Temperature Distributions and Solid Phase Reaction Rates Produced by Scanning CW Beams

A. Leitoila and J. F. Gibbons, Applications of CW Beam Processing to Ion Implanted Crystalline Silicon

N. M. Johnson, Electronic Defects in CW Transient Thermal Processed Silicon

K. F. Lee, T. J. Stultz, and J.F. Gibbons, Beam Recrystallized Polycrystalline Silicon: Properties, Applications, and Techniques

T. Shibata, A. Wakita, T. W. Sigmon and J. F. Gibbons, Metal-Silicon Reactions and Silicide

Y. I. Nissim and J. F. Gibbons, CW Beam Processing of Gallium Arsenide

Volume 18 Mercury Cadmium Telluride

P. W. Kruse, The Emergence of $(Hg_{1-x}Cd_x)Te$ as a Modern Infrared Sensitive Material

H. E. Hirsch, S. C. Liang, and A. G. White, Preparation of High-Purity Cadmium, Mercury, and Tellurium

W. F. H. Micklethwaite, The Crystal Growth of Cadmium Mercury Telluride

P. E. Petersen, Auger Recombination in Mercury Cadmium Telluride

R. M. Broudy and V. J. Mazurczyck, (HgCd)Te Photoconductive Detectors

M. B. Reine, A. K. Soad, and T. J. Tredwell, Photovoltaic Infrared Detectors

M. A. Kinch, Metal-Insulator-Semiconductor Infrared Detectors

Volume 19 Deep Levels, GaAs, Alloys, Photochemistry

G. F. Neumark and K. Kosai, Deep Levels in Wide Band-Gap III–V Semiconductors

D. C. Look, The Electrical and Photoelectronic Properties of Semi-Insulating GaAs

R. F. Brebrick, Ching-Hua Su, and Pok-Kai Liao, Associated Solution Model for Ga-In-Sb and Hg-Cd-Te

Y. Ya. Gurevich and Y. V. Pleskon, Photoelectrochemistry of Semiconductors

Volume 20 **Semi-Insulating GaAs**

R. N. Thomas, H. M. Hobgood, G. W. Eldridge, D. L. Barrett, T. T. Braggins, L. B. Ta, and S. K. Wang, High-Purity LEC Growth and Direct Implantation of GaAs for Monolithic Microwave Circuits

C. A. Stolte, Ion Implantation and Materials for GaAs Integrated Circuits

C. G. Kirkpatrick, R. T. Chen, D. E. Holmes, P. M. Asbeck, K. R. Elliott, R. D. Fairman, and J. R. Oliver, LEC GaAs for Integrated Circuit Applications

J. S. Blakemore and S. Rahimi, Models for Mid-Gap Centers in Gallium Arsenide

Volume 21 **Hydrogenated Amorphous Silicon**

Part A

J. I. Pankove, Introduction

M. Hirose, Glow Discharge; Chemical Vapor Deposition

Y. Uchida, di Glow Discharge

T. D. Moustakas, Sputtering

I. Yamada, Ionized-Cluster Beam Deposition

B. A. Scott, Homogeneous Chemical Vapor Deposition

F. J. Kampas, Chemical Reactions in Plasma Deposition

P. A. Longeway, Plasma Kinetics

H. A. Weakliem, Diagnostics of Silane Glow Discharges Using Probes and Mass Spectroscopy

L. Gluttman, Relation between the Atomic and the Electronic Structures

A. Chenevas-Paule, Experiment Determination of Structure

S. Minomura, Pressure Effects on the Local Atomic Structure

D. Adler, Defects and Density of Localized States

Part B

J. I. Pankove, Introduction

G. D. Cody, The Optical Absorption Edge of a-Si: H

N. M. Amer and W. B. Jackson, Optical Properties of Defect States in a-Si: H

P. J. Zanzucchi, The Vibrational Spectra of a-Si: H

Y. Hamakawa, Electroreflectance and Electroabsorption

J. S. Lannin, Raman Scattering of Amorphous Si, Ge, and Their Alloys

R. A. Street, Luminescence in a-Si: H

R. S. Crandall, Photoconductivity

J. Tauc, Time-Resolved Spectroscopy of Electronic Relaxation Processes

P. E. Vanier, IR-Induced Quenching and Enhancement of Photoconductivity and Photoluminescence

H. Schade, Irradiation-Induced Metastable Effects

L. Ley, Photoelectron Emission Studies

Volume 24 Applications of Multiquantum Wells, Selective Doping, and Superlattices

Volume 25 Diluted Magnetic Semiconductors

Volume 26 III–V Compound Semiconductors and Semiconductor Properties of Superionic Materials

Volume 27 **High Conducting Quasi-One-Dimensional Organic Crystals**

Volume 28 **Measurement of High-Speed Signals in Solid State Devices**

Volume 29 **Very High Speed Integrated Circuits: Gallium Arsenide LSI**

Volume 30 **Very High Speed Integrated Circuits: Heterostructure**

Volume 31 **Indium Phosphide: Crystal Growth and Characterization**

J. P. Farges, Growth of Discoloration-Free InP

M. J. McCollum and G. E. Stillman, High Purity InP Grown by Hydride Vapor Phase Epitaxy

I. Inada and T. Fukuda, Direct Synthesis and Growth of Indium Phosphide by the Liquid Phosphorous Encapsulated Czochralski Method

O. Oda, K. Katagiri, K. Shinohara, S. Katsura, Y. Takahashi, K. Kainosho, K. Kohiro, and R. Hirano, InP Crystal Growth, Substrate Preparation and Evaluation

K. Tada, M. Tatsumi, M. Morioka, T. Araki, and T. Kawase, InP Substrates: Production and Quality Control

M. Razeghi, LP-MOCVD Growth, Characterization, and Application of InP Material

T. A. Kennedy and P. J. Lin-Chung, Stoichiometric Defects in InP

Volume 32 **Strained-Layer Superlattices: Physics**

T. P. Pearsall, Strained-Layer Superlattices

F. H. Pollack, Effects of Homogeneous Strain on the Electronic and Vibrational Levels in Semiconductors

J. Y. Marzin, J. M. Gerárd, P. Voisin, and J. A. Brum, Optical Studies of Strained III–V Heterolayers

R. People and S. A. Jackson, Structurally Induced States from Strain and Confinement

M. Jaros, Microscopic Phenomena in Ordered Superlattices

Volume 33 **Strained-Layer Superlattices: Material Science and Technology**

R. Hull and J. C. Bean, Principles and Concepts of Strained-Layer Epitaxy

W. J. Shaff, P. J. Tasker, M. C. Foisy, and L. F. Eastman, Device Applications of Strained-Layer Epitaxy

S. T. Picraux, B. L. Doyle, and J. Y. Tsao, Structure and Characterization of Strained-Layer Superlattices

E. Kasper and F. Schaffer, Group IV Compounds

D. L. Martin, Molecular Beam Epitaxy of IV–VI Compounds Heterojunction

R. L. Gunshor, L. A. Kolodziejski, A. V. Nurmikko, and N. Otsuka, Molecular Beam Epitaxy of I–VI Semiconductor Microstructures

Volume 34 **Hydrogen in Semiconductors**

J. I. Pankove and N. M. Johnson, Introduction to Hydrogen in Semiconductors

C. H. Seager, Hydrogenation Methods

J. I. Pankove, Hydrogenation of Defects in Crystalline Silicon

J. W. Corbett, P. Deák, U. V. Desnica, and S. J. Pearton, Hydrogen Passivation of Damage Centers in Semiconductors

S. J. Pearton, Neutralization of Deep Levels in Silicon

J. I. Pankove, Neutralization of Shallow Acceptors in Silicon
N. M. Johnson, Neutralization of Donor Dopants and Formation of Hydrogen-Induced Defects in *n*-Type Silicon
M. Stavola and S. J. Pearton, Vibrational Spectroscopy of Hydrogen-Related Defects in Silicon
A. D. Marwick, Hydrogen in Semiconductors: Ion Beam Techniques
C. Herring and N. M. Johnson, Hydrogen Migration and Solubility in Silicon
E. E. Haller, Hydrogen-Related Phenomena in Crystalline Germanium
J. Kakalios, Hydrogen Diffusion in Amorphous Silicon
J. Chevalier, B. Clerjaud, and B. Pajot, Neutralization of Defects and Dopants in III–V Semiconductors
G. G. DeLeo and W. B. Fowler, Computational Studies of Hydrogen-Containing Complexes in Semiconductors
R. F. Kiefl and T. L. Estle, Muonium in Semiconductors
C. G. Van de Walle, Theory of Isolated Interstitial Hydrogen and Muonium in Crystalline Semiconductors

Volume 35 **Nanostructured Systems**

M. Reed, Introduction
H. van Houten, C. W. J. Beenakker, and B. J. Wees, Quantum Point Contacts
G. Timp, When Does a Wire Become an Electron Waveguide?
M. Búttiker, The Quantum Hall Effects in Open Conductors
W. Hansen, J. P. Kotthaus, and U. Merkt, Electrons in Laterally Periodic Nanostructures

Volume 36 **The Spectroscopy of Semiconductors**

D. Heiman, Spectroscopy of Semiconductors at Low Temperatures and High Magnetic Fields
A. V. Nurmikko, Transient Spectroscopy by Ultrashort Laser Pulse Techniques
A. K. Ramdas and S. Rodriguez, Piezospectroscopy of Semiconductors
O. J. Glembocki and B. V. Shanabrook, Photoreflectance Spectroscopy of Microstructures
D. G. Seiler, C. L. Littler, and M. H. Wiler, One- and Two-Photon Magneto-Optical Spectroscopy of InSb and $Hg_{1-x}Cd_xTe$

Volume 37 **The Mechanical Properties of Semiconductors**

A.-B. Chen, A. Sher, and W. T. Yost, Elastic Constants and Related Properties of Semiconductor Compounds and Their Alloys
D. R. Clarke, Fracture of Silicon and Other Semiconductors
H. Siethoff, The Plasticity of Elemental and Compound Semiconductors
S. Guruswamy, K. T. Faber, and J. P. Hirth, Mechanical Behavior of Compound Semiconductors
S. Mahajan, Deformation Behavior of Compound Semiconductors
J. P. Hirth, Injection of Dislocations into Strained Multilayer Structures
D. Kendall, C. B. Fleddermann, and K. J. Malloy, Critical Technologies for the Micromatching of Silicon
I. Matsuba and K. Mokuya, Processing and Semiconductor Thermoelastic Behavior

Volume 38 **Imperfections in III/V Materials**

U. Scherz and M. Scheffler, Density-Functional Theory of sp-Bonded Defects in III/V Semiconductors
M. Kaminska and E. R. Weber, E12 Defect in GaAs
D. C. Look, Defects Relevant for Compensation in Semi-Insulating GaAs
R. C. Newman, Local Vibrational Mode Spectroscopy of Defects in III/V Compounds
A. M. Hennel, Transition Metals in III/V Compounds
K. J. Malloy and K. Khachaturyan, DX and Related Defects in Semiconductors
V. Swaminathan and A. S. Jordan, Dislocations in III/V Compounds
K. W. Nauka, Deep Level Defects in the Epitaxial III/V Materials

Volume 39 **Minority Carriers in III–V Semiconductors: Physics and Applications**

N. K. Dutta, Radiative Transition in GaAs and Other III–V Compounds
R. K. Ahrenkiel, Minority-Carrier Lifetime in III–V Semiconductors
T. Furuta, High Field Minority Electron Transport in p-GaAs
M. S. Lundstrom, Minority-Carrier Transport in III–V Semiconductors
R. A. Abram, Effects of Heavy Doping and High Excitation on the Band Structure of GaAs
D. Yevick and W. Bardyszewski, An Introduction to Non-Equilibrium Many-Body Analyses of Optical Processes in III–V Semiconductors

Volume 40 **Epitaxial Microstructures**

E. F. Schubert, Delta-Doping of Semiconductors: Electronic, Optical and Structural Properties of Materials and Devices
A. Gossard, M. Sundaram, and P. Hopkins, Wide Graded Potential Wells
P. Petroff, Direct Growth of Nanometer-Size Quantum Wire Superlattices
E. Kapon, Lateral Patterning of Quantum Well Heterostructures by Growth of Nonplanar Substrates
H. Temkin, D. Gershoni, and M. Panish, Optical Properties of Ga_{1-x} In_xAs/InP Quantum Wells

Volume 41 **High Speed Heterostructure Devices**

F. Capasso, F. Beltram, S. Sen, A. Pahlevi, and A. Y. Cho, Quantum Electron Devices: Physics and Applications
P. Solomon, D. J. Frank, S. L. Wright and F. Canora, GaAs-Gate Semiconductor-Insulator-Semiconductor FET
M. H. Hashemi and U. K. Mishra, Unipolar InP-Based Transistors
R. Kiehl, Complementary Heterostructure FET Integrated Circuits
T. Ishibashi, GaAs-Based and InP-Based Heterostructure Bipolar-Transistors
H. C. Liu and T. C. L. G. Sollner, High-Frequency-Tunneling Devices
H. Ohnishi, T. More, M. Takatsu, K. Imamura, and N. Yokoyama, Resonant-Tunneling Hot-Electron Transistors and Circuits

Volume 42 **Oxygen in Silicon**

F. Shimura, Introduction to Oxygen in Silicon

W. Lin, The Incorporation of Oxygen into Silicon Crystals

T. J. Schaffner and D. K. Schroder, Characterization Techniques for Oxygen in Silicon

W. M. Bullis, Oxygen Concentration Measurement

S. M. Hu, Intrinsic Point Defects in Silicon

B. Pajot, Some Atomic Configuration of Oxygen

J. Michel and L. C. Kimerling, Electrical Properties of Oxygen in Silicon

R. C. Newman and R. Jones, Diffusion of Oxygen in Silicon

T. Y. Tan and W. J. Taylor, Mechanisms of Oxygen Precipitation: Some Quantitative Aspects

M. Schrems, Simulation of Oxygen Precipitation

K. Simino and I. Yonenaga, Oxygen Effect on Mechanical Properties

W. Bergholz, Grown-in and Process-Induced Effects

F. Shimura, Intrinsic/Internal Gettering

H. Tsuya, Oxygen Effect on Electronic Device Performance

Volume 43 **Semiconductors for Room Temperature Nuclear Detector Applications**

R. B. James and T. E. Schlesinger, Introduction and Overview

L. S. Darken and C. E. Cox, High-Purity Germanium Detectors

A. Burger, D. Nason, L. Van den Berg, and M. Schieber, Growth of Mercuric Iodide

X. J. Bao, T. E. Schlesinger, and R. B. James, Electrical Properties of Mercuric Iodide

X. J. Bao, R. B. James, and T. E. Schlesinger, Optical Properties of Red Mercuric Iodide

M. Hage-Ali and P. Siffert, Growth Methods of CdTe Nuclear Detector Materials

M. Hage-Ali and P. Siffert, Characterization of CdTe Nuclear Detector Materials

M. Hage-Ali and P. Siffert, CdTe Nuclear Detectors and Applications

R. B. James, T. E. Schlesinger, J. Lund, and M. Schieber, $Cd_{1-x}Zn_xTe$ Spectrometers for Gamma and X-Ray Applications

D. S. McGregor, J. E. Kammeraad, Gallium Arsenide Radiation Detectors and Spectrometers

J. C. Lund, F. Olschner, and A. Burger, Lead Iodide

M. R. Squillante and K. S. Shah, Other Materials: Status and Prospects

V. M. Gerrish, Characterization and Quantification of Detector Performance

J. S. Iwanczyk and B. E. Patt, Electronics for X-ray and Gamma Ray Spectrometers

M. Schieber, R. B. James and T. E. Schlesinger, Summary and Remaining Issues for Room Temperature Radiation Spectrometers

Volume 44 **II–IV Blue/Green Light Emitters: Device Physics and Epitaxial Growth**

J. Han and R. L. Gunshor, MBE Growth and Electrical Properties of Wide Bandgap ZnSe-based II–VI Semiconductors

Volume 45 **Effect of Disorder and Defects in Ion-Implanted Semiconductors: Electrical and Physiochemical Characterization**

Volume 46 **Effect of Disorder and Defects in Ion-Implanted Semiconductors: Optical and Photothermal Characterization**

Volume 47 **Uncooled Infrared Imaging Arrays and Systems**

R. G. Buser and M. P. Tompsett, Historical Overview
P. W. Kruse, Principles of Uncooled Infrared Focal Plane Arrays
R. A. Wood, Monolithic Silicon Microbolometer Arrays
C. M. Hanson, Hybrid Pyroelectric-Ferroelectric Bolometer Arrays
D. L. Polla and J. R. Choi, Monolithic Pyroelectric Bolometer Arrays
N. Teranishi, Thermoelectric Uncooled Infrared Focal Plane Arrays
M. F. Tompsett, Pyroelectric Vidicon
T. W. Kenny, Tunneling Infrared Sensors
J. R. Vig, R. L. Filler, and Y. Kim, Application of Quartz Microresonators to Uncooled Infrared Imaging Arrays
P. W. Kruse, Application of Uncooled Monolithic Thermoelectric Linear Arrays to Imaging Radiometers

Volume 48 **High Brightness Light Emitting Diodes**

G. B. Stringfellow, Materials Issues in High-Brightness Light-Emitting Diodes
M.G. Craford, Overview of Device Issues in High-Brightness Light-Emitting Diodes
F. M. Steranka, AlGaAs Red Light Emitting Diodes
C. H. Chen, S. A. Stockman, M. J. Peanasky, and C. P. Kuo, OMVPE Growth of AlGaInP for High Efficiency Visible Light-Emitting Diodes
F. A. Kish and R. M. Fletcher, AlGaInP Light-Emitting Diodes
M. W. Hodapp, Applications for High Brightness Light-Emitting Diodes
I. Akasaki and H. Amano, Organometallic Vapor Epitaxy of GaN for High Brightness Blue Light Emitting Diodes
S. Nakamura, Group III–V Nitride Based Ultraviolet-Blue-Green-Yellow Light-Emitting Diodes and Laser Diodes

Volume 49 **Light Emission in Silicon: from Physics to Devices**

D. J. Lockwood, Light Emission in Silicon
G. Abstreiter, Band Gaps and Light Emission in Si/SiGe Atomic Layer Structures
T. G. Brown and D. G. Hall, Radiative Isoelectronic Impurities in Silicon and Silicon-Germanium Alloys and Superlattices
J. Michel, L. V. C. Assali, M. T. Morse, and L. C. Kimerling, Erbium in Silicon
Y. Kanemitsu, Silicon and Germanium Nanoparticles
P. M. Fauchet, Porous Silicon: Photoluminescence and Electroluminescent Devices
C. Delerue, G. Allan, and M. Lannoo, Theory of Radiative and Nonradiative Processes in Silicon Nanocrystallites
L. Brus, Silicon Polymers and Nanocrystals

Volume 50 **Gallium Nitride (GaN)**

J. I. Pankove and T. D. Moustakas, Introduction

Volume 51A **Identification of Defects in Semiconductors**

Volume 51B **Identification Defects in Semiconductors**

Volume 52 **SiC Materials and Devices**

Volume 53 **Cumulative Subjects and Author Index Including Tables of Contents for Volumes 1–50**

Volume 54 **High Pressure in Semiconductor Physics I**

Volume 55 **High Pressure in Semiconductor Physics II**

Volume 56 **Germanium Silicon: Physics and Materials**

Volume 57 **Gallium Nitride (GaN) II**

Volume 58 **Nonlinear Optics in Semiconductors I**

Volume 59 **Nonlinear Optics in Semiconductors II**

Volume 60 **Self-Assembled InGaAs/GaAs Quantum Dots**

Kohki Mukai, Mitsuru Sugawara, Mitsuru Egawa, and Nobuyuki Ohtsuka, Metalorganic Vapor Phase Epitaxial Growth of Self-Assembled InGaAs/GaAs Quantum Dots Emitting at 1.3 μm
Kohki Mukai and Mitsuru Sugawara, Optical Characterization of Quantum Dots
Kohki Mukai and Mitsuru Sugawara, The Photon Bottleneck Effect in Quantum Dots
Hajime Shoji, Self-Assembled Quantum Dot Lasers
Hiroshi Ishikawa, Applications of Quantum Dot to Optical Devices
Mitsuru Sugawara, Kohki Mukai, Hiroshi Ishikawa, Koji Otsubo, and Yoshiaki Nakata, The Latest News

Volume 61 **Hydrogen in Semiconductors II**

Norbert H. Nickel, Introduction to Hydrogen in Semiconductors II
Noble M. Johnson and Chris G. Van de Walle, Isolated Monatomic Hydrogen in Silicon
Yurij V. Gorelkinskii, Electron Paramagnetic Resonance Studies of Hydrogen and Hydrogen-Related Defects in Crystalline Silicon
Norbert H. Nickel, Hydrogen in Polycrystalline Silicon
Wolfhard Beyer, Hydrogen Phenomena in Hydrogenated Amorphous Silicon
Chris G. Van de Walle, Hydrogen Interactions with Polycrystalline and Amorphous Silicon–Theory
Karen M. McManus Rutledge, Hydrogen in Polycrystalline CVD Diamond
Roger L. Lichti, Dynamics of Muonium Diffusion, Site Changes and Charge-State Transitions
Matthew D. McCluskey and Eugene E. Haller, Hydrogen in III–V and II–VI Semiconductors
S. J. Pearton and J. W. Lee, The Properties of Hydrogen in GaN and Related Alloys
Jörg Neugebauer and Chris G. Van de Walle, Theory of Hydrogen in GaN

Volume 62 **Intersubband Transitions in Quantum Wells: Physics and Device Applications I**

Manfred Helm, The Basic Physics of Intersubband Transitions
Jerome Faist, Carlo Sirtori, Federico Capasso, Loren N. Pfeiffer, Ken W. West, Deborah L. Sivco, and Alfred Y. Cho, Quantum Interference Effects in Intersubband Transitions
H. C. Liu, Quantum Well Infrared Photodetector Physics and Novel Devices
S. D. Gunapala and S. V. Bandara, Quantum Well Infrared Photodetector (QWIP) Focal Plane Arrays

Volume 63 **Chemical Mechanical Polishing in Si Processing**

Frank B. Kaufman, Introduction
Thomas Bibby and Karey Holland, Equipment
John P. Bare, Facilitization
Duane S. Boning and Okumu Ouma, Modeling and Simulation
Shin Hwa Li, Bruce Tredinnick, and Mel Hoffman, Consumables I: Slurry
Lee M. Cook, CMP Consumables II: Pad
François Tardif, Post-CMP Clean

Volume 64 **Electroluminescence I**

Volume 65 **Electroluminescence II**

Volume 66 **Intersubband Transitions in Quantum Wells: Physics and Device Applications II**

Volume 67 **Ultrafast Physical Processes in Semiconductors**

Roland E. Allen, Traian Dumitrică, and Ben Torralva, Electronic and Structural Response of Materials to Fast, Intense Laser Pulses
E. Gornik and R. Kersting, Coherent THz Emission in Semiconductors

Volume 68 **Isotope Effects in Solid State Physics**

Vladimir G. Plekhanov, Elastic Properties; Thermal Properties; Vibrational Properties; Raman Spectra of Isotopically Mixed Crystals; Excitons in LiH Crystals; Exciton–Phonon Interaction; Isotopic Effect in the Emission Spectrum of Polaritons; Isotopic Disordering of Crystal Lattices; Future Developments and Applications; Conclusions

Volume 69 **Recent Trends in Thermoelectric Materials Research I**

H. Julian Goldsmid, Introduction
Terry M. Tritt and Valerie M. Browning, Overview of Measurement and Characterization Techniques for Thermoelectric Materials
Mercouri G. Kanatzidis, The Role of Solid-State Chemistry in the Discovery of New Thermoelectric Materials
B. Lenoir, H. Scherrer, and T. Caillat, An Overview of Recent Developments for BiSb Alloys
Citrad Uher, Skutterudities: Prospective Novel Thermoelectrics
George S. Nolas, Glen A. Slack, and Sandra B. Schujman, Semiconductor Clathrates: A Phonon Glass Electron Crystal Material with Potential for Thermoelectric Applications

Volume 70 **Recent Trends in Thermoelectric Materials Research II**

Brian C. Sales, David G. Mandrus, and Bryan C. Chakoumakos, Use of Atomic Displacement Parameters in Thermoelectric Materials Research
S. Joseph Poon, Electronic and Thermoelectric Properties of Half-Heusler Alloys
Terry M. Tritt, A. L. Pope, and J. W. Kolis, Overview of the Thermoelectric Properties of Quasicrystalline Materials and Their Potential for Thermoelectric Applications
Alexander C. Ehrlich and Stuart A. Wolf, Military Applications of Enhanced Thermoelectrics
David J. Singh, Theoretical and Computational Approaches for Identifying and Optimizing Novel Thermoelectric Materials
Terry M. Tritt and R. T. Littleton, IV, Thermoelectric Properties of the Transition Metal Pentatellurides: Potential Low-Temperature Thermoelectric Materials
Franz Freibert, Timothy W. Darling, Albert Miglori, and Stuart A. Trugman, Thermomagnetic Effects and Measurements
M. Bartkowiak and G. D. Mahan, Heat and Electricity Transport Through Interfaces

Volume 71 **Recent Trends in Thermoelectric Materials Research III**

M. S. Dresselhaus, Y.-M. Lin, T. Koga, S. B. Cronin, O. Rabin, M. R. Black, and G. Dresselhaus, Quantum Wells and Quantum Wires for Potential Thermoelectric Applications

Volume 72 **Silicon Epitaxy**

Volume 73 **Processing and Properties of Compound Semiconductors**

Volume 74 **Silicon-Germanium Strained Layers and Heterostructures**